Ni-Co 2013

TMS2013

142nd Annual Meeting & Exhibition

Ni-Co 2013

Proceedings of symposium sponsored by
the TMS Extraction & Processing Division, TMS High
Temperature Alloys Committee, and the Metallurgy and
Materials Society of the Canadian Institute of Mining,
Metallurgy and Petroleum (CIM)

Held during the
TMS 2013 Annual Meeting & Exhibition
San Antonio, Texas, USA
March 3-7, 2013

Edited by

Thomas Battle
Michael Moats • Violina Cocalia
Harald Oosterhof • Shafiq Alam
Antoine Allanore • Rodney Jones
Nathan Stubina • Corby Anderson
Shijie Wang

Editors
Thomas Battle
Michael Moats
Violina Cocalia
Harald Oosterhof
Shafiq Alam
Antoine Allanore
Rodney Jones
Nathan Stubina
Corby Anderson
Shijie Wang

ISBN 978-3-319-48581-2 ISBN 978-3-319-48147-0 (eBook)
DOI 10.1007/978-3-319-48147-0

Chemistry and Materials Science: Professional

Published by Springer International Publishers, Switzerland, 2016
Reprint of the original edition published by John Wiley & Sons, Inc., 2013, 978-1-11860-575-2

Printed on acid-free paper

This Springer imprint is published by Springer Nature
The registered company is Springer International Publishing AG
The registered company address is: Gewerbestrasse 11, 6330 Cham, Switzerland

TABLE OF CONTENTS
Ni-Co 2013

Preface ix
About the Editors xi

Plenary

Laterites - Still a Frontier of Nickel Process Development 3
A. Taylor

Cobalt ('the Technology Enabler'): CDI Review & Future Challenges 25
D. Weight

The Recycling of Cobalt from Alloy Scrap, Spent Batteries or Catalysts and Metallurgical Residues–An Overview 53
C. Ferron

Ni Laterite Hydrometallurgy

Acid Leaching of Nickel Laterites with Jarosite Precipitation 75
D. White, and J. Gillaspie

Extraction of Nickel, Cobalt and Iron from Laterite Ores by Mixed Chloride Leach Process 97
V. Lakshmanan, R. Sridhar, R. DeLaat, J. Chen, M. Halim, and R. Roy

Reductive Leaching of Limonitic Laterites Using Ferrous Sulphate 107
M. Zuniga, and E. Asselin

Sulfuric Acid Leaching Characteristics of Ni-Doped Goethite 117
G. Li, W. Cai, M. Rao, Q. Zhi, and T. Jiang

Characteristics of Nickel Laterite Crushed Ore Agglomerates 125
A. Janwong, N. Dhawan, T. Vethsodsakda, and M. Moats

Electrometallurgy

Acid Mist Abatement in Base Metal Electrowinning 143
T. Robinson, D. White, and R. Grassi

Boleo Cobalt Electrowinning Development 155
J. Lu, D. Dreisinger, and T. Gluck

Comparison of Intercell Contact Bars for Electrowinning Plants 177
C. Boon, R. Fraser, T. Johnston, and D. Robinson

High Current Density Electrowinning of Nickel in EMEW® Cells 191
J. Robinson, I. Ewart, M. Moats, and S. Wang

Helm Tracker™ Cathode Current Sensing Technology 201
R. Fraser, T. Johnston, J. Yesberg, S. Nolet, and C. Boon

Pyrometallurgy: Solid-State Processing

Cobalt Recovery through Sulphating Roast of Cu/Co Concentrate of Katanga Mining 213
K. Adham

Experimental Study on Reduction-Magnetic Separation Process of Low-Grade Nickel Laterite Ore 221
F. Chen, B. Zhang, W. Li, Q. Wang, and X. Hong

State of the Art Refractory Corrosion Test Work for the Nonferrous Metals Industry 231
D. Gregurek, A. Ressler, V. Reiter, A. Franzkowiak, A. Spanring, B. Drew, and D. Flynn

Ores and Processing

Mineralogical Characterization of Cobaltic Oxides from the Democratic Republic of Congo 243
Y. Vanbrabant, C. Burlet, and P. Louis

PolyMet Mining Corporation's NorthMet Process Development 255
D. Dreisinger, M. Ounpuu, D. Imeson, and N. Verbaan

Talvivaara Nickel Mine - from a Project to a Mine and Beyond......................269
M. Riekkola-Vanhanen, and L. Palmu

The Sintering Character of Limonitic Nickel Laterite 279
C. Wu, H. Li, Y. Chen, C. Li, and Z. Zhang

Pyrometallurgy: Smelting

Nickel, Cobalt and Copper Recovery from Sea Nodules by Direct Smelting Process 291
K. Sahu, S. Agarwal, D. Mishra, A. Agrawal, N. Randhawa, K. Godiwalla, and R. Jana

Alternative Coolants and Cooling System Designs for Safer Freeze Lined Furnace Operation 299
M. Kennedy, P. Nos, M. Bratt, and M. Weaver

Outotec's Ausmelt Top Submerged Lance (TSL) Technology for the Nickel Industry...315
R. Andrews, R. Matusewicz, L. Aspola, and S. Hughes

Processing of PGM Containing Ni/Cu Bulk Concentrates in a Sustainable Way by Outotec Direct Nickel Flash Smelting Process....................................325
S. Jyrkonen, K. Haavanlammi, M. Luomala, J. Karonen, and P. Suikkanen

Nickel-Chromium-Boron Alloys Production by Aluminothermic Processes ...335
O. Yilmaz, M. Alkan, and O. Yücel

Applications & Recycling

Numerical Simulation of Temperature Field in Directional Solidification of Turbine Blade by Liquid Metal Cooling Method 343
N. Tang, Q. Xu, and B. Liu

Influence of Thermomechanical Treatment on Structure and Properties of a Cobalt Based Superalloy...353
P. Sarkar, S. Prasad, M. Chatterjee, and M. Rao

Effect of Processing Conditions on Structure, Properties and Performance of a Nickel Base Cast Superalloy for High Temperature Applications 357
M. Chatterjee, A. Kishore, P. Sarkar, and M. Rao

Development of Nickel Boron Alloys for Brazing Materials 365
K. Tasyürek, M. Alkan, and O. Yücel

Ni and Co Hydrometallurgy

Continuous Co-Precipitation Behaviour and Stability of Arsenic(V) from Fe(II,III)-Al(III)-Ni(II) Sulphate Effluent Solutions.......................................371
C. Doerfelt, and G. Demopoulos

Hydrochloric Acid Regeneration via Calcium Sulfate Crystallization for Non-Ferrous Chloride Leaching Processes..379
T. Feldmann, and G. Demopoulos

Hydrometallurgical Nickel Laterite Processing: A Review of Current SX Flowsheets and Industry Trends 391
A. Fischmann, S. Wiggett, T. Bednarski, V. Cocalia, and C. Bourget

Studies on Refining Cobalt Salt Solution by Extraction Chromatography to Prepare High Purity Cobalt...405
C. Song, Z. Li, L. Shuling, C. Zhenping, and W. Lijun

The Starved Acid Leaching Technology (SALT) for Recovery of Nickel and Cobalt from Saprolites and Caron Plant Residues 413
D. Dreisinger, and J. Clucas

Selective Production of Co and Ni Powders Through Hydrothermal Reduction of Leach Solutions of a Synthetic Matte Containing Cu-Ni-Co-Fe-S...............425
D. Mishra, K. Park, K. Sahu, A. Agrawal, and C. Nam

Author Index...435

Subject Index 437

Preface

The physical and extractive metallurgy of nickel and cobalt have been a focus of much activity in TMS over the years. For many years, indeed, a Cu-Ni-Co Committee existed in the Extraction and Processing Division, before their efforts were taken over by the Pyrometallurgy and Hydrometallurgy/Electrometallurgy Committees. These committees have conducted extensive programming in Ni-Co metallurgy, including physical metallurgy in cooperation with the High Temperature Alloys Committee and other committees in other divisions of the society. Many of the larger symposia in this area were organized in collaboration with our partners in The Metallurgy and Materials Society of the Canadian Institute of Mining, Metallurgy and Petroleum (MetSoc). TMS-led symposia included Extractive Metallurgy of Ni and Co in 1988, and the International Laterite Ni Symposium in 2004. MetSoc has organized successful symposia on Ni Co in 1997 and 2005.

Soon after the 2005 conference, the two societies decided to collaborate on future programming in this area. The contract we've signed guarantees a series of symposia in coming years, organized by both societies, with the two organizations alternating in the lead. The first of this new sequence was held by MetSoc as the keystone of their Conference of Metallurgists in 2009 in Sudbury, Ontario. The 2013 symposium in San Antonio is the first in the sequence with TMS as the lead organizer.

Following are presentations and papers representing all eight of the sessions organized in this symposium: a four-paper plenary session, one session on Ores and Processing, two sessions each on Pyrometallurgy and Hydrometallurgy, one on Electrometallurgy, and one on Applications and Recycling. We also have a short course on Ni-Co, and a number of exhibitors to support the success of this conference.

Special thanks are due to MetSoc for their assistance and advice, including three members of the organizing committee: Nathan Stubina, Shafiq Alam, and Ron Schonewille. In addition, there is a strong core of TMS organizers representing all aspects of Ni-Co metallurgy: Violina Cocalia, Corby Anderson, Mike Moats, Harald Oosterhof, Phil Mackey, Xingbo Liu, and Shijie Wang, with assistance from long-time members Norbert Piret and Ram Ramachandran. The affiliations of all organizers, as well as logos of the many co-sponsoring organizations, can be found elsewhere in this book's frontmatter.

I also wish to acknowledge the co-sponsoring organizations for this symposium: MEI (Minerals Engineering International), GDMB (German Society for Mining, Metallurgy, Resource and Environmental Technology), MMIJ (The Mining and Materials Processing Institute of Japan), SF2M (Societe Francaise de Metallurgie et de Materiaux), SME (Society for Mining, Metallurgy, and Exploration), SAIMM (Southern African Institute of Mining and Metallurgy), High Temperature Materials Committee of the Chinese Society for Metals (CSM), ALTA Metallurgical Services,

ABM (Brazilian Association for Metallurgy, Mining, and Materials), and the Cobalt Development Institute.

We hope you find this proceedings volume a useful review of previous work, summary of the state-of-the-art, and reference for the future.

Thomas Battle, lead organizer
Senior Metallurgist
Midrex Technologies
Pineville, North Carolina

Editors/Organizers

Thomas Battle is currently a Senior Metallurgist at Midrex Technologies in Pineville, North Carolina. He has undergraduate degrees in Materials Engineering and Astronomy from the University of Michigan, a Master's in Metallurgical Engineering from the Colorado School of Mines, and a Doctorate in Materials and Metallurgical Engineering from the University of Michigan. After a time as a Post-Doctoral Research Fellow in the Centre for Numerical Modelling and Process Analysis at Thames Polytechnic, he spent 18 years at various positions with the White Pigments and Mineral Products business at DuPont (now known at DuPont Titanium Technologies). He has spent the last four years as a senior metallurgist at Midrex Technologies, focusing on iron ore pelletizing technology and the direct reduction of iron.

Tom has been active with TMS for more than 20 years, holding a number of volunteer positions, both technical and administrative, mainly for the Extraction and Processing Division. This culminated in a three-year term as chair of the division, and a position on the society's Board of Directors. He is a founding member of the North American Extractive Metallurgy Council.

Michael Moats is an associate professor of Metallurgical Engineering within the Department of Materials Science and Engineering at the Missouri University of Science and Technology. He has published more than 60 papers on various hydro- and electro-metallurgical topics with a particular emphasis on industrial surveys of copper and zinc operations, copper electrowinning and electrorefining and gold leaching. He is a co-author of *The Extractive Metallurgy of Nickel, Cobalt and Platinum Group Metals.* He has been employed in industry, academia, and consulting.

Violina Cocalia holds a Ph.D. in Inorganic Chemistry from the University of Alabama. She has experience in coordination chemistry, solvent extraction, hydrometallurgy, and ionic liquids. She is currently leading the R&D Metal Extraction Products Group at Cytec Industries Inc.

Harald Oosterhof graduated as chemical engineer from the group of Prof. Van Swaaij at Twente University (The Netherlands) in 1994. In the same year, he assumed a position as researcher at Delft University of Technology (The Netherlands) where he worked in the Laboratory for Process Equipment that was headed by Prof. Van Rosmalen. The research that focussed on anti solvent crystallization of well-soluble salts was rewarded with two patents and a dozen publications.

After receiving his Ph.D. in Industrial Crystallization and Precipitation from Delft University in 1999, Harald started working at Umicore, a global materials technology group that is based in Belgium. During his first assignment as Project Leader Hydrometallurgy, he focussed on the refining of cobalt, nickel, and germanium.

Since 2011, Harald has worked as Scientist in the group "Recycling and Extraction" at Umicore's Central R&D department. His main competence areas are:

- Special metals hydrometallurgy (In, Se, Te)
- Recycling and refining of rare earth metals
- Base metals recycling and hydrometallurgy
- Recycling and refining of spent rechargeable batteries (NiMH and Li-ion)

In his current job, Harald is frequently involved in business development of scarce metals recycling and he is heading a team of hydrometallurgical specialists.

Shafiq Alam is an Associate Professor at Memorial University of Newfoundland, Canada. In 1998, he received his Ph.D. degree in Chemical Engineering from Saga University, Japan. From 1999–2001, he was appointed as a post-doctoral research fellow at the University of British Columbia and the University of Toronto, Canada.

Before joining Memorial University in 2006, he has worked with many different companies, such as, Shell, Process Research ORTECH Inc. (Mississauga, Ontario), Fluor Canada Ltd. (Vancouver), and the National Institute of Advanced Industrial Science and Technology (AIST), Japan. Dr. Alam is highly experienced in the area of extractive metallurgy. He is the lead researcher of the Hydrometallurgy lab at the INCO Innovation Centre (IIC) at Memorial University. He possesses two patents and has over 70 publications in the area of hydrometallurgy.

Dr. Alam is a registered professional engineer and has worked on projects with many different mining companies including Xstrata, Phelps Dodge, INCO Ltd. (Vale), and Barrick Gold Corporation. He is an Executive Committee Member of the Hydrometallurgy Section of the Canadian Institute of Mining, Metallurgy and Petroleum (CIM), and is actively involved with the committee in organizing different international conferences, such as, the Conference of Metallurgists (COM) in Canada; Ni-Co Symposium at TMS 2013 in Texas, USA; and the Copper 2013 Conference in Chile.

Antoine Allanore is Assistant Professor of Metallurgy at the Massachusetts Institute of Technology (Cambridge, Massachusetts), in the Department of Materials Science and Engineering where he currently holds the Thomas B. King Chair. He earned a chemical engineering degree from the Ecole Nationale Superieure des Industries Chimiques de Nancy and a M.Sc. and Ph.D. in chemical engineering from the Institut National Polytechnique de Lorraine. Prior joining MIT, he worked as a research engineer at ArcelorMittal R&D on the development of new electrolytic processes for primary steel production.

Dr. Allanore was a TMS Extraction and Processing Division Young Leader Professional Development Award winner in 2011, and co-recipient of the Vittorio de Nora Prize awarded at TMS 2012.

Rodney Jones has worked in the Pyrometallurgy Division at Mintek since 1985. He holds a B.Sc. (Eng) degree in chemical engineering from the University of the Witwatersrand (Wits) in Johannesburg, a B.A. degree in logic and philosophy from the University of South Africa, and a M.Sc. (Eng) degree in metallurgy from Wits University. He is a registered Professional Engineer, a Fellow and Vice President of the Southern African Institute of Mining and Metallurgy (SAIMM), a Fellow of the South African Institute of Chemical Engineers (SAIChE), and a full member of the Computer Society of South Africa (CSSA). He was a Visiting Professor at the Center for Pyrometallurgy, University of Missouri-Rolla, during July and August 1996, and in 2002 and 2003 also lectured in pyrometallurgy at Murdoch University, Perth, as an Adjunct Associate Professor. The National Research Foundation rated him in 2009 as an "Internationally Acclaimed Researcher." SAIMM awarded him an Honorary Life Fellowship in 2010. Rodney is also a member of the Board of Trustees for OneMine.org.

Nathan Stubina earned a B. Eng in Metallurgical Engineering from McGill University and a Ph.D. in Metallurgy and Materials Science from the University of Toronto. He started working for Falconbridge in 1987 and joined Noranda in 1999. While with those companies he worked at several sites including Sudbury, Norway, Sweden, Belgium and Rouyn-Noranda. Nathan has a mix of plant and R&D working experience.

Nathan joined Barrick Gold in 2005 where he is the Manager of the Barrick Technology Centre. He is a member of the Professional Engineers of Ontario. Nathan is currently VP International for the CIM, is a member of the CIM's Board of Directors, was Chair of the 2009 Conference of Metallurgists (COM) in Sudbury and is a MetSoc Past-President. He is a founding member of NAEMC (North American Extractive Metallurgy Council), is a trustee for the Harold A. Steane Memorial Scholarship, and sits on the board of the Strathcona BIA (Business Improvement Association) for Vancouver's Downtown Eastside. Nathan is the author of over 30 technical papers and patents, and he is a co-editor of the MetSoc commemorative book titled *The Canadian Metallurgical & Materials Landscape* (*1960 to 2011*).

Corby Anderson has over 33 years of global experience in industrial operations, management, engineering, design, consulting, teaching, research, and professional service. His career includes positions with Morton Thiokol, Key Tronic Corporation, Sunshine Mining and Refining Company, H. A Simons Ltd., and Montana Tech. He holds a B.Sc. in Chemical Engineering and an M.Sc. and Ph.D. in Metallurgical Engineering. He is a Fellow of the Institution of Chemical Engineers and the Institute of Materials, Minerals and Mining. He currently serves as the Harrison Western Professor in the Kroll Institute for Extractive Metallurgy as part of the George S. Ansell Department of Metallurgical and Materials Engineering at the Colorado School of Mines. In 2009 he was honored by SME with the Milton E. Wadsworth Extractive Metallurgy Award for his contributions in hydrometallurgical research.

Shijie Wang received his B.Sc. in Mineral Processing from China and his Masters and Ph.D. in Metallurgical Engineering from the University of Nevada at Reno. Dr. Wang has experience working at the Beijing General Research Institute for Non-Ferrous Metals, ASARCO Inc., and Phelps Dodge Corporation. He is currently Principal Advisor at Rio Tinto Kennecott Utah Copper. Dr. Wang has been active in extractive metallurgy and has experience in metallurgical process development and existing operation optimization as well as troubleshooting. He is copper and precious metals refining subject matter expert.

Dr. Wang has been a TMS member since 1991 and is currently Chair of the Hydrometallurgy and Electrometallurgy Committee of TMS.

Ni-Co 2013

Plenary

Session Chair

Thomas Battle

Ni-Co 2013
Edited by: Thomas Battle, Michael Moats, Violina Cocalia, Harald Oosterhof, Shafiq Alam, Antoine Allanore, Rodney Jones, Nathan Stubina, Corby Anderson, and Shijie Wang
TMS (The Minerals, Metals & Materials Society), 2013

LATERITES – STILL A FRONTIER OF NICKEL PROCESS DEVELOPMENT

By

Alan Taylor
ALTA Metallurgical Services

Presented by
Alan Taylor
alantaylor@altamet.com.au

INTRODUCTION

- Nickel production has been historically based on sulphide resources & high grade saprolitic laterites.
- Based on currently known resources, future production will increasingly have to come from low grade limonitic & saprolitic laterites with <1.5% Ni.
- Commercially applied processes for low grade laterites are:
  - Caron Process: reduction roast – ammonia leach)
  - PAL (or HPAL) Process: high pressure sulphuric acid leaching
  - EPAL Process: enhanced pressure acid leaching

INTRODUCTION

Note: There is a also single known application of ferronickel smelting operated by LARCO, Greece.

However, ferronickel smelting is generally considered to be uneconomic for low grade laterites.

CARON PROCESS

- It is not a new process and was first proposed by Professor Caron, Delft Univ. Netherlands, in the 1920s.
- It was pioneered commercially by Freeport at Nicaro, Cuba, in 1944, then taken over by the Cuban Government in 1960. It is still in operation.
- A further five plants were constructed in the 1970s-1990s, one of which was closed (Nonoc in the Philippines) and one never completed (Los Camariocas in Cuba).

TYPICAL CARON FLOWSHEET

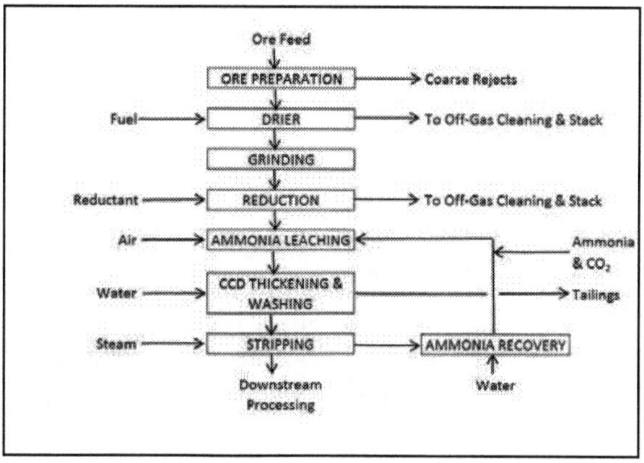

CURRENT CARON INSTALLATIONS

Plant	Start-up Date	Builder	Location
Nicaro	1944	Freeport	Cuba
Yabulu	1974	Freeport	Qld, Australia
Tocantins	1982	Votorantim	Brazil
Punta Gorda	1986	Cuban-Russian	Cuba

YABULU OPERATION, AUSTRALIA

ISSUES WITH CARON FLOWSHEET

- It has s high energy consumption for the initial ore drying step.
- Nickel recovery is only moderately high.
- Cobalt recovery is relatively low.
- It is generally limited to limonite and mixed ores with >35% Fe.
- Recovery falls off with saprolite ores.
- It is sensitive to mineral composition and requires careful mining and blending.

PAL PROCESS

- The process is also not new. The first plant was commissioned by Freeport at Moa Nay, Cuba, in 1959. It was also taken over by the Cuban Government in 1960 and remains in operation today.
- The next major action was not until the development of the AMAX and Nical Processes in the 1970s and 1980s for the Prony Project in New Caledonia and the Gasquet Project in Northern California respectively. Neither reached commercialization due mainly to low nickel price.
- In fact there were no further commercial PAL operations till the 1990s when 3 plants were built in Western Australia, namely Bulong, Cawse & Murrin Murrin (Bulong & Cawse are now closed).

PAL PROCESS

- Since then, 5 PAL plants (not including the Ravensthorpe EPAL plant) have been constructed and are either operating or in commissioning; a further two are under construction.
- The main driving forces for the PAL projects include:
  - Increased nickel demand and higher price
  - Availability of numerous low grade laterite deposits
  - Higher energy efficiency and higher nickel and cobalt recoveries than the Caron Process
  - Longevity of the Moa Bay operation.
  - Successful commercial application of horizontal autoclaves and related facilities for pressure leaching of other ores such as gold and zinc.

TYPICAL PAL FLOWSHEET

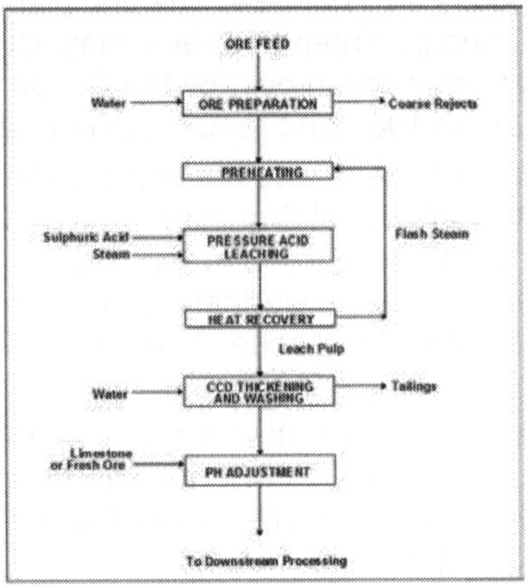

CURRENT PAL INSTALLATIONS (EXCLUDING RAVENSTHORPE EPAL PLANT)

Plant	Country	Nickel (t/y)	Status
Moa Bay	Cuba	33,000	Operating
Murrin Murrin	Australia	40,000	Operating
Coral Bay	Philippines	20,000	Operating
Goro	New Caledonia	60,000	Operating
Ambatovy	Madagascar	60,000	Commissioning
Ramu	PNG	31,000	Commissioning
Taganito	Philippines	30,000	Construction
Gordes	Turkey	10,000	Construction

MURRIN MURRIN OPERATION, AUSTRALIA

ISSUES WITH PAL PROCESS

- The capital cost is high.
- It has a high acid consumption for saprolitic ores with high magnesium content.
- The process conditions are highly corrosive.
- The maintenance cost is relatively high.
- Autoclave descaling and maintenance involves significant pant downtime.
- Sophisticated control/safety systems are needed.
- Downstream processing is complex.
- There has been a lengthy ramp-up time for the majority of projects.

EPAL PROCESS

- It was developed by BHP Billiton and installed at Ravensthorpe, Western Australia. Commissioning commenced in late 2007, and operation was suspended in early 2009, attributed to a fall in nickel price.
- It was acquired by First Quantum in Dec. 2009 and restarted after modifications, with commercial production achieved in late 2011.
- Limonite ore is treated by PAL and saprolite ore by atmospheric tank leaching.

EPAL PROCESS FLOWSHEET
(Ref: Ravensthorpe Presentation at ALTA 2005)

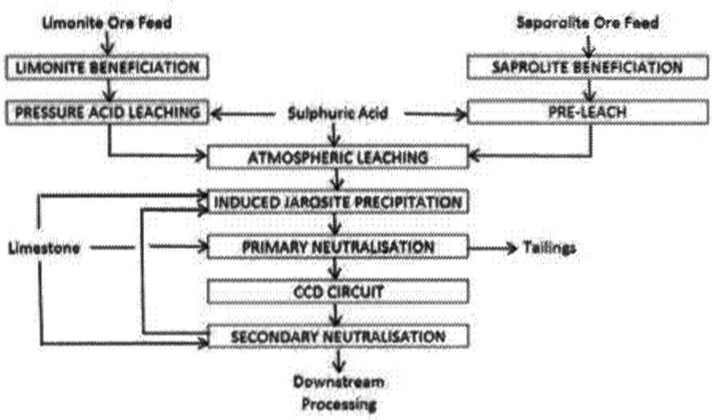

ISSUES WITH EPAL PROCESS

- It retains the disadvantages of PAL for the limonite ore portion of the flowsheet.
- It is relatively complex, with two different leach processes.

DEVELOPING PROCESSES USING SULPURIC ACID

- Pressure acid leaching (PAL) combined with atmospheric pressure agitated tank leaching (AL).
- Stand-alone atmospheric pressure agitated leaching tank (AL).
- Heap leaching (HL)

PAL COMBINED WITH AL

- These combine PAL for limonite with AL for saprolite in a parallel circuit and/or use saprolite to neutralize the residual acid from PAL.
- Objectives include:
  - treating the whole orebody
  - reducing the net acid consumption for saprolite
  - eliminating or reducing the limestone needed to neutralize residual acid from limonite PAL
  - reducing capital cost.
- It has been piloted by a number of companies as far back as AMAX and Nical in the 1970s, but not yet commercialized with the exception of EPAL.

STAND-ALONE AL

- Stand-alone AL has also been tested by a number organizations. A recent example is Eramet who have adopted AL for the Weda Bay Project in Indonesia.
- The advantage of stand-alone AL is the absence of a PAL circuit which should translate into capital cost savings, reduced down time and lower maintenance costs.
- However, this may be offset by increased sulphuric acid consumption.
- Higher acid consumptions lead to higher iron and other impurities to downstream processing.

PROPOSED WEDA BAY FLOWSHEET
(Ref: Eramet Paper at ALTA 2012)

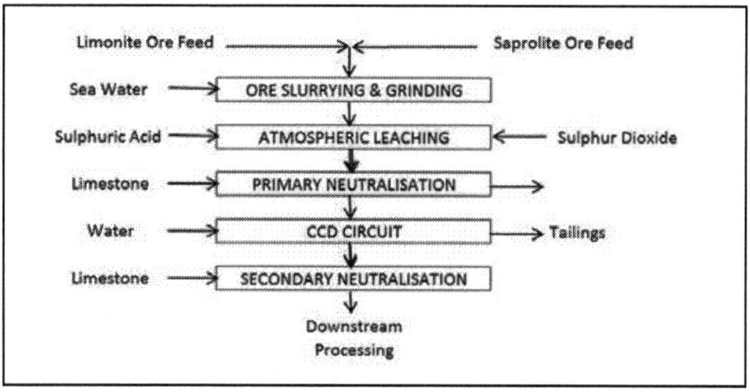

ISSUES WITH AL

- It generally has a high acid consumption, especially for saprolite ore, leading to higher operating cost which offsets the saving in capital cost versus PAL.
- It produces high levels of iron and other impurities in the leach solution which makes downstream processing more challenging.
- It is generally less effective for leaching limonite ore and sulphur dioxide injection may be needed to achieve high cobalt extraction.
- It typically operates close to 100C so that expensive materials of construction are needed.

HEAP LEACHING

- Development work has been carried out by numerous organizations since the 1990's, but no stand-alone commercial operation has yet been established.
- A commercial satellite operation treating grinding circuit scats was established in 2007 by Minara at Murrin Murrin in Western Australia. Fresh ore treatment was also successfully demonstrated, but is not currently practiced as it is said to be more profitable to send ore through the existing PAL circuit.
- A stand-alone HL demonstration plant was operated by European Nickel at Caldag in Turkey, in 2004; but the project was held up by permitting issues. It was sold in 2011 and is now owned by VTG Nikel of Turkey who are aiming for commercial operation.

CALDAG HL DEMONSTRATION PLANT
(Ref: European Nickel Paper at ALTA 2006)

CALDAG DEMONSTRATION PLANT HL FLOWSHEET (Ref: European Nickel Paper at ALTA 2006)

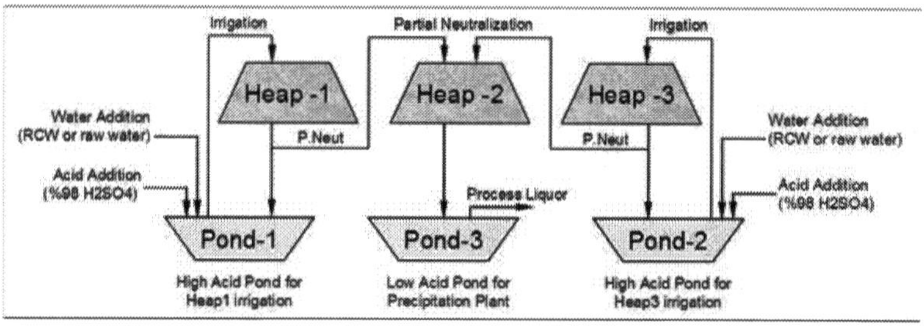

ISSUES WITH HL

- Laterites frequently have a high fines content and agglomeration with a polymer is typically required.
- Heap stability is a problem, especially in high rainfall locations. High rainfall can also cause breakdown of agglomerates and cause dilution of the leach solution.
- It generally has a high acid consumption, especially for saprolite ore, leading to higher operating cost which offsets the saving in capital cost versus PAL.
- It typically produces higher levels of iron and other impurities in the leach solution than PAL, which makes downstream processing more difficult.

ISSUES WITH HL

- Heap leaching of limonite ore typically yields low recoveries.
- Decommissioning can be challenging especially at high rainfall sites.
- Environmental issues can be greater that for PAL or AL.

ADVANTAGES OF SULPHURIC ACID BASED LEACHING PROCESSES

Key advantages of sulphuric acid for leaching laterites include:

- It is the best known and most widely applied leaching reagent in the mining and metallurgical industry.
- it is widely available at moderate cost.
- It is commonly produced in the same industry – eg smelters.
- On-site production yields very useful energy.
- Intermediate products can be produced which significantly reduces cost and simplifies operation.

ISSUES WITH SULPHURIC ACID BASED PROCESSES

- Downstream processing is typically complex and is a major component of overall capital cost, especially if separate nickel and cobalt products are produced at site.
- The leach solution is generally dilute with regard to nickel and cobalt, which adversely affects capital and operating costs.
- The acid not regenerated and recycled.

DEVELOPING PROCESSES USING CHLORIDE LEACHING SYSTEMS

- Chloride leaching processes for laterites have been tested by a number of organizations.
- Recent examples include Neomet in Canada, Anglo Research in South Africa, Nichromet in Canada, Process Research Ortech in Canada and SMS Siemag in Austria.
- There are no commercial plants for laterites as yet; however, there are a number of long established commercial operations refining nickel matte in chloride conditions.

POTENTIAL ADVANTAGES OF CHLORIDE LEACHING OF LATERITES

- It is an agitated tank leach at atmospheric pressure.
- It is applicable to both limonite and saprolite ores.
- It has lower capital and operating costs than PAL.
- Nickel and cobalt recoveries are high.
- The lixiviant is regenerated and recycled.
- Separation and recovery of nickel and cobalt products is comparatively straightforward and commercially proven methods are available.
- Secondary neutralization to remove residual iron is not needed.
- Solid/separation properties are superior to sulphuric acid processes.
- There are opportunities to recover saleable by-products.

NEOMET PROCESS FLOWSHEET (Ref: Neomet Paper at ALTA 2011)

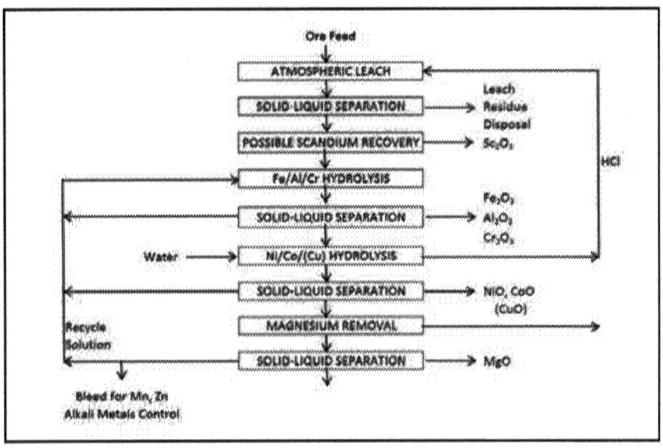

ISSUES WITH CHLORIDE SYSTEMS

- Chloride solutions are highly corrosive, especially at high temperatures.
- The handling of HCl vapours involves personnel safety issues.
- Depending on the overall water balance, there can be environmental problems due to the release of chloride containing solutions.

DEVELOPING PROCESSES USING NITRIC LEACHING SYSTEMS

- Nitric leaching systems have attracted limited interest for laterites.
- A recent example is the DNi Process being developed by Direct Nickel in Western Australia.
- There are no commercial plants as yet, however nitric systems have been used for other leaching and refining applications.

POTENTIAL ADVANTAGES OF NITRIC LEACHING OF LATERITES

- It is an agitated tank leach at atmospheric pressure.
- It is applicable to both limonite and saprolite ores.
- Lower capital and operating costs than PAL.
- Nickel and cobalt recoveries ate high.
- The lixiviant is regenerated and recycled .
- The is no secondary neutralization to remove residual iron.
- Solid/liquid separation properties are superior to sulphuric acid processes.
- There are opportunities to recover saleable by-products.

DNi PROCESS FLOWSHEET (Ref: Direct Nickel Paper at ALTA 2011)

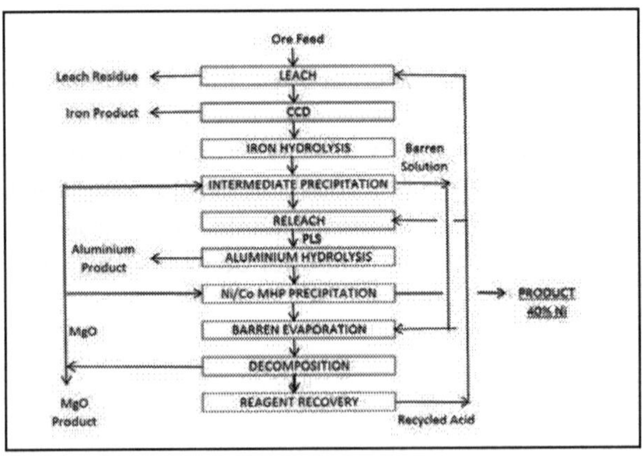

ISSUES WITH NITRIC SYSTEMS

- Nitric acid solutions are highly corrosive, especially at high temperatures.
- The handling of vapours involves personnel safety issues.
- Depending on the overall water balance, there can be environmental problems due to the release of chloride containing solutions.

SUMMARY OF PRESENT STATUS FOR PROCESSING OF LOW GRADE LATERITES

- PAL dominates for recent projects. No recent Caron plants have been built.
- AL is being pursued for a number of projects, but there are no stand-alone operations as yet.
- HL is still being pursued , though Interest has waned, and there are no stand-alone operations as yet.
- A number of chloride leaching processes are being developed, but there are no operations as yet.
- A nitric acid leaching system is being developed by Direct Nickel, but there is no commercial plant as yet.

THE IDEAL LATERITE PROCESS

- High recoveries of nickel and cobalt.
- Lower capex and opex than existing processes
- No initial drying and low overall energy requirement.
- Atmospheric pressure operation.
- Low net reagent consumption.
- Straight forward downstream processing.
- Ability to provide separate nickel and cobalt products.
- Potential for saleable by-products.
- Suitable for large and small projects.
- Low to moderate corrosivity.
- Low short and long term environmental impact.

The field is still wide open

There is no clear winner as yet

Laterites are still a very much a frontier of nickel process development

REFERENCES

- 1. Nickel Laterite Processing Technologies – Where To Next?, Jim Kyle, ALTA 2010, Perth, Western Australia.
- 2. Treatment of Nickel-Cobalt Laterites Short Course, Alan Taylor, ALTA 2011, Perth, Western Australia.
- 3. Developments and Trends in Hydrometallurgical Processing of Nickel Laterites, Boyd Willis, ALTA 2012, Perth, Western Australia.
- 4. Nickel Laterite and Three Mineral Acids, Mike Dry and Bryn Harris, ALTA 2012, Perth, Western Australia

Ni-Co 2013
Edited by: Thomas Battle, Michael Moats, Violina Cocalia, Harald Oosterhof, Shafiq Alam, Antoine Allanore, Rodney Jones, Nathan Stubina, Corby Anderson, and Shijie Wang
TMS (The Minerals, Metals & Materials Society), 2013

Cobalt Development Institute

Promoting the Sustainable and Responsible use of Cobalt in all Forms

Cobalt

('the technology enabler')

CDI Review & Future Challenges

By

David Weight

General Manager

TMS2013

DISCLAIMER

You are solely responsible for evaluating the accuracy and completeness of any content appearing in this presentation. Whilst the Cobalt Development Institute (CDI) has endeavoured to provide accurate and reliable information, it does not make any representations or warranties in relation to the content of this presentation. In particular, the CDI does not make any representations or warranties regarding the accuracy, timeliness or completeness of the content of the presentation or in respect of its suitability for any purpose. No action should be taken without seeking independent professional advice. The CDI will not be responsible for any loss or damage caused by relying on the content contained in this presentation.

TMS2013

55 Years of History

- **1957**
  - **Started as the 'Cobalt Information Centre'**
  - **Largely Patronised by Union Minière, later Gecamines**
  - **Closed in the 70s**
- **1982**
  - **Restarted as the CDI**
  - **Gecamines and ZCCM dominant**
- **Today**
  - **non-profit organisation for primary and secondary producers, for converters and recyclers, for users and traders of cobalt and for the scientific community**

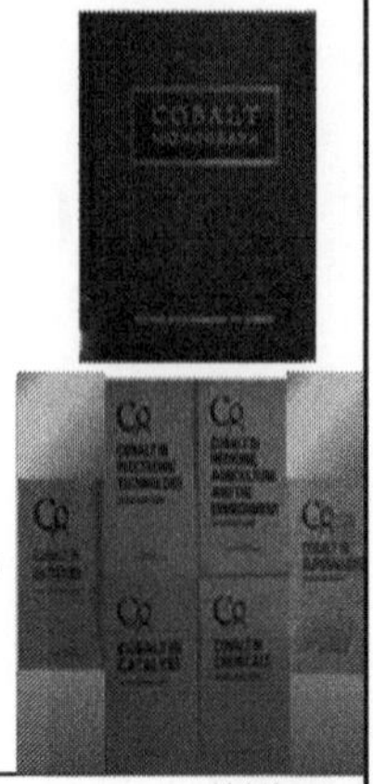

TMS2013

Mission Statement

"Promoting the Sustainable and Responsible Use of Cobalt in all its Forms"

❖ **Promote the use of and knowledge about Cobalt in all its forms**

❖ **Support the sustainable development and use of Cobalt, Cobalt compounds and products**

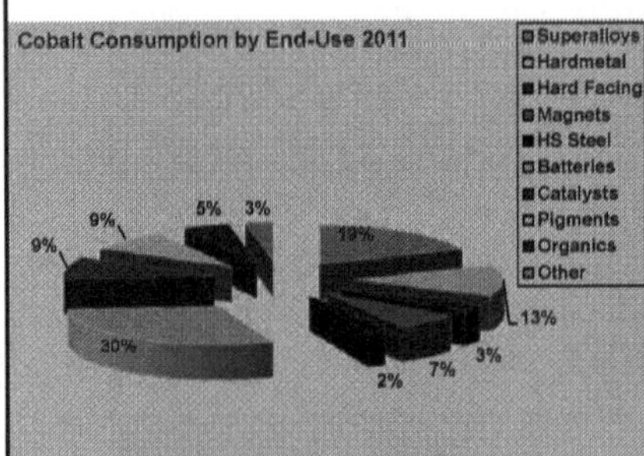

TMS2013

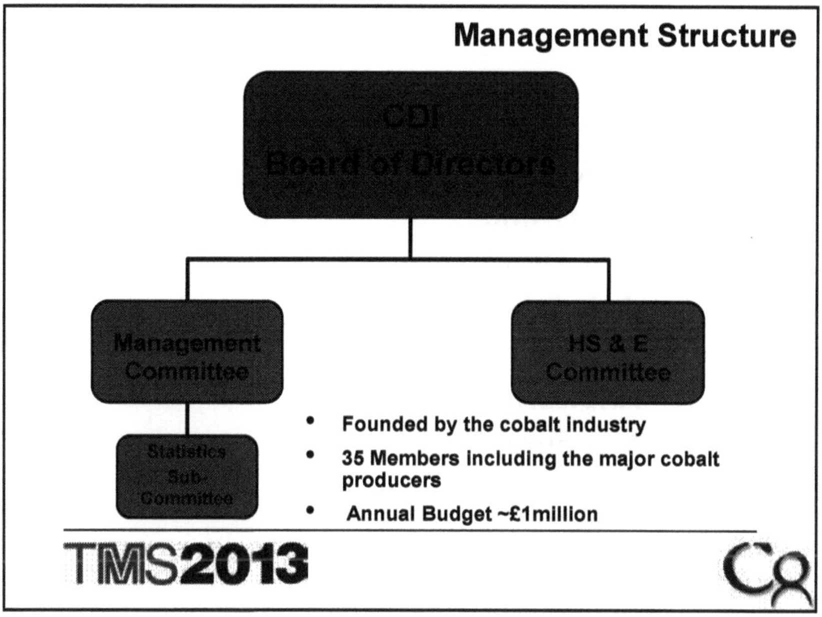
Management Structure
CDI
Board of Directors
Management
Committee
HS & E
Committee
Statistics
Sub-
Committee
• Founded by the cobalt industry
• 35 Members including the major cobalt producers
• Annual Budget ~£1million
TMS2013

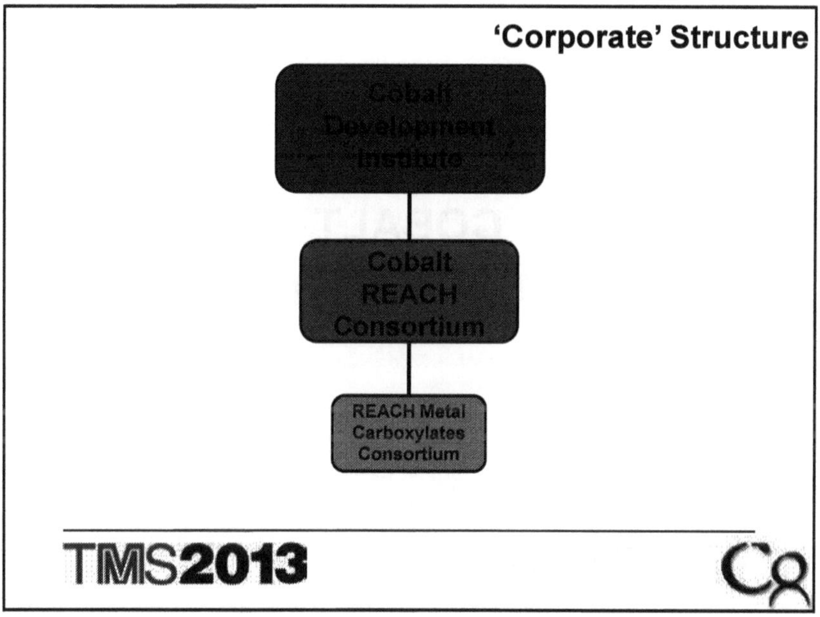
'Corporate' Structure
Cobalt
Development
Institute
Cobalt
REACH
Consortium
REACH Metal
Carboxylates
Consortium
TMS2013

CDI Foundations

- **HS&E - Science/regulatory monitoring/informing**
- **Advocacy and PR**
- **REACH – and beyond (Authorisation/Restriction…)**
- **Conference**
- **Statistics**
- **Cobalt News**
- **Website**
- **Answering enquiries**
- **Monographs**
- **Global Interaction**

TMS2013

COBALT APPLICATIONS

TMS2013

Cobalt End-Use

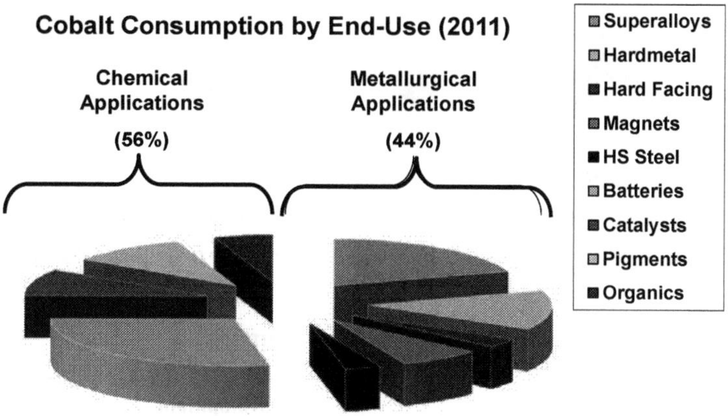

TMS2013

Rechargeable Battery

Typical raw material for LIB

- Cobalt metal or oxide/hydroxide
- Converted to lithiated cathode material

2011 cobalt consumption

- >20,000 MT

Use

- Active cathode material
- Application requirements dictate the chemistry and physical properties

Comment

- New chemical systems developed during high price period
- Long-term growth

TMS2013

Hardmetal & Diamond Tool

Typical raw material

- Cobalt oxalate
- Converted to fine cobalt powders

Use

- Binder for cemented carbides (WC)
- Powder metallurgical parts processing

2011 cobalt consumption

- ~10,000 MT

Comment

- New alloy alternatives in development
- Recycle plays an important role

TMS2013

Pigment

Typical raw material

- Predominantly cobalt oxide
- Some cobalt sulphate, carbonate, and hydroxide

Use

- Colourant additive
- Ceramic tile, glass, and plastic

2011 cobalt consumption

- ~6,500 MT

Comment

- High cobalt price limits demand
- Had seen strong growth in China, Asia, and Eastern Europe

TMS2013

Catalyst – Polymerisation/Oxidation

Typical raw material

- Cobalt acetate and hydroxide
- Converted to carboxylates in some cases

Use

- Increases polymerisation rates in:
  - PTA – Purified Terephthalic Acid
  - UPR – Unsaturated Polyester Resin

2011 cobalt consumption

- ~6,000 MT Cobalt (PTA the driver)

Comment

- Centered in Asia/Pacific
- Recycle is important

TMS2013

Rubber Adhesion

Typical raw material

- Cobalt hydroxide and chloride
- Converted to cobalt carboxylates

Use

- Wet blended adhesion additive
- Strengthens bond between rubber and steel cord

2011 cobalt consumption

- >2,000 MT

Comment

- Competitive market – move of activity to Asia

TMS2013

Coatings

Typical raw material

- Cobalt hydroxide and metal
- Converted to cobalt carboxylates

2011 cobalt consumption

- ~1,500 MT

Use

- Wet blended drier additive
- Accelerates drying of alkyd based paint, ink, and varnish

Market trends

- Movement towards low VOC formulations or water emulsified systems

TMS2013

Catalyst – Hydroprocessing

Typical raw material

- Cobalt carbonate and hydroxide
- Cobalt nitrate solution

2011 cobalt consumption

- ~1,500 MT Cobalt

Use

- Synthesised with other metals (typically molybdenum)
- Promoter in Hydro - desulphurisation (HDS) Catalyst

Market trends

- Driven by tighter sulphur legislation
- Cyclical with catalyst change-outs

TMS2013

Catalyst – Fischer Tropsch for GTL

Typical raw material

- Cobalt metal converted to cobalt nitrate/carbonate

2011 cobalt consumption

- Several hundred tonnes

Use

- Proprietary synthesis
- Key catalyst for gas-to-liquid conversion at large plants in Qatar (Shell, Sasol-Chevron)

Market trends

- GTL projects are more attractive with high oil prices
- Recycle will be important
- Long term growth potential

TMS2013

PW 7000 Gas Turbine Series. As used in the A380

Superalloys

>14,000 tonnes in these applications.

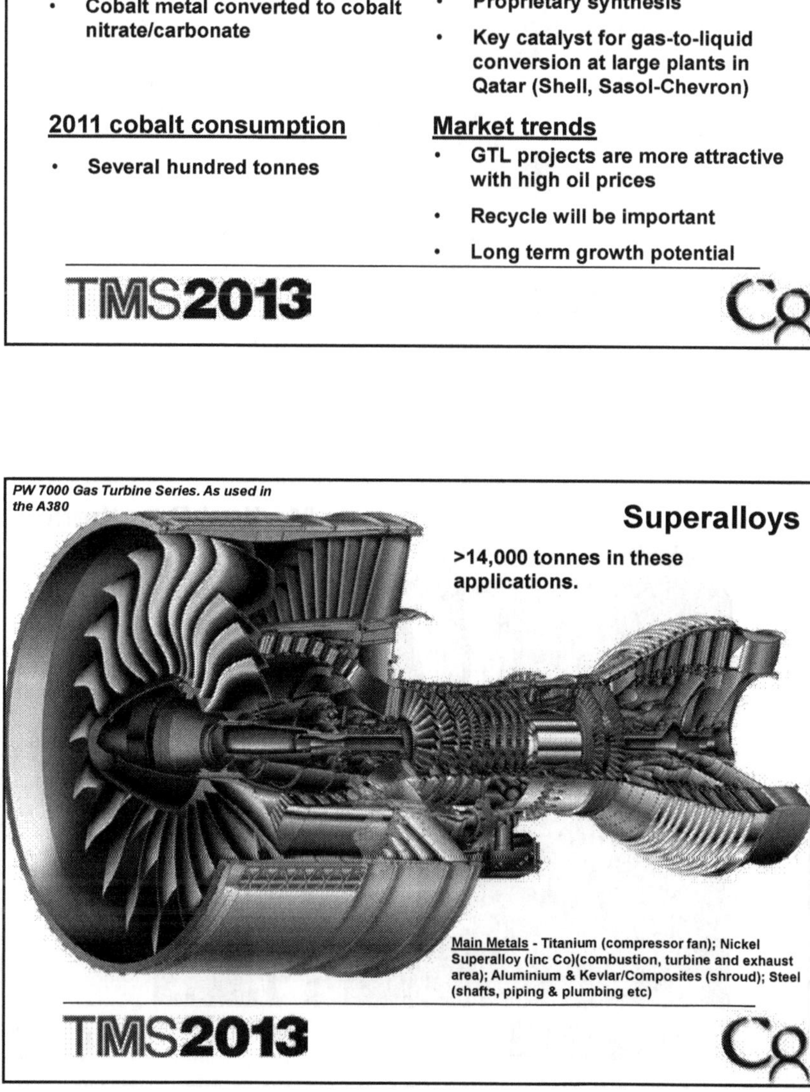

Main Metals - Titanium (compressor fan); Nickel Superalloy (inc Co)(combustion, turbine and exhaust area); Aluminium & Kevlar/Composites (shroud); Steel (shafts, piping & plumbing etc)

TMS2013

Casting, Hard Facing and High Speed Steel

~3,000 tonnes in these applications.

TMS2013

Medical Prosthetics

>1,000 tonnes in these applications

TMS2013

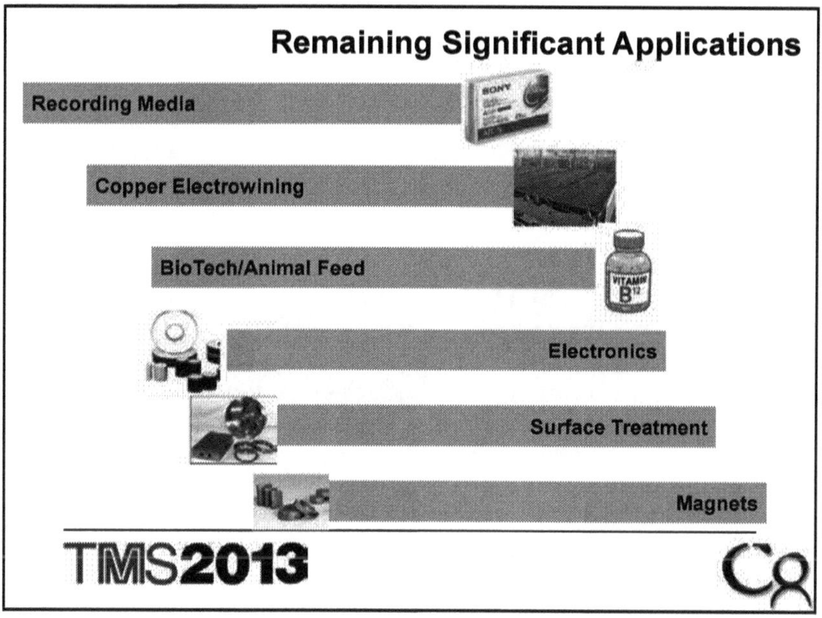

Cobalt Market

TMS2013

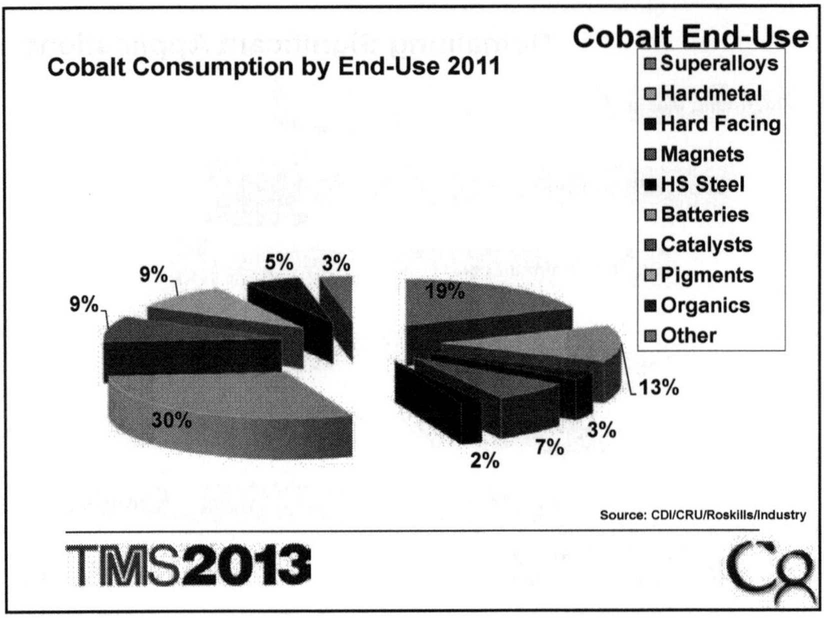
Cobalt End-Use
Cobalt Consumption by End-Use 2011
Superalloys
Hardmetal
Hard Facing
Magnets
HS Steel
Batteries
Catalysts
Pigments
Organics
Other
9%
9%
5%
3%
30%
19%
13%
3%
7%
2%
Source: CDI/CRU/Roskills/Industry
TMS2013

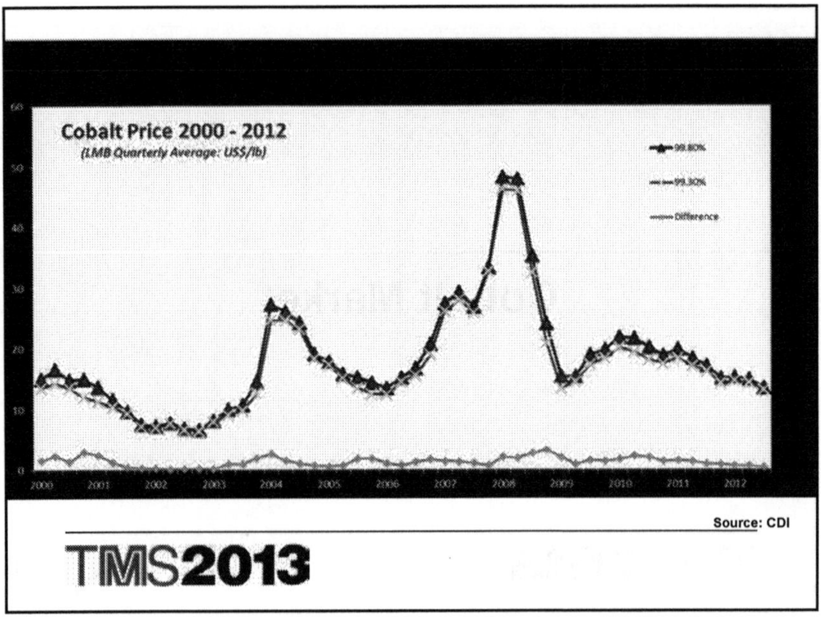
Cobalt Price 2000 - 2012
(LMB Quarterly Average: US$/lb)
Difference
Source: CDI
TMS2013

Cobalt Production

Current Production Sources:

Nickel Industry	~	57%
Copper Industry & Other	~	37%
Primary Cobalt Operations	~	6%

Source: CDI estimate

TMS2013

Refined Production by Geographical Area

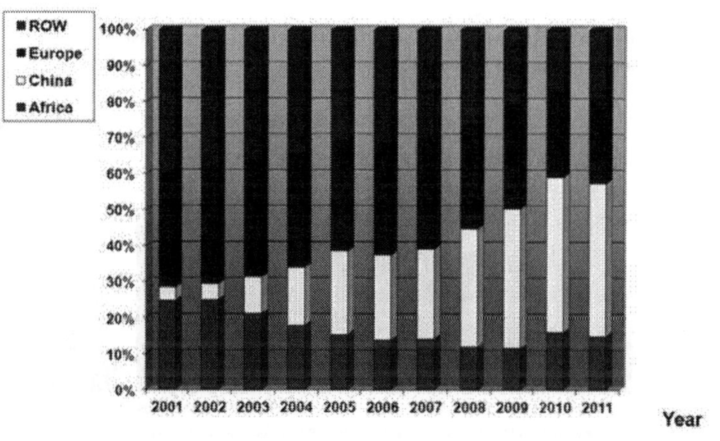

Source: CDI

TMS2013

Refined Cobalt Apparent Demand by Region

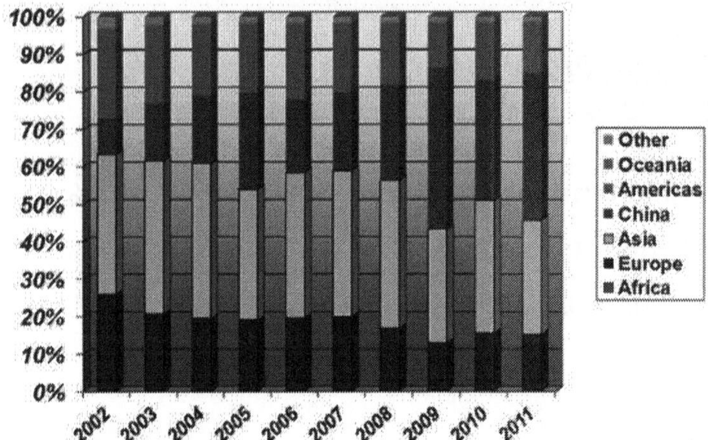

•Source: CDI/WBMS

TMS2013

Refined Cobalt Demand

Historically strong demand – dampened by 08 and 09

(Cobalt demand by sector, 1999-2011 tonnes)

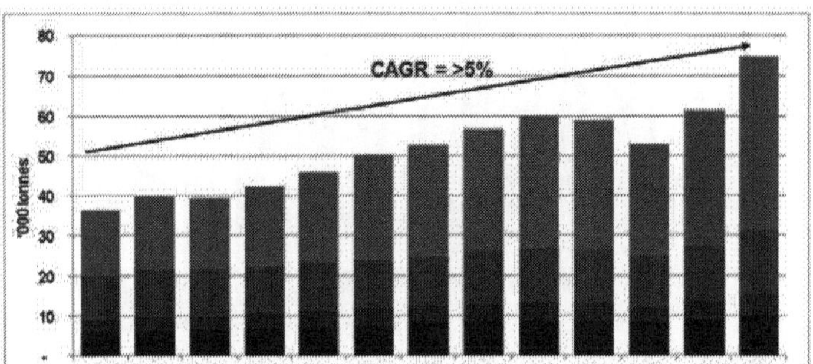

•Source: CRU Analysis

TMS2013

Supply & Demand

	2008	2009	2010	2011
Supply *(CDI)*	*55,256*	*59,851*	*76,363*	*82,247*
Demand *(Apparent*)*	*60,040*	*58,736*	*65,677*	*75,100*

Historically the CAGR has been ~5% per annum

NB: Availability affected by Stocks in China - ~3000mt in 2011

* Source WBMS/CRU/Antaike

TMS2013

Projects

PROJECT		OPERATING COMPANY
• Luita, DRC	-	ENRC
• Tenke Fungurume, DRC	-	FreeportMcMoran/Lundin
• Katanga/KOV, DRC	-	Katanga Mining
• Kolwezi Tailings, DRC	-	ENRC
• Ruashi, DRC	-	Metorex (Jinchuan)
• Ravensthorpe, Australia	-	First Quantum
• El Boleo, Mexico	-	Baja
• Nkamouna, Cameroon	-	Geovic
• Ramu, PNG	-	CMCC/Highlands Pacific
• Ambatovy, Madagascar	-	Sherritt/Sumitomo/Kore Res…
• Goro, New Caledonia	-	Vale
• Vermelho, Brazil	-	Vale
• Idaho, USA (Primary Co)	-	Formation Capital
• Koniambo, New Caledonia	-	Xstrata/SMSP
• Halmahera, Indonesia	-	Eramet
• Nama, Zambia	-	Caledonia Mining
• Muliashi, Zambia	-	CNMC
• Kara-Kul, Russia/Mongolia Boarder	-	Global Cobalt Corp
• Werner Lake, Canada (Primary Co)	-	Global Cobalt Corp
• New Caledonia/Pacific Region	-	Gladstone Pacific Nickel (QNI?)
• Mumi, Kolwezi DRC	-	Mutanda Mining

TMS2013

Global Reserves

Africa	~50%
Americas	~10%
Australasia	~24%
Russia	~ 3%
ROW	~13%
	100%

(Based on reserves of about 7,500,000 tonnes and Identified world cobalt resources are 15million tonnes)*

*Source: USGS (figures for 2011) & CDI

TMS2013

Hazard & Risk

TMS2013

Cobalt – An Essential Trace Element

Who needs cobalt?

Cobalt is essential to various bacteria, which utilize cobalt to improve plant growth and human health.

- **Bacteria need cobalt to aid plants in obtaining nutrients for growth.**
- **Bacteria transform cobalt into cobalamin – the form of cobalt needed by animals and humans, known as vitamin B12.**

TMS2013

Cobalt Natural Life Cycle

...occurs naturally in soil → ...is ingested by ruminants → Rumen bacteria synthesize cobalt-containing vitamin B12 → Healthy lifestock

...is needed by bacteria to obtain macronutrients (nitrogen)

Bacteria, aided by cobalt, deliver macronutrients to legumes and other plants

Good growth of agricultural plants for human nutrition → Delivery of healthy plant food, cobalt and Vit B12 traces to humans → Healthy blood formation and cell division (growth)

Vit B12 in dairy products, meat

Delivery of Vit B12 to humans

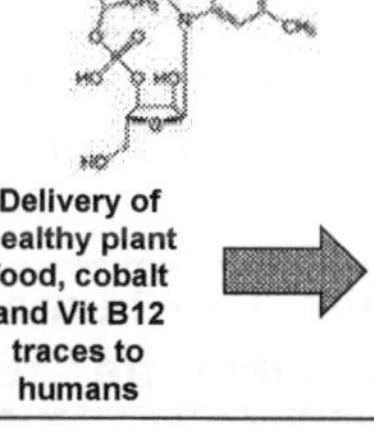

Healthy Human Being

TMS2013

Natural occurrence of cobalt

Cobalt is a natural component of the earth's crust and therefore it is present in:

- Sea
- Surface and ground water
- Soil, dusts, rocks, volcanoes
- Plants

Cobalt occurs naturally in the environment as "Cobalt-59" and is NOT radioactive

Cobalt levels in the environment can -and must- never be "zero".

TMS2013

Natural occurrence of cobalt

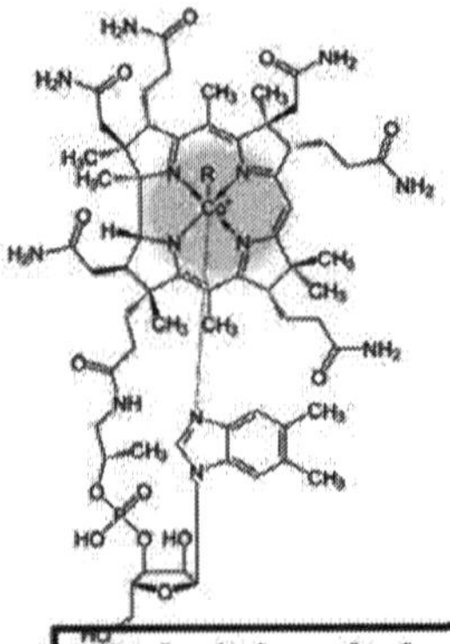

Cobalt, as part of cobalamin (vitamin B12), occurs naturally in the tissues of animals and humans

Cobalt, as part of vitamin B12, is essential for life

Cobalt levels in animal- or human tissue can -and must- never be "zero"

TMS2013

Deficiency is a concern?

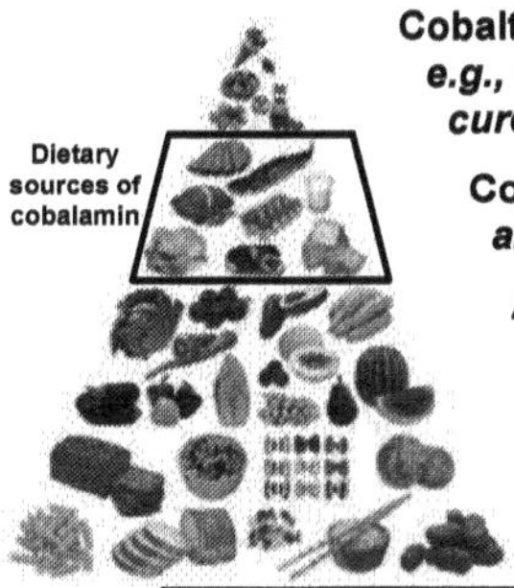

Cobalt deficiency: ***Historically a concern in farm animals, e.g., "bush sickness" in New Zealand cured by adding cobalt to soil fertilizers.***

Cobalamin deficiency: ***a realistic concern for vegans and vegetarians.***

A major concern ***in patients who have undergone weight-loss surgery or with stomach issues.***

Easily bioavailable ***forms of cobalamin need to be provided, e.g. as nasal sprays.***

Deficiency symptoms:

Cobalamin deficiency hallmarks: anaemia, gastrointestinal problems and neurological symptoms

TMS2013

Too Much of a Good Thing!

Toxicity

- **Cobalt toxicity is a concern in the workplace.**
- **It is addressed by adequate risk-control measures such as personal protective equipment of the workers.**
- **The CDI and the cobalt industry are actively addressing health and environmental concerns related to cobalt, such as inhalation, skin sensitisation and systemic effects.**

"The dose makes the poison."

Paracelsus (1493–1541)

German-Swiss Renaissance genius physician and scientist. Many of his ground-breaking advances in the field of medicine are relevant to this day.

TMS2013

Toxicity, Hazard and Risk

TMS2013

Toxicity, Hazard and Risk

Toxicity: *an adverse effect to a human, animal or environment*

Hazard: *the intrinsic property of a substance to cause harm; the maximum toxic effect*

Exposure: *contact by various pathways (ingestion, inhalation, touching,...)*

Risk: *The probability that harm will occur under a specified set of exposure circumstances*

All above aspects of risk assessment are studied by the CDI and the cobalt industry for cobalt and its various substances.

TMS2013

Toxicity / Hazard

- **Asthma**
- **Allergic dermatitis**
- **Hard metal lung disease *(Co + tungsten carbide)***
- **Occupational lung cancer *(animal data)***
- **Thyroid hyperplasia *(goiter)***
- **Heart effects including cardiomyopathy**
- **Polycythemia *(> red blood cells)***
- **Reproductive toxicity *(animal data)***

TMS2013

Human Exposure

Exposure

- **Man-made sources:** *burning of fossil fuels, sewage sludge, phosphate fertilizers, mining/smelting of cobalt containing ores, industrial processing of cobalt substances and alloys, consumer products.*
- **Most significant source of exposure to cobalt for the general public is from food:** *Humans require vitamin B12 which contains cobalt as the central co-factor. Many animals and plants require small amounts of divalent cobalt directly for growth and vitality.*

TMS2013

Risk

Hazard + Exposure = Risk

- Cobalt has human health and environmental hazards
  - → ***these are communicated by the classification and appropriate labelling for each cobalt substance***
- Cobalt exposure does occur
  - → ***Exposure to each cobalt substance is described and documented in "Exposure Scenarios"***
- Based on both hazard and exposure,
  - → ***Risk is assessed under "Chemical Safety Assessment" and risks are controlled***

TMS2013

Risk Regulation

Regulation requires the process of:

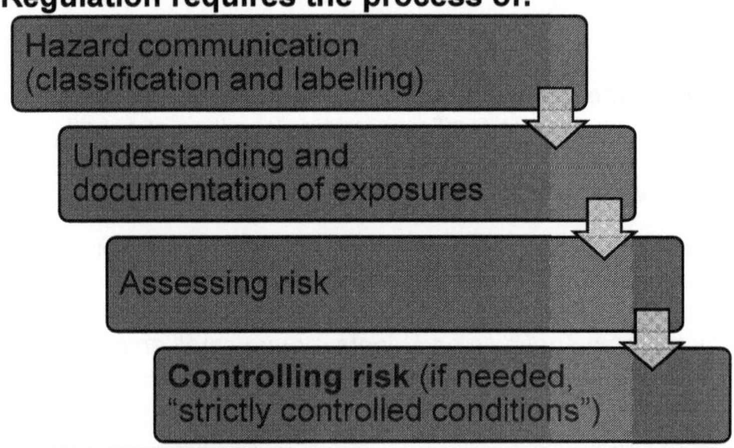

TMS2013

Relevant Regulations - Examples

European Union

- REACH – *Self classification (hazard) and risk assessment*
- Restriction of some uses of cobalt - *controlling/eliminating risk*
- Toys – *no CMR (carcinogenic, mutagenic, reproductive toxin)*
- Water Framework Directive

North America

- USA: EPA – *Safe Drinking Water Act, IRIS*
- *California: State Proposition 65 relating to 'safe' levels of substances (Co included!)*
- *Maine: Toxic Chemicals in Children's Products (Co included!)*

TMS2013

Relevant Regulations – Examples cont/-

North America

- Canada – *Domestic Substance List evaluating environmental and health effects. Also at Provincial level (Ontario)*

Globally Harmonised System (GHS):

- Harmonization of classification and labelling of chemicals (GHS), focus on intrinsic hazards
- Issue of "disharmony" between regions of the world
- Different regions implement different versions and different "building blocks" of GHS
- Language and translation issues, esp. Asia

TMS2013

Relevant Regulations – Examples cont/-

Asia, South America, other

- **"moving targets"**

TMS2013

Finding the Appropriate Balance

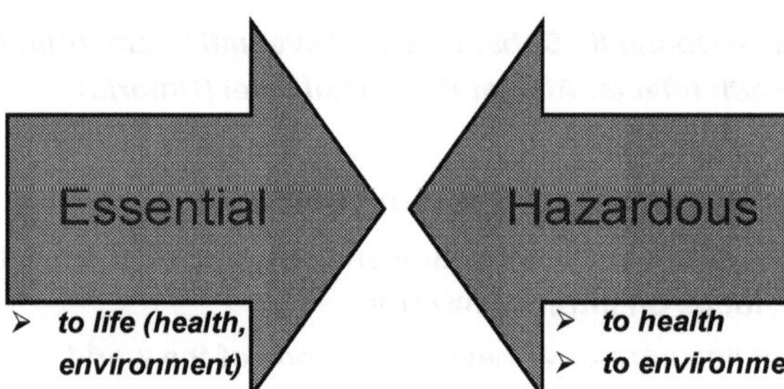

- ***to life (health, environment)***
- ***to industry***
- ***to sustainable development***

- ***to health***
- ***to environment***

TMS2013

Cobalt benefits and risks -

need to be communicated proactively by industry!

TMS2013

Challenges Going Forward

- Market Factors
- Hazards and Risks
- Regulation – market access
- Sustainability
- Strategic Issues

TMS2013

Sustainability – EU Perspective

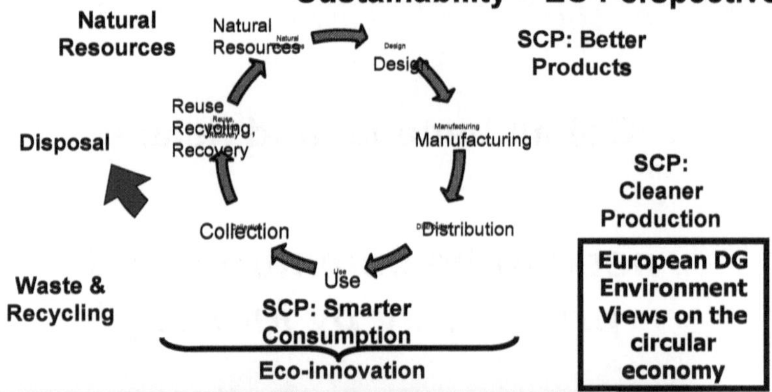

•Long-term trend: everything that will be produced and used in EU (and later also worldwide) will have to demonstrate acceptable 'sustainability footprint'!

SCP = Sustainable Consumption and Production

Source: Eurometaux

TMS2013

Sustainable Mining, Mineral & Metals

- **Innovative** – hence attracts talent
- **Resource efficient** – doing more with less
- **Energy efficient** – energy management + technology
- Eco-efficient Production – **Best Available Technology**
- Investing in the future – **high-tech driven/R&D**
- Products are safe, intelligent, recyclable & effectively recycled – the **'circular economy' ⟹ Sustainable Footprint!**
- The use of minerals and metals is **environmentally friendly**
- **Image** of mining, minerals and metals ⟹ <u>good</u>
- **Risks** associated with operations and products are fully under control

*Source Eurometaux/CDI

TMS2013

Challenges

- **Strategic/Critical Raw Materials**
  - EU with Raw material Initiative
  - UK with Business Risk Initiative
  - USA with Resource Risk Initiative
  - EU with Development of Company rules on Financial Transparency
  - France promoting exploration and recycling
- **Other Challenges**
  - Dodd-Frank *(Sledge hammer approach?)*
  - EICC/GeSI: Transparency in the supply chain

TMS2013

CDI Proposition

Representing Cobalt's Interests Globally

TMS2013

THANK YOU FOR YOUR ATTENTION

Contact Details:
David Weight – General Manager
The Cobalt Development Institute Ltd
167 High Street, Guildford, Surrey, GU1 3AJ, UK.
+44 (0) 1483 510471
david.weight@thecdi.com
Website: www.thecdi.com

TMS2013

Ni-Co 2013
Edited by: Thomas Battle, Michael Moats, Violina Cocalia, Harald Oosterhof, Shafiq Alam, Antoine Allanore, Rodney Jones, Nathan Stubina, Corby Anderson, and Shijie Wang
TMS (The Minerals, Metals & Materials Society), 2013

THE RECYCLING OF COBALT FROM ALLOY SCRAP, SPENT BATTERIES OR CATALYSTS AND METALLURGICAL RESIDUES – AN OVERVIEW

C. J. Ferron

HydroProc Consultants
1373 Wildlark Dr., Peterborough, Ontario, K9K 2J4, Canada

Keywords: Recycling cobalt, spent batteries, spent catalysts, cobalt slag.

Abstract

In 2011, the US Department of Energy declared cobalt a critical metal, one of 15 such metals deemed to be indispensable but in a risky supply situation in the near future for the production of clean energy.

Because of the uncertainty surrounding the supply of primary cobalt from Central Africa, it is more and more imperative to improve the recycling of cobalt from its various uses.

This paper reviews methods to recycle cobalt from the scraps of various alloys, from spent rechargeable batteries or spent catalysts and from metallurgical residues, presenting both actual and suggested processes.

Introduction

In 2011, the US Department of Energy/DOE published the Critical Materials Strategy, in which it examined the role of selected materials in the clean energy economy (1). It highlighted the importance of certain critical materials to wind turbines, electric vehicles, photovoltaic thin films and energy-efficient lighting. Cobalt for its use in batteries was one of the 15 metals examined. Cobalt was presented to approach the near-critical level both in supply risk and importance to clean energy, in the short (to 2015) and medium (2015-2025) terms. DOE's strategy to manage this crisis included improving extraction processes, substitution, and recycling/reusing materials.

Similarly, the European Union (EU) has listed cobalt as one of the 40 critical metals for future development (2). These facts clearly indicate that cobalt is badly needed for further world development.

What about the supply of cobalt? Irrespective of where the cobalt reserves are located, today and in the near future, a significant portion of the world primary cobalt is still produced from the Cu-Co deposits of Central Africa, the remainder originating from Ni laterites and Ni-Co sulphides and sulpho-arsenide ores (Morocco, Sudbury, Norils'k, Australia).

In 2005, the British Geological Survey published data that indicated that 45% of the world primary cobalt originated from the Shaba province, in the DRC (3). Moreover, no primary cobalt production exists in the USA, which is completely dependent on foreign imports for their cobalt usages. The announced additional supply of cobalt from the HPAL laterite leaching plants has not yet met with expectations. To minimize the risks of supply due to world competition and political instability, it is therefore essential to maximize the recycling of cobalt from metallurgical scrap generated during the fabrication of cobalt products, and from spent cobalt compounds. The term "recycling" should be extended to the recovery of cobalt containing metallurgical wastes to further lower reliance on external import.

This paper first gives a brief overview of the major cobalt uses, potential feeds to recycling operations; the paper then reviews recycling processes applied commercially and/or proposed for application. The paper will finally review potential or actual processes to "recycle" cobalt from metallurgical wastes.

Overview of Cobalt Uses

To identify areas of potential recycling applications, the easiest is to examine the various uses of cobalt. The amount of cobalt used in various applications obviously varies with time with new applications being developed and other becoming obsolete.

The following table presents a 2007 survey of cobalt demand by end uses (4).

Table I. Cobalt Demand by end use (2005)

End use	% Usage
Batteries	23
Recording Materials	6.5
Tire adhesives	10.3
Catalysts	10.2
Pigments	12.0
Magnets	10.0
Hardfacing + other alloys	7.0
Hard materials (Carbides)	15.0
Superalloys (Ni/Co/Fe)	26.0

Recycling of a material can be carried out at the source, for example, when an alloy is cast and machined, or at the EOL (end-of-life) of the compound (for example when the catalyst/battery is spent). Some end use of the metal does not lend itself to recycling because the cobalt in it is very dilute and/or very difficult to collect (Pigments). Other end uses appear more favorable to the collection and the recycling.

Examples of end uses (far from exhaustive) are presented below to provide a range of cobalt concentration, and also to indicate the accompanying metals that could add to the value of the metals to recycle, or add to the complexity of the recycling process.

<u>Cobalt as alloying element</u>

Overall, this is the most significant use of cobalt, as shown in Table I (48% total use). Metallic cobalt is an important alloying element in numerous applications, such as: super alloys (cobalt, nickel, iron based), magnetic alloys (soft, AlNiCo, SmCo), wear-resistant alloys, high speed steels, prosthetic alloys, and cemented carbide.

Table II (5) below presents an example of alloy compositions for a variety of cobalt-containing alloys.

Table II. Examples of Cobalt-Containing Alloy Compositions

Type/Name	% Element															
	Co	Ni	Fe	Al	Cr	Mn	Ta	Sm	Mo	W	Nb	Si	Ti	V	Cu	C
Ni-base Superalloys IN-100	15	60	0	5.5	10	0	0	0	3	0	0	0	4.7	0	0	
Co-base Superalloys HAYNES 188	39.2	22	3	0	22	0	0	0	0	14	0	0	0	0	0	
Fe-base Superalloys INCOLOY 903	15	38	41	0.7	0	0	0	0	0	0	3	0	1.4	0	0	
Soft Magnetic alloys Rotelloy 3	49	0	49	0	0	0	0	0	0	0	0	0	0	2	0	
AlNiCo Alloys Alcomax 3	24.5	13.5	50.4	8	0	0	0	0	0	0	√	0	0	0	√	
Sm.Co Alloys $SmCo_5$	66.2	0	0	0	0	0	0	33.8	0	0	0	0	0	0	0	
High-speed Steels M42	8	0	72	0	3.9	0	0	0	9.4	1.5	0	0	0	3.1	0	1.08
Prosthetic alloys Vitallium	62	0	0	0	30	0	0	0	5	0	0	0	0	0	0	
Wear-resistant alloys Stellite 314	40.1	6	5	0	32	0	0	0	0	6	6	1	0	0	2	1.9
Cemented Carbide (for cutting)	10.7	0	0	0	0	0	5.4	0	0	Base	1.6	0	6.7	0	0	base

Cobalt in Catalysts
Cobalt-containing catalysts (5) are used in the petroleum industry (hydroprocessing) and desulphurisation: COMOX (3-5% Co_3O_4, 14% MoO_3, remainder z- Al_2O_3), and in the chemical industry for OXO synthesis and the production of TPA (terephthalic acid) and DMT (di-methylterephthalate) where cobalt acetate or cobalt naphtenate/octoate are used. Finally, cobalt-containing catalysts are also used during the Fischer Tropsch synthesis of fuels, oils and waxes from natural gas (GTL gas-to-liquid conversion). These catalysts also contain non-trivial amounts of precious metals (Pt,Ru,Rh,...).

Cobalt in Batteries
Cobalt is used in various types of rechargeable batteries. In the Ni-Cd batteries, cobalt represents 1-5% of the weight of nickel hydroxide at the cathode. In the nickel metal hydride battery (Ni/MH), cobalt is moreover employed at the anode as a hydrogen storage alloy such as mischmetal (NiCoAlMn (5-15% Co). Finally, in the Li-ion batteries, the cathode could be $LiCoO_2$ where cobalt could represent 50% of the weight (6, 7).

Recycling of Cobalt

Generalities
Similarly to other recycling concepts, the success of cobalt recycling will depend on several factors such as the recycling process itself but also the concentration of cobalt in the material to be recycled and the ability to collect the material. Some cobalt-containing material has the cobalt in such a dispersed form that the recycling economics are not favourable (pigments). Other materials are difficult to collect or are lost (cobalt in animal feed), or the time to reach their end-of-life is very long.

Recycling of Cobalt from Alloys
Alloy "scrap" potentially recyclable is generated in various locations, at the casting shop, at the machining shop, or as EOL (scrap dealer). These are very localized areas, which makes the gathering of scrap for recycling easier.

A. Internal Recycle Or Recycle At The Source.The most convenient method of recycling cobalt (and other valuable constituents) from alloys is to re-melt the scrap directly (without significant treatment) where it is produced, at the caster (rejected pieces) or the machine shop (turnings), provided the composition can be controlled and no harmful elements have been added (polishing grit, cutting oils,…).
When simple, direct re-melting is not feasible, internal recycle can involve a more complex treatment process, such as for example the recycle of carbides, as shown in the following Figure 1.

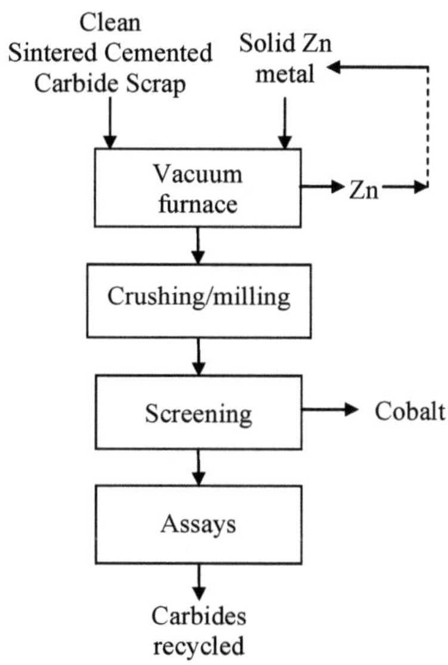

Figure 1. Simplified process to recycle cemented carbides

In such a process, zinc metal is added and melted with clean cemented carbide scrap; cobalt is dissolved in the molten zinc, zinc is distilled off, leaving a spongy cake easily crushed, milled and screened to separate the cobalt metal powder and the hard carbide (tungsten, tantalum and other carbides) grains.

B. Recycling Cobalt at a Custom Sulphide Smelter. Cobalt-containing alloy scrap have been for many years industrially recycled as custom feeds, for example at the Falconbridge (now Xstrata) smelter in Sudbury, Ontario, together with primary cobalt (and nickel) sulphide concentrates and other scrap (8). In fact, a significant proportion of Sudbury smelter cobalt production derives from cobalt originating from custom feeds, in particular alloy scrap.
There are significant advantages in recycling alloy scrap at a smelter like Sudbury:

- Economy of scale (large existing smelter)
- The scrap can be used as a heat sink
- Co-recovery of other valuable elements, such as nickel, precious metals, if present in the scrap
- Capability for large tonnages, metal separation and refining

In a smelter like Xstrata Sudbury, cobalt-containing custom feed could be added at various stages of the process to sulphide concentrates. A simplified process diagram is shown in Figure 2.

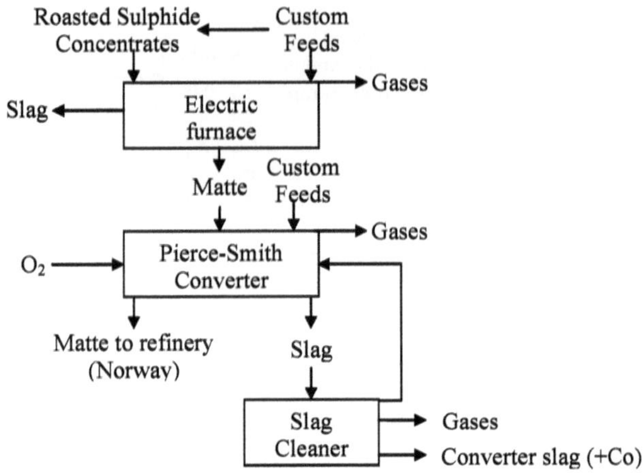

Figure 2. Simplified process diagram of Xstrata Sudbury smelter

Various types and shapes of cobalt (+ nickel) scrap (large pieces, grindings, turnings of superalloys, Ni-Co alloys (and also Li-ion batteries, catalysts) are processed.

In 2011, out of about 5000T of cobalt produced by Xstrata, about 75% originated from custom feeds (8).

Because of the large variety of scrap (types and compositions), a dedicated network is usually in place to collect/procure the custom feeds and great care must be taken not to introduce harmful impurities to the smelter. Other elements in the scrap easily oxidizable (Ta, W, Re,...) will typically be lost to the slag and will therefore not contribute to payment terms. More specifically, cobalt itself is relatively easy to oxidize, and in the process of converting to remove most of the iron, a significant portion of cobalt is also slagged off and would be lost if it were not for a slag cleaning stage to recycle some of the cobalt from the original converter slag.

Other sulphide smelters (OMG in Kokkola, Finland) are also equipped to take cobalt containing custom feeds.

C. Recycling Cobalt Alloy Scrap At A Custom Smelter Recycler. Various companies are concerned with "simply" re-melting alloys (i.e. no sulphides). Such a company, as an example, is Greenville Metals (9).

Greenville can process various shapes (solid scrap, spills, grindings, turnings, swarfs, sludges, oxides in the form of dust and pellets) and types of alloys, in particular, cobalt based alloy and miscellaneous Ni/Co reverts. Such a plant depends heavily on assaying and sorting the various types of compositions. The process used is not complex and involves melting (arc or induction furnaces) under non-oxidizing or reducing conditions, some refining and casting.

D. Recycling Cobalt Alloy Scrap By Hydrometallurgical Methods. Several companies recycle cobalt alloys (usually Ni-Co alloys) using hydrometallurgical techniques. These are typically smaller operations, intended at recovering and separating nickel and cobalt, and/or recovering valuable products otherwise lost in a smelter (such as Ta, W, Re,...). Depending on the complexity of the alloy composition, the process used can become increasingly complex. Figure 3 presents the simplest case, while Figure 4 is applied for the recovery of valuable by-products.

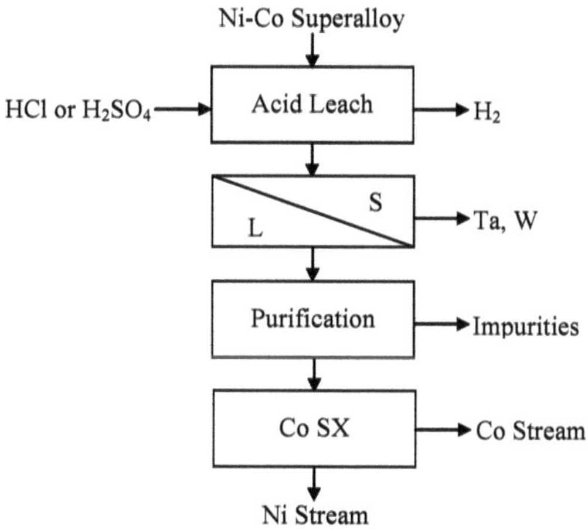

Figure 3. Recovery of cobalt and valuable by-products (Ni, Ta, W) using hydrometallurgical techniques

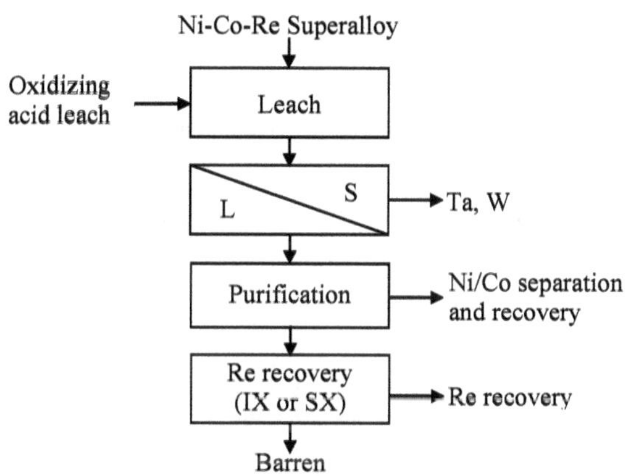

Figure 4. Recovery of cobalt and valuable by-products (Ni, Ta, W, Re) using hydrometallurgical techniques (10)

A lot of research was carried out by the USBM on the recycling of superalloy scrap. Some of the routes they investigated relied on electrochemical techniques to separate the components of super alloy scrap, as illustrated in Figure 5.

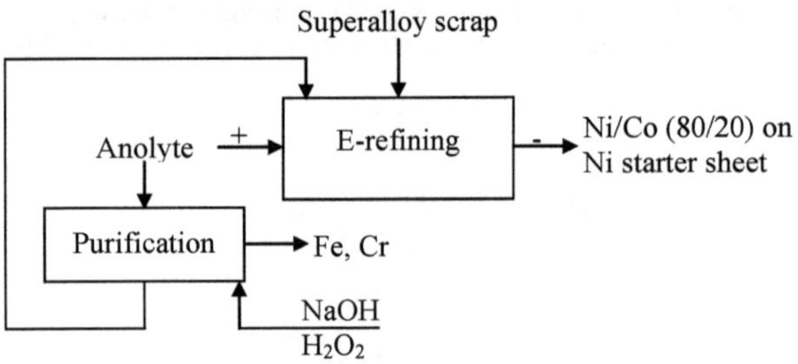

Figure 5. Simplified electro-dissolution process to recover nickel and cobalt from superalloy scrap (11)

A significantly more complex process making use of a novel double membrane electrochemical cell (DMEC) was also proposed by the USBM in 1990 (12) to recover separate cobalt and nickel metal from the superalloy scrap.

- Very many other processes have been developed to recycle valuable constituents from superalloys.

Recycling Cobalt From Spent Catalysts

A. Recycling Cobalt Catalysts to Custom Smelter. Most of the large custom smelters are equipped to recycle cobalt containing catalysts: Ni-Co catalysts are processed at nickel-cobalt custom smelters such as Sudbury (Xstrata) and Kokkola (OMG), while catalysts containing high values of precious metals are recycled at precious metals custom smelters such as Umicore's.

B. Recycling Cobalt Catalysts By Regeneration. Some petroleum catalysts can be regenerated by burning off the carbon and sulphur deposited on the catalyst during operation that inhibit the catalytic activity and reduce the yield of hydrocarbon product manufactured. After regeneration, the catalyst recovers 70-80% of its original capacity and can be used more than once. Regeneration can be carried out in-situ or (preferably) ex-situ by companies such as Tricat Inc., Eurecat, Porocel (in North America). Moreover, the React™ catalyst rejuvenation technology was developed by Akzo Nobel to restore Co/Mo and Ni/Mo "stars" or "Type II" catalysts to >95% of their original activity (13).

C. Recycling Cobalt Catalysts On Their Own. Methods to treat spent catalysts are known. One such method developed by Gulf Chemical and Metallurgical Corporation has been described to recover Mo, V, Ni, Co and Al_2O_3 from spent catalysts (14).
The simplified process flowsheet is illustrated in Figure 6.

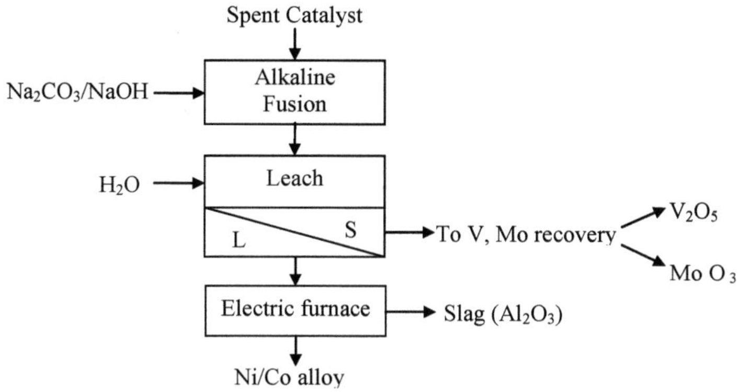

Figure 6. Simplified process flow sheet to treat spent catalysts

Simple hydrometallurgical processes can also be used to treat specific cobalt catalysts. Figure 7 below illustrates such a simple process developed to treat Co-Mn catalysts.

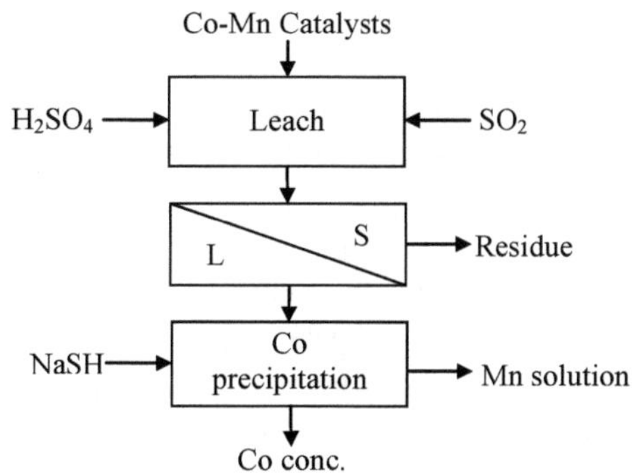

Figure 7. Simplified process flow sheet to recover cobalt from a cobalt catalyst

Using a relatively simple process, greater than 95% of the cobalt in the catalyst was dissolved and most of it precipitated as a high grade cobalt sulphide concentrate for further treatment.

Recycling Cobalt From Spent Batteries
Here again, depending on the type of cobalt-containing battery, various options are available to recycle the cobalt within.

A. Recycling cobalt from batteries to custom smelters. Large Ni-Co custom smelters are usually equipped to recycle also cobalt (and nickel) from batteries. As an example, Xstrata Nickel

Sudbury smelter has been, since 2006, processing Li-ion manufacturing scrap (OEM, 15-50% Co) and spent Li-ion batteries (EOL, 12-17% Co). (15).

B. Recycling cobalt from batteries in dedicated plants. Umicore, through its Umicore Battery Recycling (UBR) division, has recently installed a pilot plant capable of treating 7000 tonnes per year of batteries (Ni-MH and Li-ion) (16). The plant makes use of the patented ultra high temperature process (>3000°C plasma torch UHT), as illustrated in Figure 8 below.

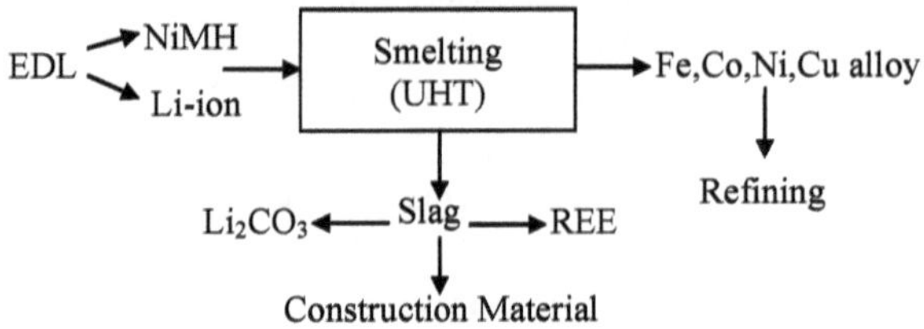

Figure 8. Simplified process flow sheet of Umicore UBR process

During the high temperature melting, all the base metals are produced as an alloy that could be refined and separated into individual constituents. Elements like Al, Ca and Li are slagged off. If the economics allow it, lithium can be recovered from the slag as lithium carbonate (Val'Eas process), and if Ni-MH batteries are processed, the REE in the slag could be recovered and processed into a concentrate.

TOXCO Inc., in their Trail, B.C. Canada recycling plant operates an unique hydrometallurgical process for rechargeable Li-ion batteries (and others). Their process is illustrated in Figure 9 below (17).

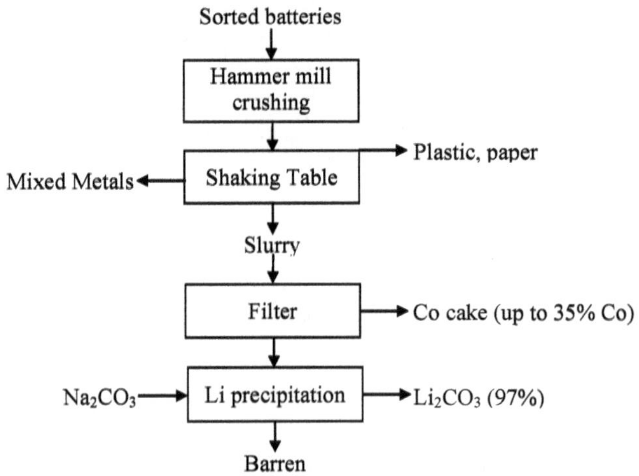

Figure 9. Simplified process flowsheet to recycle Co from Li-ion battery

Other fully hydrometallurgical processes have been developed to treat Ni-MH (Figure 10) or Ni-MH plus Ni-Cd batteries (Figure 11).

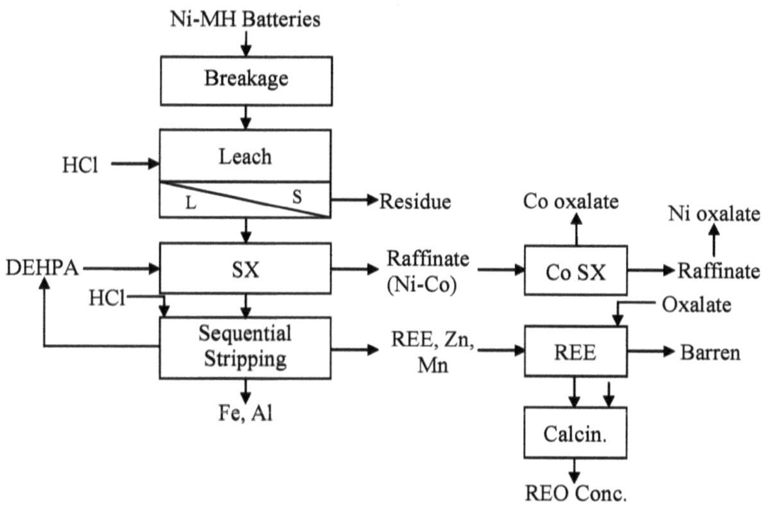

Figure 10. Simplified process flow sheet to recycle cobalt from Ni-MH batteries (18).

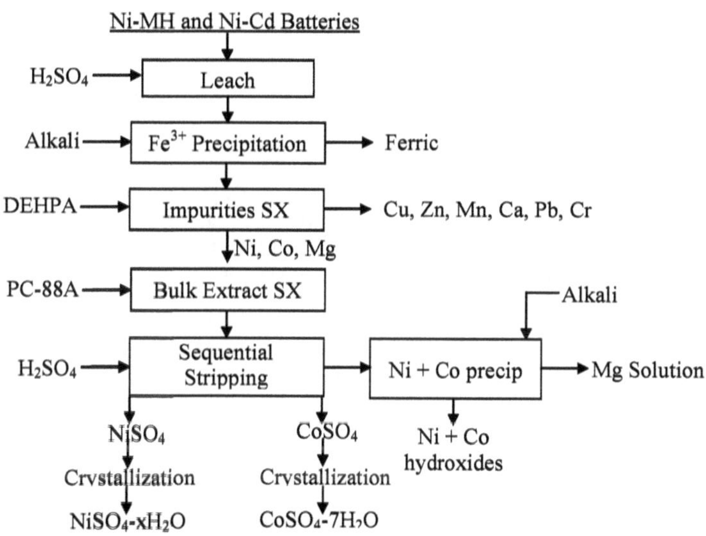

Figure 11. Simplified process flow sheet to recover cobalt from Ni-MH and Ni-Cd batteries (19)

Recent research has also been focusing on electrochemical recovery of cobalt from spent Li-ion cathodes. Researchers in Italy (20) and Brazil (21) investigated the recycling of cobalt from the cathodic material of Li-ion spent batteries, where the final cobalt recovery step relied on electrochemical techniques. In one example, the cathodic material (also containing nickel) was dissolved in acid, cobalt and nickel separated by solvent extraction, and the cobalt electrowon at 250 A/M^2, pH ~ 4.1 and 50°C. In the other example, cathodic $LiCoO_2$ was acid leached, and the metallic cobalt deposited directly on 430 steel. The deposit was then oxidized to Co_3O_4 in air at 850°C.

Recycling Cobalt From Metallurgical Wastes and Residues

If we extend the meaning of the term "recycling" to include also the cobalt losses generated (due to inefficiency of some sort) during the metallurgical operations of base metals (copper, cobalt, nickel, zinc), the potential to recover large quantities of cobalt increases substantially. Potentials exist where cobalt ores are produced, smelted or where cobalt is removed as impurity or used as reagent.

Recycling Cobalt Lost During the Mineral Processing of Cobalt Ores – During Flotation

Oxidized cobalt (together with oxidized copper) is processed by flotation in Central Africa, mostly in DRC. For various reasons, the recovery of cobalt during flotation lags that of copper by several tens of percentage points, and the lost cobalt ends up in the flotation tailings. Over the years, several attempts/investigations have been made to improve cobalt recovery during the flotation of oxidized cobalt, and in particular, hydroxamate collectors had been shown to be promising (22), but were never implemented (in DRC). Recently, hydroxamates have again been suggested for the flotation of oxidized copper (23).

Recycling Cobalt Lost During the Mineral Processing of Cobalt ores – Historic Flotation Tailings

As mentioned above, flotation tailings generated over the years in DRC during the processing of cobalt oxide ores contain high tenors of cobalt. During the last decade, companies have attempted to recover the values (cobalt and copper) from those flotation tailings in DRC, usually by hydrometallurgical techniques. Examples of projects dealing with flotation tailings are the Kakanda tailings and the Kolwezi tailings, from which the original plans would have extracted large quantities of cobalt (600T Co/yr at Kakanda, and 3000 T/y at Kolwezi). Both projects appeared technically successful, both piloted, and the Kolwezi tailings projects reached the feasibility (Adastra) and construction (First Quantum) stages.

The project undertaken by Reminex at their Bou Azzer operation in Morocco met with industrial success: in 1997, a plant was started to re-process flotation tailings assaying 0.4% Co (24). The process flow sheet is illustrated in Figure 12 below.

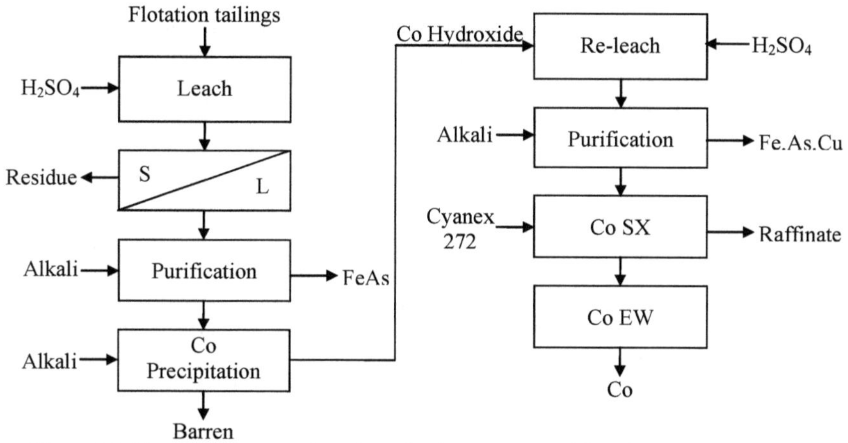

Figure 12. Simplified process flow sheet for the treatment of flotation tailings at Bou Azzer

<u>Recycling Cobalt from Slags Generated During Smelting of Cu-Co Ores in Central Africa</u>

As briefly discussed earlier, cobalt losses to the slag during oxidizing smelting/converting are high. Over the years, large slag dumps have been accumulated in the DRC (Lubumbashi) and Zambia (Rkona), that contain large quantities of cobalt. OMG have operated since 1999 in Lubumbashi a plant to treat slags from historical Gecamines waste dump ("The Big Hill"). It involved reducing smelting of the slag in an electric arc furnace to generate the white alloy containing copper and cobalt, this white alloy being then processed further in the company Kokkola refinery in Finland.

The process used in Zambia (Chambishi) is illustrated (25) in Figure 13 below.

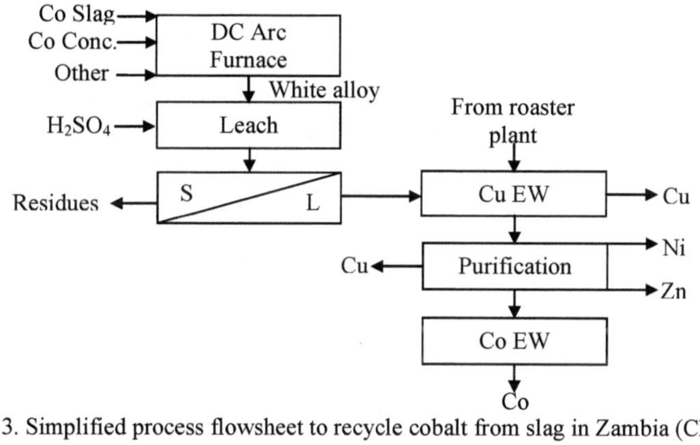

Figure 13. Simplified process flowsheet to recycle cobalt from slag in Zambia (Chambishi)

In 2001, the plant to re-process the slag dump was commissioned at Kitwe. The slag, assaying 0.7% Co, was melted under reducing conditions (with other feed materials) in a DC electric arc furnace to produce the cobalt white alloy. After granulation, the white alloy was leached with sulphuric acid and then blended with other streams to be directed to the copper electrowinning plant where cobalt was recovered as metal from a bleed stream using traditional techniques (No SX).

Recycling Cobalt From Slags Generated During Smelting of Ni-Co Sulphide Ores
In the last decade, work has been undertaken at the University of Toronto to examine the feasibility of recovering values (i.e. nickel, cobalt, copper) from Vale (ex Inco) Sudbury smelter slag (26, 27), in particular the converter slag assaying 1.07% Ni, 0.68% Cu and 0.67% Co, and the EAF slag assaying 0.23% Ni, 0.27% Cu and 0.12% Co. Results have indicated that a high temperature (250°C) acidic (40-60 g/L H_2SO_4) process with oxygen overpressure could dissolve greater than 95% of the values (including cobalt) in 45 minutes, provided the slag had been slowly cooled. To follow on Xstrata Sudbury lead, slag cleaning technology is being installed on Ni-Co sulphide smelters, and that will significantly improve cobalt recovery.

Recycling of Cobalt From Waste Streams of the Copper Industry
Interesting cobalt values are present in waste solution form copper hydrometallurgical or electrometallurgical industries. Cobalt values in the 12-20 ppm range (shooting sometimes as high as 80-100 ppm) have been measured in Arizona copper leaching operations. Moreover, bleed streams in copper electrowinning plants range from 50 to 150 ppm cobalt, when cobalt is added to prevent excessive corrosion of the Pb-Ca-Sn anodes, likely by reducing the anodic over-potential for oxygen evolution (28). From these solutions containing dilute cobalt with varying amounts of copper, nickel, iron, aluminum, nickel, processes have been developed that rely on combinations of ion exchange and solvent extraction. For example, the USBM has developed a process that can be illustrated (29) as in Figure 14.

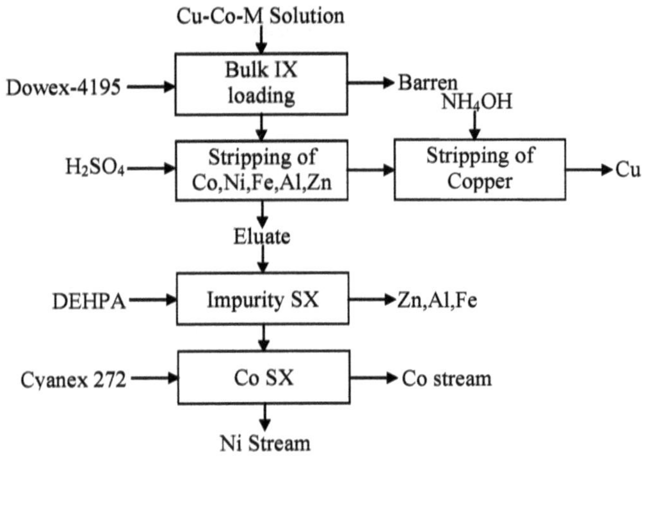

Figure 14. Simplified process flow sheet to recover cobalt from copper waste streams

The economics of such a process must be examined but they were favourable when the process was developed, and large volumes are involved.

Recently a plant was commissioned in China to recycle cobalt from a low grade waste solution from a copper open pit mine in Dexing (Jiangxi Copper) consisting of a blend of ARD, raffinate, heap leach drainage and ground water assaying typically ~150 mg/L Cu, 900 mg/L Fe^{3+}, 390 mg/L Fe, 6 mg/L Ni, and 7.5 mg/L Co. BioTeq Environmental Technologies installed from 2008 to 2011 a series of 3 water treatment plants, as illustrated below in Figure 15 (30).

After neutralization and copper precipitation as a sulphide, Ni+Co are extracted by ion exchange and precipitation as a 10% Ni, 10% Co carbonate for sale. The final effluent is further treated in a high-density-sludge (HDS) plant. Forecast annual metal recovery amounts to 33 tonnes of cobalt.

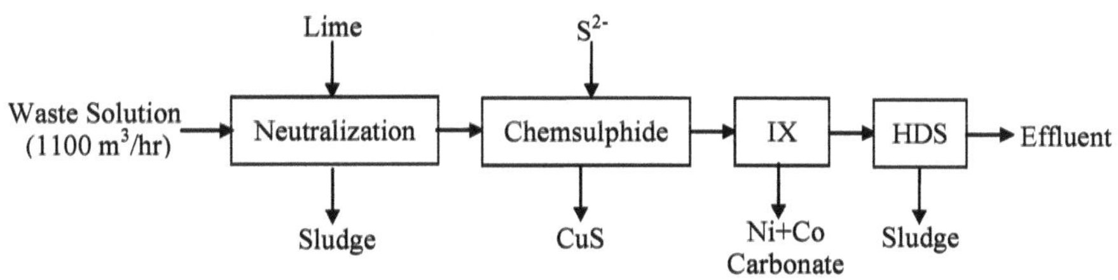

Figure 15. Simplified process flowsheet to recycle cobalt from a Cu open pit/heap leach operation in China

<u>Recycling Cobalt From Zinc Smelter Waste Streams</u>

Prior to electrowinning zinc, all zinc plants must purify the electrolyte, and in particular cobalt has to be removed, usually by cementation with zinc powder. Such a cement can be re-processed to recover its metal values, in particular cobalt. Hydrometal S.A. a subsidiary of Jean Goldschmidt Int. S.A. (JGI) processes such cobalt-containing cements at its plant in Engis, Belgium (31). Cements from various sources are treated, with typical content of 15-20% Cu, 2-5% Ni, 20-30% Zn and 5-7% Co. A simplified process flowsheet is presented in Figure 16.

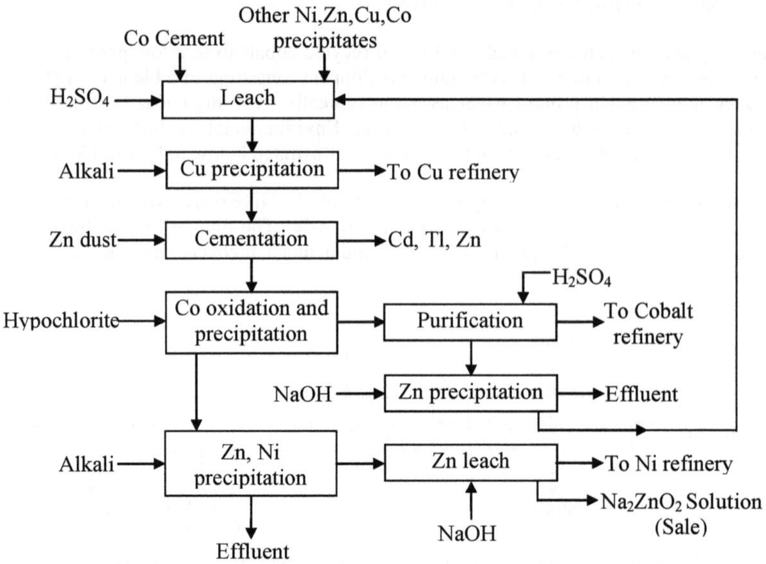

Figure 16. Simplified Process Flowsheet to Recovery Cobalt From Zinc Plant Cements

Typical cobalt concentrates sold to cobalt refineries assay >65% Co, 3% Zn, 3% Fe, 3.8% Mn, <0.15% Cd, <1.5% Cu, <0.45% Ni (dry basis).

In the cases where cobalt has been removed from the electrolyte using napthol derivatives (32), a process has been suggested to recover cobalt from the precipitate by sulphation roasting it, followed by water leaching the calcine, precipitating out iron and manganese, removing other impurities (Zn Cd, Cu) by ion adsorption, and solvent extracting the cobalt with P507.

<u>Recycling Cobalt From Nickel Refinery Sludges Residues</u>

A process has been developed to recover cobalt from stockpiled sludges and residues from a large nickel refinery (33). The feed material assayed 41.8% Cu, 5.3% Ni, and 1.48% Co. The process consisted in acid leaching, followed by Cu electrowinning (99.9% Cu cathode), and carbonate precipitation of Ni + Co from the spent electrolyte to a product assaying 31% Ni, 13.6% Co and 0.49% Cu.

Conclusions

Although it is by no means exhaustive, this paper clearly indicates that the recycling of cobalt from various manufacturing and end-of-life products is technically feasible and already practiced widely. Further extension of this recycling and additional recovery of cobalt from metallurgical wastes would further lower our dependence on primary cobalt. Expansion of cobalt recycling to other areas is usually not constrained by technical issues but by economics (collection) and the will to do so.

References

1. U.S. Department of Energy "2011 Critical Materials Strategy - Summary".

2. D. Van Der Linde, "Recycling Cobalt-Sustainable and Affordable" (Cobalt Investing News, June 22, 2011).

3. British Geological Survey (http://www.bgs.uc.uk/mineralsdownloads/african_mp_01_05.pdf)

4. J.P.T. Kapusta, "Cobalt Production and Markets: A Brief Overview" *Cobalt News, 2007, Vol. I*, pp 9-13).

5. Cobalt Facts, Cobalt Development Institute (CDI), Guilford England, 1992

6. L.A. Dominey, "Cobalt (and) Nickel in Advanced Rechargeable Batteries. The 1997 Commercial and Technical Update" (Paper presented at the Cobalt Conference CDI, Hong Kong, 1997).

7. G. Whiting, J. Currie, B. Dewar, "Batteries and Cobalt" (New Cobalt Facts, 1995, Cobalt Development Institute (CDI).

8. Xstrata Nickel, B. Gutauskas Communication to Colorado School of Mines, July 2012

9. N. Wilson, "Greenville Metals- A unique resource for metallic recycling", *Recycling Metals from Industrial Waste,* Colorado School of Mines, Golden, June 2009

10. C.J. Ferron, L.E. Seeley, " Rhenium Recovery" PCT Application PCT/CA/2010/000333, Sept 16, 2010

11. L.J. Lutz, S.A. Parker, J.B. Stephenson, " Recycling of Contaminated Superalloy Scrap via Electrochemical Processes" in The Paul E. Queneau International Symposium, Extractive Metallurgy of Copper, Nickel and Cobalt, Reddy and Weizenbach Ed., Vol 1, TMS, Denver, 1993, 1211-1220.

12. S. Wang, "Cobalt-Its Recovery, Recycling and Application," *Journal of Metals,* Vol 58, # 10, October 2006, 47-50,

13. J.P. Rosso, "TRICAT Triage: Innovative Management of Spent HDS Catalyst," *Recycling Metals from Industrial Waste,* Colorado School of Mines, Golden, June 2009.

14. Z.R. Llanos, G.F. Provoost, W.G. Deering, F.J. Debaene, "Integrated Process for the Recovery of Metals and fused Alumina from Spent Catalysts," US Patent 5,702,500, Dec. 30, 1997.

15. Cobalt News, "The Recycling of Li-ion Batteries and GTL Catalysts," (Cobalt News, 2007, Vol. 1), 14-15.

16. Umicore Battery Recycling (http://www.batteryrecycling.umicore.com)

17. S. Thompson,"Large Format EV Battery Recycling Challenges and Opportunities," *Recycling Metals from Industrial Waste, Colorado School of Mines, Golden,* June 2011.

18. P. Zhang et al, "Hydrometallurgical Process for Recovery of Metal Values from Spent Nickel-Metal Hydride Secondary Batteries," *Hydrometallurgy,* 50, 1998, 61-75.

19. F. Wu et al., "Separation of Nickel, Cobalt and Magnesium from the Spent MH-Ni and Cd-Ni Battery anode Materials," (Presented at EPD Congress 2009), 1119-1125.

20. C. Lupi, M. Pasquali, A. Dell'Era, "Nickel and Cobalt Recycling from Lithium-ion-batteries by Electrochemical Processes," *Waste Management,* 2005, 25(2), 215-220.

21. E.M. Garcia et al, "Electrochemical Recycling of Cobalt from Spent Cathodes of Lithium-ion Batteries: Its Application as Coating on SOFC Interconnects," *Journal of Applied Electrochemistry, Vol. 41, # 11,* 2011, 1373-1379.

22. T. Shungu, N. Vermaut, C.J. Ferron, "Recent Trends in the Gecamines Copper-Cobalt Flotation Plants," *Minerals and Metallurgical Processing,* (August, 1988), 163-170.

23. G.A. Hope, A. Numprasanthai, A.N. Buckley, G.K. Parker G. Sheldon, "Bench-scale flotation of chrysocolla with n-octanohydroxamate" Minerals Engineering, 36-38 (2012), 12-20

24. I. Akalay,"Moroccan Cobalt", in The Cobalt Conference, CDI, Toronto, May 2001

25. R. T. Jones et al.," Recovery of Cobalt from Slag in a DC Arc Furnace at Chambishi, Zambia", in The Journal of The South African Institute of Mining and Metallurgy, Vol 102, # 1, 2002, 5-9.

26. W. Curlook, V.G. Papangelakis, "Pressure and Leaching of Non-Ferrous Smelter Slags for the Recovery of their Base Metal Values," *Pressure Hydrometallurgy 2004,* Collins, Papangelakis Ed., CIM, Montreal 2004, 823-837.

27. I. Perederly, V.G. Papangelakis, I. Mihaylov, "Nickel Smelter Slag Microstructure and Its Effect on Slag Leachability," *T.T. Chen Symposium, Wang, Dutrizac, Free, HWang, Kun,* Ed., TMS, Orlando, 2012, 225-237.

28. K.C. Sole, J.B. Hiskey, "The Potential for Recovery of Cobalt from Copper Leach Solutions," *Copper 91 - Cobre 91, Vol III, Hydrometalluragy and Electrometallurgy of Copper,* Cooper, Kemp, Lagos, Tan. Ed., Pergamon Press, CIM, Ottawa 1991, 229-243

29. T.H. Jeffers, M.R. Harvey, "Cobalt Recovery from Copper Leach solutions," USBM, RI#8927, 1985.

30. D. Kratochvil, " Plant Commissioning in China for Ni-Co Recovery from Low Grade Waste Solution" in *Recycling Metals from Industrial Waste, Colorado School of Mines, Golden,* June 2011.

31. A. de Kandelaer, P. Henry, "Recycling of Non-Ferrous Metals at Hydrometal, Belgium," (Paper presented at European Metallurgical Conference, EMC 2007).

32. Y. Wang, C. Zhou," Hydrometallurgical Process for the Recovery of Cobalt from Zinc Plant Residues", Hydrometallurgy, 63, 2002, 225-234

33. T. Xue, A Singhal, I. Mihaylov, D. Wrana, " Treatment of Thompson Copper/Arsenic Residue" in Pressure Hydrometallurgy 2012, 42nd Annual Hydrometallurgy Meeting, COM 2012, Collins, Filippou, Harlamovs Ed, METSOC, Niagara Falls, 199-210.

Ni-Co 2013

Ni Laterite Hydrometallurgy

Session Chair

Shafiq Alam

Ni-Co 2013
Edited by: Thomas Battle, Michael Moats, Violina Cocalia, Harald Oosterhof, Shafiq Alam, Antoine Allanore, Rodney Jones, Nathan Stubina, Corby Anderson, and Shijie Wang
TMS (The Minerals, Metals & Materials Society), 2013

ACID LEACHING OF NICKEL LATERITES WITH JAROSITE PRECIPITATION

David T White[1], James D Gillaspie[2]

[1]45 Mudlark Way, Perth, WA 6164, Australia

[2]Freeport-McMoran Copper & Gold, 333 N. Central Avenue, Phoenix, AZ 85004-3014

Keywords: Nickel Laterite, Atmospheric Leaching, Jarosite Precipitation, Pressure Acid Leaching

Abstract

Alkali addition for iron precipitation, associated with sulfuric acid leaching of nickel laterites, has been considered periodically since the late 1960s, shortly after the development of the jarosite process for iron control in zinc refining.

In the late 1990's a series of flowsheets were proposed and tested involving simultaneous atmospheric leaching, iron precipitation and acid regeneration. These were considered particularly suited to sequential leaching of limonite and saprolite ores. A number of laterite sulfuric acid leaching processes utilising this chemistry have been put into commercial operation.

Introduction

By its nature this topic is somewhat of a summary of the development of nickel laterite acid leaching with jarosite and alunite precipitation, as iron and to some extent aluminum are significant acid consumers and can be problematic for downstream metal recovery.

Therefore this paper will firstly describe some of the important historical developments, then provide descriptions of the commercialized processes involving alkali jarosite/alunite precipitation, including discussion of the downstream impurity removal requirements necessary for efficient recovery of nickel and cobalt from solution.

History

Early History

The history of sulfuric acid leaching of nickel laterite ores dates back a century. A patent specification dated 1913 (granted 1914), in the name of Madagascar Minerals Syndicate discusses leaching of low iron, high magnesium "garnierite" ore, followed by solid-liquid separation and precipitation of a nickel-cobalt intermediate with magnesium hydroxide (or oxide) [1]. A further specification in 1914 (granted 1916) by the same group included sulfide precipitation using sodium sulfide (Na_2S), with subsequent recycling of the barren solution [2]. Given the leaching temperature of greater than 90°C, it is likely some jarosite precipitation would have occurred if this process was ever operated as specified.

Freeport Moa Bay

Freeport considered atmospheric leaching for the Moa Bay project, but noted that iron dissolution was high [3]:

"Under normal laboratory conditions, the great bulk of this ore is soluble in mineral acids. No selectivity is apparent in the rates of solution of the various components up to the normal boiling point."

They therefore took advantage of the iron hydrolysis – precipitation and acid regeneration process that occurs con-currently with leaching at elevated temperatures in pressure acid leaching (PAL). It was noted that PAL residues contained significant quantities of sulfate, consistent with precipitation of hydronium alunite, $H_3O.Al_3(SO_4)_2(OH)_6$.

Jarosite Precipitation

Development of the Jarosite process in the zinc industry in the 1960s led to an understanding that iron could be precipitated from nickel laterite atmospheric leach solutions, at moderate acid levels, by the addition of an alkali such as potassium, sodium, or ammonia, whilst maintaining a temperature in excess of 90°C [4].

Work in the late 1960s led to a number of patents being issued concerning the sulfuric acid leaching of nickel laterites with addition of an alkali to precipitate iron [5, 6, 7]. The abstract of one of the patents summarizes the process intent well:

"A process for the leaching of nickel bearing laterites, beneficiated laterites or other nickel bearing products wherein the leach is conducted at a temperature below the atmospheric boiling point of the pulp and nickel and cobalt are effectively dissolved while iron is prevented from remaining in solution through the use of precipitating agents capable of introducing alkali metal ion or ammonia ion to the pulp."

Ores tested were predominantly limonitic in character, although some contained in excess of 2% magnesium. Leaching and precipitation reactions took place over an extended period of up to 70 hours from addition of acid. Nickel extractions were in excess of 80%, at acid addition rates of around 500 kg/t. Terminal iron solution concentrations of less than 0.5 g/L were typically achieved. Both sodium and potassium additions were made to promote iron precipitation. The potential for using seawater to slurry the ore was noted. Potassium, in the form of a carbonate was also added on occasion to raise the solution pH to around 1.5 to 2.

Jarosite precipitation was further discussed in several papers in the following two decades. Canterford noted in 1975 that [8]:

"Processes requiring sulphation require the use of aqueous or concentrated sulphuric acid ... as the sulphating agent. In many cases the leach liquors produced are highly contaminated and recovery procedures play a vital part. Fortunately, it has recently become easier to remove dissolved iron using either the jarosite process or the goethite process."

Researchers at the National Technical University of Athens published a number of papers in the 1980s and early 1990s, where jarosite precipitation of iron from nickel laterite acid leach solutions was considered [9, 10, 11, 12].

In particular, their papers presented at the 1988 TMS Extractive Metallurgy of Nickel and Cobalt Conference complement each other, with leach solutions described in one paper [10] treated to precipitate iron as jarosite as discussed in the second [11]. They also make reference to the various techniques utilized for iron precipitation in the zinc industry and options considered for their own studies:

"Iron removal by precipitation has been extensively studied for the refining of solutions in the zinc industry, where the techniques of jarosite, goethite and hematite precipitation have found extensive industrial application.

"In our previous papers iron removal from sulphate solutions has been studied by precipitation as goethite and by solvent extraction in D2EHPA solutions. In the present paper, jarosite precipitation has been studied."

<u>AMAX Pressure Acid Leaching</u>

AMAX researchers added an alkali to nickel laterite PAL slurry, primarily to improve settling properties [13]. However it was noted that iron and aluminum precipitation occurs:

"...the alkali metal and ammonium cations react with solubilized iron and aluminum to form highly insoluble species of jarosites and alunites..."

AMAX separately considered atmospheric leaching of saprolite ore, after PAL of the limonite ore, with the aim of fully utilizing the laterite ore body in a single flowsheet [14]. One variant, as shown in Figure 1 consisted of leaching raw saprolite ore in PAL discharge liquor, followed by a second leach of roasted saprolite, with both atmospheric leach residues being subjected to PAL. It is apparent that this flowsheet serves mainly to utilize (and neutralize) the free acid remaining after PAL to dissolve magnesium and associated nickel from saprolite, with the partially leached saprolite further treated by PAL to recovery the remaining nickel values.

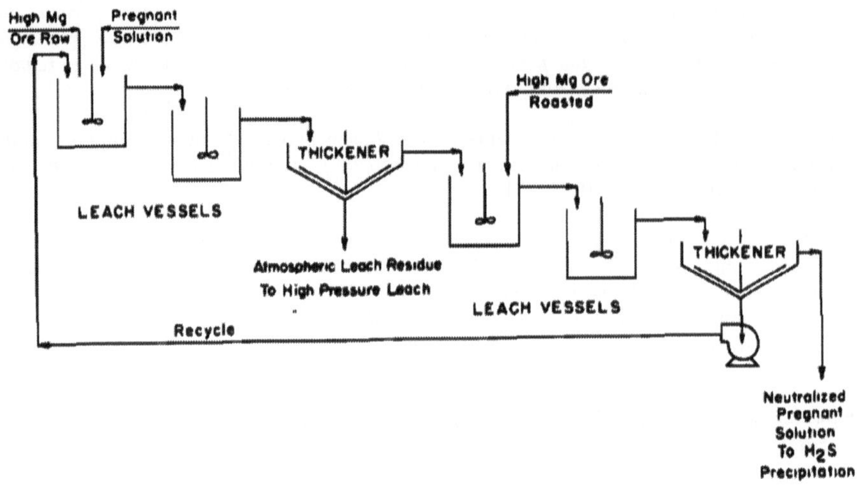

Figure 1. AMAX Saprolite Atmospheric Leach Flowsheet [14].

Australian Nickel Laterite Pressure Acid Leaching Operations

In the late 1990s, the Cawse and Bulong nickel laterite PAL operations utilized saline waters, resulting in similar chemistry to that of the AMAX alkali addition work. The Murrin Murrin operation subsequently added ammonium sulfate to achieve a similar result. A number of laboratory studies by the CSIRO identified that the precipitation products are mixed alunite-jarosite compounds, with the iron progressively converting from jarosite to hematite as the leach progresses [15, 16].

BHP Minerals

In the 1990s, BHP Minerals developed a number of hydrometallurgical technologies with the potential to improve on perceived shortcomings with PAL, to support development of the Gag Island nickel laterite project. The aims included being:

1. Cheaper to build than a PAL facility;
2. Simpler to operate than PAL in a remote environment;
3. Able to process saprolite as well as limonite ore.

Amongst the options considered were heap leaching [17], agitated tank atmospheric leaching [18, 19, 20, 21] and combinations of PAL and atmospheric leaching [22, 23, 24].

It is apparent from reading papers published on the various leaching technologies [20, 21], that precipitation of iron with regeneration of (at least some of the) acid during the leach was a desirable goal, for both chemistry control and minimizing acid consumption, emulating the same technique inherent in PAL.

In mid-2001, BHP merged with Billiton, thereby making the BHP Minerals' technologies available for the Ravensthorpe nickel project. There was potential to use jarosite atmospheric leaching technology because Ravensthorpe was already using seawater for its process and had substantial saprolite as well as limonite resources.

Major Process Options Developed

Whilst BHP Billiton developed a range of technologies in the area of nickel laterite acid leaching, two technologies stand out in the area of atmospheric leaching with jarosite precipitation. These are:

1. Sequential atmospheric leaching of limonite followed by leaching of saprolite with simultaneous jarosite precipitation and acid regeneration, also known as Jarosite Atmospheric Leaching (JAL) [18, 19, 20].
2. Pressure acid leaching of limonite, followed by atmospheric leaching of saprolite, which subsequently became known as Enhanced pressure acid leaching, or EPAL® [22, 23, 24].

Although outside the scope of this paper, it is worth noting that a sequential Goethite Atmospheric Leach process was also developed and has been described in detail [21].

The Laterite Profile

An idealized nickel laterite profile is shown in Figure 2. The areas containing commercial nickel values are usually the goethite section of the limonite zone and the clay and saprolite zones. As indicated by its name, the goethite dominant section is high in iron and low in magnesium. Iron content can be in excess of 35%, with magnesium as low as 1%. Conversely, the saprolite zone ores are low in iron and high in magnesium. Iron grades of less than 10% and magnesium grades in excess of 15% are typical.

It is inevitable that in practice some cross contamination will occur in mining, so therefore a broader definition would apply to each ore type for the purpose of hydrometallurgical processing. Thus ores of limonitic character could be defined by having an iron content of greater than 25% and a magnesium content of less than 6%. Those of a saprolitic nature would have an iron content less than 20% and a magnesium content greater than 8%. The relative ratio of the limonitic and saprolitic ore types is variable among deposits and will vary as mining progresses through any given deposit.

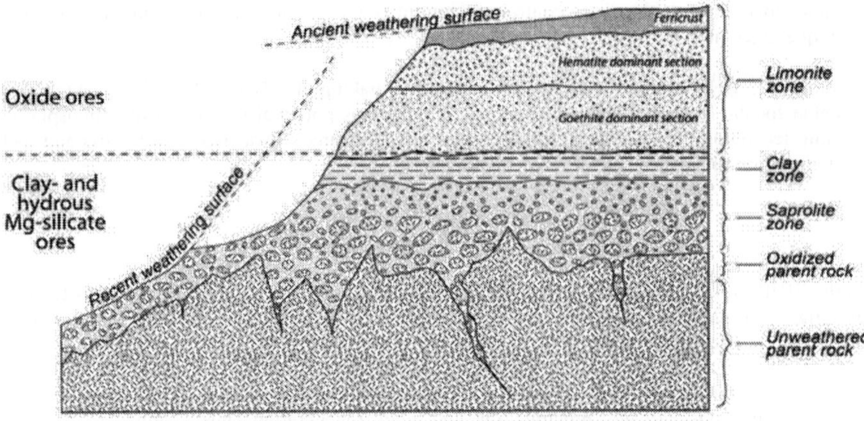

Figure 2. Idealized Cross Section through a Nickel Laterite Profile [25]

Jarosite Atmospheric Leaching

Flowsheet and Basic Leach Chemistry

A flowsheet for the jarosite atmospheric leach is shown in Figure 3. Both limonite and saprolite are slurried in an alkali containing water, such as seawater, to ensure there is enough for jarosite formation.

Concentrated sulfuric acid is added directly to limonite ore slurry. This substantially dissolves the nickel containing goethite matrix causing over 90% of the nickel and cobalt to leach along with substantial amounts of iron. Leach temperatures close to the boiling point of the liquor are attained, due to the heat of dilution of the acid. The goethite dissolution reaction is represented by equation 1.

$$2FeO(OH) + 3H_2SO_4 \rightarrow Fe_2(SO_4)_3 + 4H_2O \quad (1)$$

Sulfur dioxide or other suitable reductant may be added to assist with leaching of manganese minerals, which are generally higher in cobalt. This reaction, which is represented by equation 2 also suppresses chromium(VI) formation, ensuring downstream processes are not affected.

$$MnO_2 + SO_2 \rightarrow MnSO_4 \quad (2)$$

Saprolite slurry is then added. Much of the magnesium and nickel is leached, causing the free acid to drop to a level where jarosite precipitation begins, regenerating acid and furthering the leach process. Jarosite Precipitation and acid regeneration are represented by equation 3, with the saprolite leach by equation 4.

$\frac{1}{3}Na_2SO_4 + Fe_2(SO_4)_3 + 4H_2O \rightarrow \frac{2}{3}NaFe_3(SO_4)_2(OH)_6 + 2H_2SO_4$ (3)

$3MgO*2SiO_2*2H_2O + 3H_2SO_4 \rightarrow 3MgSO_4 + 2SiO_2 + 5H_2O$
(4)

Overall acid addition rates are typically in the range of 700 to 900 kg per tonne of combined limonite plus saprolite, depending on the ratio of saprolite to limonite and the nickel extraction desired.

Combining equations 1 and 4 gives a net reaction as shown in equation 5, indicating the jarosite reaction can regenerate two thirds of the acid consumed in leaching iron.

$2FeO(OH) + 3H_2SO_4 + \frac{1}{3}Na_2SO_4 \rightarrow \frac{2}{3}NaFe_3(SO_4)_2(OH)_6 + 2H_2SO_4$
(5)

The combined leach slurry passes to induced jarosite precipitation (IJP) and primary neutralization (PN). Recycle streams from downstream processes are leached initially, further decreasing free acid and allowing more jarosite formation to occur. Limestone is then added to decrease iron levels further and neutralize free acid.

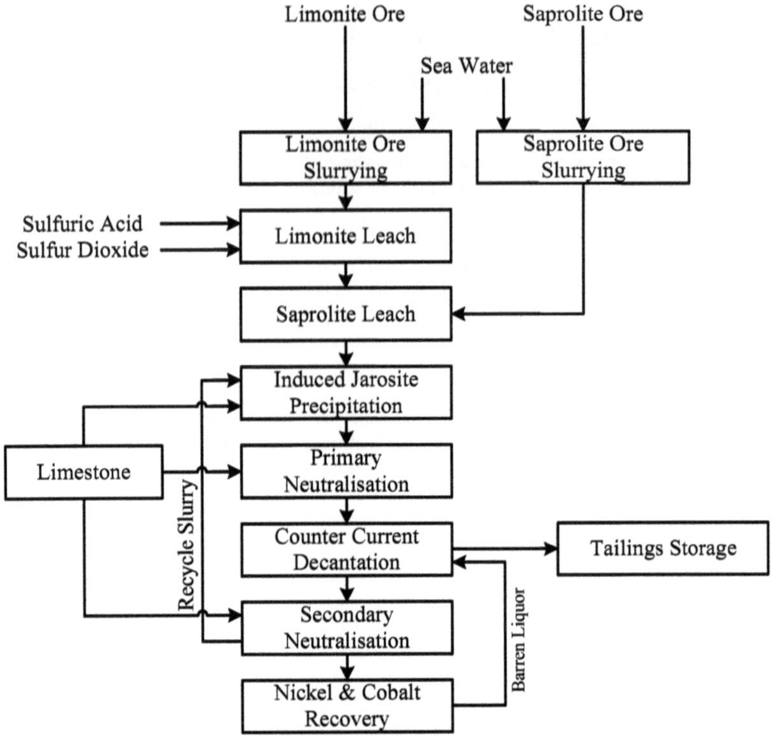

Figure 3. Jarosite Atmospheric Leach Flowsheet.

Ore Preparation

Limonite and saprolite are typically mined and processed through slurry preparation plants separately. The high iron goethite component of the ore is generally very fine, so there are ore bodies where separation can be achieved based on sizing. In any event, it is preferable to have some saprolite contamination of limonite, as saprolite will leach in the high acid conditions of the limonite leach, but the same cannot necessarily be said of limonite in the saprolite leach.

Limonite Leach

The goethite component of limonite ore is relatively difficult to leach, requiring almost total dissolution for liberation of the nickel held in the matrix.

Consequently, sulfuric acid is mixed with limonite slurry, at a rate of 1000 to 1500 kg per tonne of limonite ore. Limonite leach times are 3 to 4 hours. Peak iron concentrations are up to 100 g/L. At the end of limonite leaching free acid concentrations can range from a low of 10 g/L to over 30 g/L. If the free acid at the exit of the limonite leach is below 30 g/L, some jarosite may start to form in the limonite leach, especially where high alkali containing waters such as seawater are being used.

Table 1 Limonite Atmospheric Leach - Feed and Residue Assays [19]

Stream	Wt loss, %	Elemental Analysis (wt %)					Dissolution, %				
		Ni	Co	Fe	Mg	Mn	Ni	Co	Fe	Mg	Mn
Feed	-	1.85	0.10	35.2	4.1	0.73	-	-	-	-	-
Residue	68.5	0.43	0.02	13.5	4.5	0.04	92	93	88	65	98

Saprolite Leach

Although saprolite is a relatively high acid consumer due to the presence of magnesium, the free acid driving force required for leaching the majority of the nickel is lower than for limonite.

Sufficient saprolite ore slurry is mixed with the limonite leach discharge to reduce free acid levels below 30 g/L H_2SO_4, so that iron commences to precipitate as jarosite, simultaneously regenerating acid to maintain a driving force for continued nickel and cobalt leaching. Ferric iron in solution is reduced to less than 5 g/L, with free acid in the range of 20 to 25 g/L H_2SO_4. The free acid – iron interaction and leaching progress are well illustrated in Figure 4 and Figure 5. Saprolite leach times are in the range of 10 to 12 hours.

It has been found in practice that the reaction is self-seeding, with sufficient solids available in the agitated tanks to initiate and sustain jarosite precipitation.

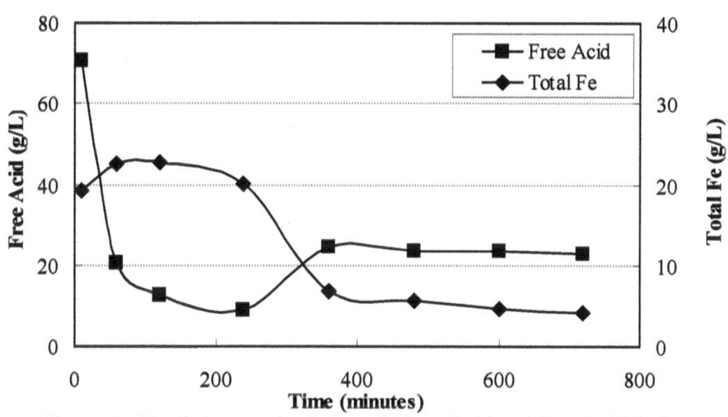

Figure 4. Batch Atmospheric Leach Free Acid and Iron in Solution

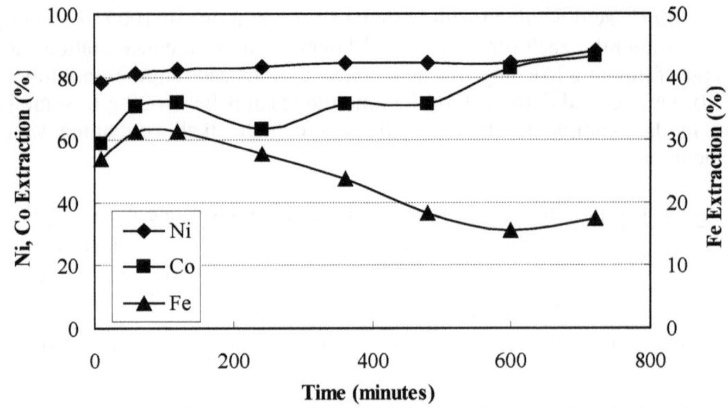

Figure 5. Batch Atmospheric Leach Nickel, Cobalt and Iron Extractions

Induced Jarosite Precipitation

The function of the IJP stage is to precipitate excess ferric iron from solution under hot, mildly acidic conditions. Some precipitation of other impurities, such as aluminum, also occurs at this stage.

Saprolite leach discharge is partially neutralized with recycled Secondary Neutralization Precipitation residue slurries, simultaneously recovering the metal values associated with those streams. The residual acid is then further reduced to approximately 5 g/L H_2SO_4 using ground limestone slurry. Temperature is maintained at 95°C.

Primary Neutralization

In Primary Neutralization (PN), further limestone slurry is added in stages to precipitate the bulk of the remaining iron(III). Solution pH is controlled to avoid unrecoverable losses of nickel and cobalt. Discharge from primary neutralization typically has a pH of 2 to 2.5 and has an iron(III) concentration of less than 0.5 g/L.

Enhanced Pressure Acid Leaching

In EPAL, much the same chemistry is applied as for the JAL process, but the sequence of leaching is changed, as illustrated in Figure 6.

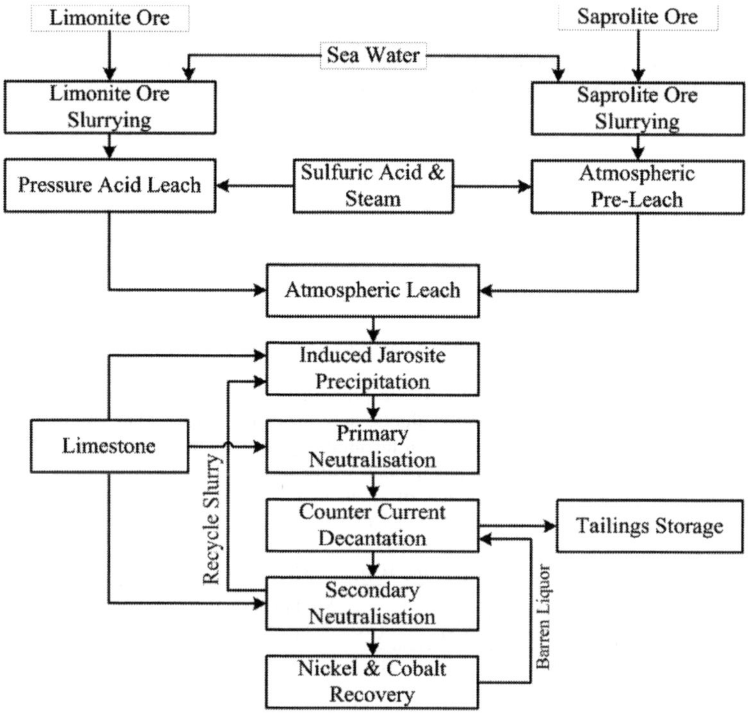

Figure 6. Enhanced Pressure Acid Leach Flowsheet.

Limonite Pressure Acid Leaching

Limonite ore is leached with sulfuric acid at around 250°C. At this temperature, acid consumption is minimised because iron is hydrolysed to hematite, kinetics are rapid and nickel and cobalt extractions typically exceed 95%.

Equation 1 remains relevant for leaching of goethite, and is repeated here for clarity.

$$2FeO(OH) + 3H_2SO_4 \rightarrow Fe_2(SO_4)_3 + 4H_2O \quad (1)$$

Hematite precipitation as shown in equation 6 results in nominally complete regeneration of acid, as shown in equation 7, although for practical purposes iron extractions of 1 to 2% are typical.

$$Fe_2(SO_4)_3 + 3H_2O \rightarrow Fe_2O_3 + 3H_2SO_4 \quad (6)$$

$$2FeO(OH) \rightarrow Fe_2O_3 + H_2O \quad (7)$$

Again a reductant may be added to aid leaching of manganese minerals, and suppress chromium(VI) formation, with elemental sulfur being quite effective at higher temperatures [26].

$$3MnO_2 + S + 2H_2SO_4 \rightarrow MnSO_4 + 2H_2O \quad (8)$$

In addition to the hematite precipitation – acid regeneration chemistry, sodium and potassium in seawater react with aluminum in solution to form alunite, resulting in low net extraction. Alunite precipitation also regenerates acid, helping to minimize acid consumption, analogous to jarosite precipitation in atmospheric leaching. Sodium alunite precipitation, as shown in equation 9, is an analogue of jarosite precipitation in equation 3.

$$\tfrac{1}{3}Na_2SO_4 + Al_2(SO_4)_3 + 4H_2O \rightarrow \tfrac{2}{3}NaAl_3(SO_4)_2(OH)_6 + 2H_2SO_4 \quad (9)$$

Aluminum to sulfur ratios in residues point to iron co-precipitation, such that an alunite-jarosite compound is formed, typically about two-thirds alunite and one-third jarosite. The sodium plus potassium to sulfur ratios indicate that 10 to 20% of the alunite-jarosite is in the hydronium form. Kyle [27] gives a useful summary of these phenomena while Johnson and co-workers [15, 16] have carried out detailed studies into the effects of alkali addition and leach liquor salinity on PAL. Some typical assays are shown in Table 2 and Table 3.

Table 2. Pressure Acid Leach in Saline Waters - Metal Extractions [24].

Stream	Acid Additon (kg/t)	Free Acid (g/L)	Metal Extractions (%)						
			Ni	Co	Fe	Al	Cr	Mg	Mn
Batch	320	34	97.4	97.6	1.6	0.7	1.3	87.6	98.0
Pilot Plant	325	26	95.7	95.8	2.3	2.2	1.7	88.7	96.1

Table 3 Pressure Acid Leach in Saline Water - Feed and Residue Assays [24]

Stream	Elemental Analysis (wt %)							
	Ni	Co	Fe	Cr	Al	K	Na	S
Feed	2.2	0.088	27.6	0.28	2.0	0.02	0.04	0.05
Residue	0.11	0.004	30.0	0.30	2.3	0.09	0.80	3.0

Saprolite Leaching

The presence of high levels of magnesium means that not only is there no particular advantage in leaching saprolite in an autoclave, but the relative solubility limits favor atmospheric leaching. If the saprolite to limonite ratio in the ore body is relatively low, the acid contained in the PAL discharge solution can be used for atmospheric leaching of saprolite. At higher saprolite to limonite ratios, there is insufficient acid available to attain high metal recoveries during saprolite leaching, so additional acid is added in a Pre-leach step. This flexibility allows processing of the entire ore profile.

Pre-Leach

In the Pre-Leach step, concentrated sulfuric acid is added to saprolite slurry. This raises the slurry temperature to greater than 95°C and leaches substantial amounts of nickel, cobalt, iron and other impurities. As with the limonite leaching in JAL, some jarosite may start to precipitate.

Atmospheric Leach

PAL and Pre-Leach discharge streams are combined and allowed to react to complete saprolite leaching.

Induced Jarosite Precipitation and Primary Neutralization

The functions of IJP and PN remain the same as those for the JAL process.

Operational Flexibility

Configuration Changes

An aspect of plant design and operation that is apparent from the above description is that the atmospheric leach section of the EPAL® configuration can be used as a JAL plant, simply by piping limonite to feed the Pre-leach and adding saprolite directly to the atmospheric leach tanks.

In the event that PAL becomes an operating constraint, operation of atmospheric leach in JAL mode can be used to fully utilize the acid plant capacity and minimize production disruptions. The comparative configurations are shown in Figure 7.

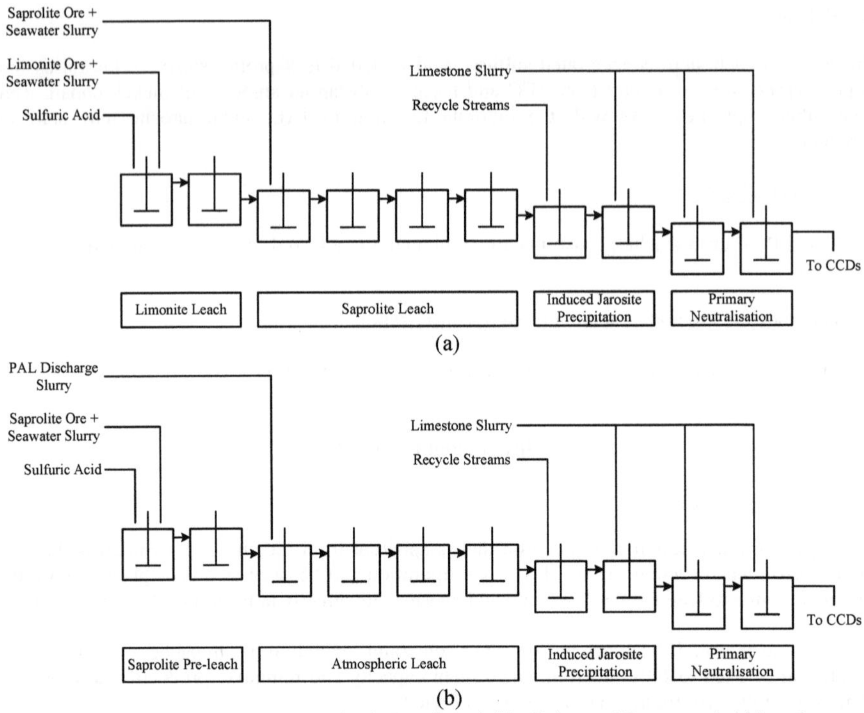

Figure 7. (a) Jarosite Atmospheric Leach and (b) Enhanced Pressure Acid Leach Configurations.

<u>Other Atmospheric Leaching Configurations</u>

It is feasible to carry out a saprolite only atmospheric leach. The main technical constraint of this configuration is the solubility limit of magnesium.

It is also possible to pre-blend limonite and saprolite slurries. However as the examples in Table 4 demonstrate, recoveries from the sequential limonite followed by saprolite leach (JAL) are at least equivalent and arguably superior, while producing end solutions of lower acidity and iron content [28]. This has material implications for the quantity of limestone required for neutralization and equipment sizing for downstream solid-liquid separation and residue storage.

Table 4. Comparative Batch Jarosite Atmospheric Leach Results for Sequential and Blended Ore Feeds [28].

Leach Type	Lim:Sap Ratio	Overall Acid:Ore Ratio	Leaching Time, hours	Ni Ext, %	Co Ext, %	Free Acid, g/L	Fe, g/L	Equivalent Free Acid, g/L*
Sequential	1:1	0.80	13	92	97	28	7.5	48
Sequential	1:1	0.80	24	93	93	34	5.4	48
Sequential	1:2	0.83	24	94	95	25	4.9	38
Combined	1:1	0.77	24	87	92	39	7.7	59
Combined	1:1	0.80	28	87	95	25	27.4	97
Combined	1:1	0.87	23	94	97	36	21.4	92
Combined	1:2	0.89	18	89	94	30	23.6	92
Combined	1:1	0.84	24	94	97	36	11.0	65
Combined	1:1	0.80	24	97	95	42	7.6	62

* Equivalent free acid is the sum of the free acid and the sulfate associated with the iron in solution, assumed to be as iron(III).

Disposition of Other Elements

Aluminum

Disposition of aluminum is of particular concern where the process calls for precipitation of a mixed hydroxide intermediate, or if a direct solvent extraction is utilized.

Re-leaching of mixed hydroxide intermediates in ammonia solutions is impeded if aluminum is present [29]. Cobalt - nickel separation using solvent extraction with Cyanex 272 as practiced at the Bulong operation from 1999 to 2003, requires low levels of aluminum to avoid crud formation [30]. Aluminum can also adversely affect the performance of the Cyanex 301 reagent, as used at Goro for extraction of nickel and cobalt [31, 32].

Pressure Acid Leaching. As discussed above, net aluminum extraction is around 1% when pressure acid leaching is carried out in saline waters. Thus aluminum removal in the PAL process step has the potential to be an "out" for aluminum, even if downstream removal steps prove to be ineffective.

Atmospheric Leach and Precipitation Processes. Aluminum is completely precipitated at the pH used for ferrous iron oxidation. The precipitated aluminum reports to the residue recycle stream and is releached in the Induced Jarosite Precipitation stage. There is therefore the risk of a recirculating load building up in the circuit between the Induced Jarosite Precipitation and Secondary Neutralization circuits. However, the pH of the intervening Primary Neutralization step is sufficiently high to precipitate excess aluminum and provide a safe bleed from the process.

Magnesium

Magnesium exhibits reverse solubility at very high temperatures and may precipitate in the PAL autoclave as kieserite [24]. A tolerable level seems to be around 20 g/L in flash discharge, although this may be conservative.

At the 95°C temperature conditions of atmospheric leaching, the magnesium solubility limit is around 50 g/L. A good working limit has been found to be around 40 g/L, as this allows for solution cooling without the risk of magnesium sulfate crystallization.

Calcium

Calcium in the form of calcium carbonate contained in limestone is added in the neutralization processes after nickel laterite leaching. Calcium sulfate solids in the form of gypsum (and possibly hemihydrate and anhydrite) precipitate and have the potential to form scale on equipment and piping surfaces.

In the induced jarosite precipitation and primary neutralization processes, this is typically done in-slurry, at temperatures in excess of 80°C. Few issues with scale formation occur, apparently due to the presence of a large surface area available from the leach residue solids, relative to equipment surfaces. It is important to provide adequate retention time for the complete reaction of the limestone, to avoid reactions continuing to occur in downstream thickeners and pipework. Such reactions can lead to both process upsets and scale formation over the longer term.

The downstream secondary neutralization stage is particularly susceptible to scaling because the feed liquor is effectively solids free, so more susceptible to heterogeneous precipitation of gypsum on plant equipment such as tank walls, pipes and agitators. Recirculation of gypsum containing precipitated solids is a viable control option, with operational improvements still occurring at a seed recycle rate of 500% of net precipitation. These include reduced scaling rates, higher settled slurry density and improved limestone consumption efficiency.

Good design practice typically includes dedicated thickener underflow recirculation pumps to deliver adequate volumes of precipitate as seed, and blending of seed slurry and feed liquor ahead of limestone addition to the first neutralization vessel.

Manganese

A major issue with manganese is that it is in the ore in the manganese(IV) state, which is a strong oxidant, so will oxidize chromium to chromium(VI). Addition of a reducing agent such as sulfur dioxide, as discussed above avoids this, while also increasing manganese and associated cobalt dissolution.

Manganese causes few problems with nickel recovery, being almost totally left in solution during sulfide precipitation and partially so during mixed hydroxide precipitation.

Excess barren liquors must generally be treated before discharge to remove manganese by a combination of pH adjustment with lime and oxidation.

Chromium

Chromium(III) behaves in a similar manner to iron(III) so causes few problems in downstream processing. It precipitates after iron but before aluminum as the pH is raised.

Chromium(VI) is quite persistent once formed and will cause contamination issues in downstream nickel and cobalt recovery. It can however be easily eliminated by the addition of a reducing agent into solution before neutralization. Suitable reagents used include sulfur dioxide, hydrogen sulfide and their alkali equivalents.

Silicon

Silicon dissolves at temperatures prevailing in PAL operations. During pressure letdown, solubility decreases considerably, giving the potential for scale formation on vessel walls. AMAX devoted considerable effort to understanding and controlling the phenomena [33]. The draft tube crystallizer concept adapted by AMAX from the geothermal industry was incorporated at the Cawse and Ravensthorpe operations.

Silica levels in seawater are less than 1 mg/L, so the source of soluble silica in Jarosite atmospheric leach solutions is from the ore.

Commercial operations seem to have had few major issues with downstream silica formation, especially if there is a secondary neutralization step, where levels are reduced to less than 30 mg/L. The magnesia based mixed hydroxide process further reduce this to less than 5 mg/L in processes analogous to lime softening [24, 29].

Leach Residue Stability

The stability of jarosite containing residues from atmospheric leaching has been studied in depth by Reynolds [34]. It was found that residues are chemically stable if kept in contact with alkali (sodium) containing liquors. Jarosite breakdown in the absence of such liquors is retarded in the presence of lime, which reacts with jarosite breakdown products and encapsulates the remaining jarosite.

Concluding Remarks

Sulfuric acid atmospheric leaching of laterites with jarosite precipitation has proven itself to be a versatile process, capable of treating both limonites and saprolites and variable ratios of these two ore types.

The EPAL® variant is well suited to orebodies which are predominantly limonite, and based on commercial experience, can also operate in full atmospheric leach mode, to the limit of acid plant capacity.

Of the purely atmospheric leach options presented, sequential limonite – saprolite atmospheric leaching gives the best extractions for a given total acid addition and generally produces a leach slurry ready for further (induced) jarosite precipitation with a minimum of limestone addition.

In-slurry acid neutralization at 95°C to pH 2.5 allows almost complete iron(III) precipitation as jarosite, without risk of significant nickel or cobalt co-precipitation. Jarosite chemistry also lends itself well to managing aluminum in solution, with partial removal achieved in the neutralization step, providing a purge without risk of nickel and cobalt losses.

Jarosite containing nickel laterite leach residues are chemically stable if kept in contact with alkali (sodium) containing liquors. Jarosite breakdown remains manageable, even in the absence of the stabilizing alkali ions, particularly where excess acid neutralizing reagent has been added, as occurs when processing via the mixed hydroxide process route.

The Jarosite Atmospheric Leach process may be particularly suited to arid climates, where the only water source readily available is saline and it is possible to operate with zero liquid discharge.

References

1. H.L. Sulman, H.F.K. Picard, Recovery of Nickel from its Ores, U.S. Patent 1,091,545, 31 March 1914.

2. H.L. Sulman, H.F.K. Picard, Recovery of Nickel from its Ores, U.S. Patent 1,193,734, 8 August 1916.

3. E.T. Carlson, C.S. Simons, Pressure Leaching of Nickeliferous Laterites with Sulfuric Acid, *Extractive Metallurgy of Copper, Nickel, and Cobalt*, P.E. Queneau ed., Interscience Publishers, New York, (1961), 363-396.

4. A.R. Gordon, R.W. Pickering, Improved Leaching Technologies in the Electrolytic Zinc Industry, *Metallurgical Transactions B, Vol 6B, March 1975*, 43-53.

5. D. Weston, "Hydrometallurgical Treatment of Nickel, Cobalt and Copper Containing Materials", United States Patent 3,793,430, 19 February 1974.

6. D. Weston, "Hydrometallurgical Treatment of Nickel Group Ores", United States Patent 3,793,432, 19 February 1974.

7. D. Weston, "Hydrometallurgical Treatment of Nickel Cobalt and Copper Containing Materials", United States Patent 3,798,304, 19 March 1974.

8. J.H. Canterford, Treatment of Nickeliferous Laterites, *Minerals Sci. Engng, vol. 7, no. 1, January, 1975*, 3-17.

9. S. Agatzini, A. Kontopoulos, P. Marboutis, A. Xenidis, Removal of iron from iron-nickel-cobalt solutions by precipitation and solvent extraction techniques, *Iron Control in Hydrometallurgy, ed. J.E. Dutrizac, A.J. Monhemius, Ellis Horwood Ltd, 1986 (CIM, Montreal, Quebec, 1996)*, 353-373.

10. N. Panagiotopoulos, A. Kontopoulos, Atmospheric Pressure Sulfuric Acid Leaching of Low-Grade Hematitic Laterites, *Extractive Metallurgy of Nickel and Cobalt, G.P Tyroler, C.A. Landolt, Eds., TMS, Warrendale, PA, U.S.A., 1988*, 447-459.

11. P. Maraboutis, A. Kontopoulos, . "Jarosite Precipitation from Iron – Nickel – Cobalt Sulfate Solutions", *Extractive Metallurgy of Nickel and Cobalt, G.P Tyroler, C.A. Landolt, Eds., TMS, Warrendale, PA, U.S.A., 1988*, 479-492.

12. A. Kontopoulos, "Sulfuric Acid Leaching of Laterites", *EPD Congress '91, ed. D.R. Gaskell, TMS, Warrendale, PA, U.S.A., 1991*, 147-163.

13. W.P.C. Duyvesteyn, "Leaching Nickeliferous Oxide Ores", United States Patent 4,195,065, 25 March 1980.

14. W.P.C. Duyvesteyn, G.R. Wicker, R.E. Doane, An Omnivorous Process for Laterite Deposits, *Chapter 28, International Laterite Symposium 1979, D.J.I. Evans, R.S. Shoemaker, H. Veltman, Eds, AIME, New York, New York, U.S.A., 1979*, 553-570.

15. J.A. Johnson, etal, Process Water Salinity Effects in the Pressure Acid Leaching of Cawse Nickel Laterite Ore, *Chloride Metallurgy 2002, E. Peek, G. Van Weert, Eds., CIM, Montreal, Quebec, Canada, vol. 1, 2002*, 339-354.

16. J.A. Johnson, B.I. Whittington, Effect of Ammonium, Sodium and Potassium Sulphates and Chlorides in Pressure Acid Leaching of Western Australian Nickel Laterite Ore, *Pressure Hydrometallurgy, M.J. Collins, V.G. Papangelakis, Eds, CIM, Montreal, Canada, 2004*, 199-214.

17. W.P.C. Duyvestyn, H. Liu, M.J. Davis, Heap Leaching of Nickel Containing Ore, United States Patent 6,312,500, 6 November 2001.

18. J.C. Arroyo, D.A. Neudorf, Atmospheric Leach Process for the Recovery of Nickel and Cobalt from Limonite and Saprolite Ores, United States Patent 6,261,527, 17 July 2001.

19. J.C. Arroyo, D.A. Neudorf, Atmospheric Leach Process for the Recovery of Nickel and Cobalt from Limonite and Saprolite Ores, United States Patent 6,680,035, 20 January 2004.

20. BHP Minerals, Nickel Laterites ... Taking the Pressure Off, *Presented at: New Caledonian Nickel Conference, June 2001*, 19pp

21. H. Liu, et al., Atmospheric Leaching of Laterites with Iron Precipitation as Goethite, *International Laterite Nickel Symposium 2004, W.P.Imrie, D.M. Lane, Eds, TMS, Warrendale, Pennsylvania, U.S.A., 2004*, 347-368.

22. J.C. Arroyo, J.D. Gillaspie, D.A. Neudorf and E.M. Weenink, Method for Leaching Nickeliferous Laterite Ores, United States Patent 6,379,636, 30 April 2002.

23. M. Adams, etal., Piloting of the Beneficiation and EPAL® Circuits for Ravensthorpe Nickel Operations, *International Laterite Nickel Symposium 2004, W.P.Imrie, D.M. Lane, Eds, TMS, Warrendale, Pennsylvania, U.S.A., 2004*, 193-202.

24. D.T White, M.J. Miller and A.C. Napier, Impurity Disposition and Control in the Ravensthorpe Acid Leaching Process, *Iron Control Technologies, J.E. Dutrizac and P.A. Riveros, Eds., CIM, Montreal, Quebec, Canada, 2006*, 591-610.

25. J-C. Samama, Ore Fields and Continental Weathering, *Van Nostrand Reinhold Co., New York, 1986*, 326p.

26. J. Johnson, et al., Effect of Eh in the High Pressure Acid Leaching of Nickel Laterite Ores, *International Laterite Nickel Symposium 2004, W.P.Imrie, D.M. Lane, Eds, TMS, Warrendale, Pennsylvania, U.S.A., 2004*, 273-288.

27. J.H. Kyle, Jarosite / Alunite in Nickel Laterite Leaching - Friend or Foe?, *ALTA 2003 Nickel / Cobalt – 9, ALTA Metallurgical Services, Castlemaine, Victoria, Australia, 2003.* 28pp

28. J. Agin, Y. Le Quesne, B. Berthomieu, Process for the Hydrometallurgical Treatment of a Lateritic Nickel/Cobalt and Process for Producing Nickel and/or Cobalt Intermediate Concentrates or Commercial Products Using It, United States Patent Application 2010/0098608, 22 April 2010.

29. D.T White, Commercial Development of the Magnesia Mixed Hydroxide Process for Recovery of Nickel and Cobalt from Laterite Leach Solutions, *Hydrometallurgy of Nickel and Cobalt 2009, J.J. Budac, R. Fraser, I. Mihaylov, V.G. Papangelakis and D.J. Robinson, Eds., CIM, Montreal, Quebec, Canada, 2009*, 351-367.

30. S. Donegan, Direct Solvent Extraction of Nickel at Bulong Operations, *Minerals Engineering 19, 2006*, 1234-1245.

31. W.G. Bacon, D. F. Colton, E. Krause, I.O. Mihaylov, A. Singhal, A. Vahed, J.P. Duterque, Development of the Goro Nickel Process, *ALTA 2005 Nickel/Cobalt Conference, ALTA Metallurgical Services, Melbourne, Australia, 2005*, 20pp

32. Y. Okita, A. Singhal, J-J. Perraud, Iron Control in the Goro Nickel Process, *Iron Control Technologies, J.E. Dutrizac, P.E. Riveros, Eds., CIM, Montreal, Quebec, Canada, 2006*, 635-651.

33. P.B. Queneau, et al., Control of Silica Deposition During Pressure Let-Down of Acidic Leach Slurries, *Hydrometallurgy – Research, Development and Plant Practice, K. Osseo-Asare, J.D. Miller Eds., 112th AIME Annual Meeting, Atlanta, Georgia, USA, 1983*, 121-137.

34. G.A. Reynolds, The Jarosite Group of Compounds – Stability, Decomposition and Conversion, *MSc Thesis, University of Newcastle, NSW, Australia*, 2007, 173pp.

Ni-Co 2013
Edited by: Thomas Battle, Michael Moats, Violina Cocalia, Harald Oosterhof, Shafiq Alam, Antoine Allanore, Rodney Jones, Nathan Stubina, Corby Anderson, and Shijie Wang
TMS (The Minerals, Metals & Materials Society), 2013

EXTRACTION OF NICKEL, COBALT AND IRON FROM LATERITE ORES BY MIXED CHLORIDE LEACH PROCESS

V.I. Lakshmanan, R. Sridhar, R. deLaat, J. Chen, M.A. Halim and R. Roy

Process Research Ortech Inc., 2350 Sheridan Park Drive, Mississauga, ON L5K 2T4

Keywords: Base Metals, Laterite, Leaching, Solvent Extraction, Mixed Chloride

Abstract

Primary nickel is produced from both lateritic and sulfidic ores. Lateritic ores are formed by prolonged weathering of ultramafic rocks. The laterite deposits exhibit graded layers of ore, with high iron, low nickel ore on top and low iron, high nickel ore below. The low iron (< 15 % Fe) ores containing 1.7- 2.3% Ni are treated by pyrometallurgical processes to produce ferronickel or matte. High iron (> 35 % Fe) limonitic ores can be treated by high pressure leach using sulfuric acid. An alternative process for the production of base metals (Ni, Co and Fe) from both high and low iron laterite ores is being developed by Process Research ORTECH Inc. (PRO), a world leader in the development of chloride metallurgy. The process flowsheet developed by PRO is efficient and environmentally friendly. Novel leaching and solvent extraction separation steps have been developed to make this alternative more economically attractive in the current market conditions.

Introduction

Primary nickel is produced from both lateritic and sulfidic ores. Although currently the majority of nickel is produced from sulfidic ores, the production of nickel from lateritic ores is increasing and may overtake nickel production from sulfidic ores [1]. Lateritic ores are formed by prolonged weathering of ultramafic rocks. Typical mineralogical profile of nickel laterite ore bodies is as follows [2, 3]:

1. Overburden with low nickel values of < 0.8% Ni, which is discarded.
2. Limonitic ore zone with 1.2-1.7% Ni, ~ 40 % Fe and 1-4% MgO. These zones are suited for hydrometallurgical processes capable of extracting nickel from limonitic ores. High iron (> 35 % Fe) limonitic ores can be treated by High Pressure acid Leach (HPAL) process using sulphuric acid.
3. Saprolitic ore zones with 1.7 to 2.3 % Ni and low iron (< 15 % Fe) are suited for pyrometallurgical smelting processes to produce ferronickel or matte.
4. Base rocks contain < 0.5% Ni and are not economic for extracting nickel.

In some of the major laterite projects possessing the typical profile as described above above typical profile, the limonitic ores are stockpiled and stored for subsequent treatment after exhausting the higher grade saprolitic ores. There are however, laterite projects in which the saprolitic ore profile is either absent not present or too trace for economical extraction, thereby necessitating treatment of the limonitic profile for nickel recovery.

Pyrometallurgical Processing of Lateritic Ores

The ores in the saprolitic zone with high nickel values are commonly treated by pyrometallurgical processing to recover ferronickel and matte with 25 to 75% Ni content. Most of these ores contain less than 20% Fe and the recovery of nickel from this ore zone exceeds 92% and is about 95% in the best operations. The smelting process continues to be the preferred process for treating this ore zone. The flowsheet for pyrometallurgical processing of lateritic ores is shown in Figure 1 [1]. The flowsheet consists of ore drying followed by calcination/reduction in a rotary kiln. The calcine from the rotary kiln is smelted in an electric furnace with carbon as reductant. This is followed by refining or conversion to produce ferro-nickel or matte. The process is energy intensive as it involves drying, calcining and smelting of entire ore body at successively higher temperatures.

More recently the upper iron zone near the saprolitic zone is treated by mini blast furnaces converted from iron producing operations in China [4] to recover lower grade ferronickel (< 20% Ni). Such technology is now banned in China as they are highly environmentally unfriendly and are only economical in periods of high nickel prices (> $6/lb Ni).

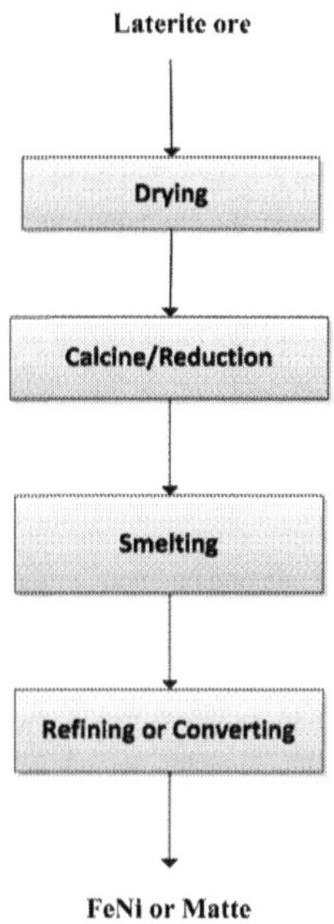

Figure 1: Pyrometallurgical Processing of Lateritic Ores

Hydrometallurgical Processing of Lateritic Ores

There are two major processes for the hydrometallurgical processing of lateritic ores: Caron process and High Pressure acid Leach (HPAL) process. The flowsheets for these two processes are shown in Figure 2 [1]. The Caron process involves ore drying followed by selective reduction of nickel. This is followed by leaching in an ammoniacal solution to bring metals in solution. Leaching is followed by precipitation of basic nickel carbonate, which is then decomposed to form nickel oxide. Nickel oxide is then reduced to form nickel metal. The Caron process is used either for limonitic ores or for mixtures of limonitic and saprolitic ores.

To treat the high iron laterite ores, High Pressure Acid Leach (HPAL) processes have been developed by many companies and are in the planning or execution stage [5]. The acid requirement for these ore bodies depends on the MgO content and is typically about 30 to 40 weight percent of the ore weight [6]. As shown in Figure 2, the HPAL process involves high pressure leaching followed by counter-current decantation (CCD) and neutralization. There are several options available for purifying nickel containing solutions and separation of nickel and cobalt. Purification can be conducted through solvent extraction, followed by electrowinning of nickel. Historically, sulfur has been available in abundance and therefore sulfuric acid was available for such processes at a cost of about \$100 per ton. When recovering 30 lbs of nickel with market price of about \$6/lb Ni, the plants could afford \$30 to \$40 per ton of ore for acid costs.

A few years ago the sulfur and acid prices saw a peak of about \$200 per ton in many parts of the world. At these prices the economics of the new plants became questionable. The market has however corrected itself. Nevertheless, when growth in the developing world is considered along with the requirement for fertilizer for their food security, it is projected that there will be corrections in the coming years for the acid price. Therefore it was of interest to develop an alternate process less dependent on sulfuric acid, the demand of which is expected to increase in the developing economies. This new innovative process is based on Chloride hydrometallurgy, which will be described in the next section. This process is applicable to both high and low iron laterite ores.

Choride Hydrometallurgy

Chloride chemistry is complex and quite different from corresponding sulfate systems. In concentrated magnesium chloride salt solutions, the activity of water is <<1, and the activity of the hydrogen ion, H^+ (or H_3O^+) increases rapidly with chloride concentration as shown in Figure 3 [7].

In addition, many metal chlorides are considerably more soluble than their corresponding sulphate salts, thus allowing the use of more concentrated solutions [8]. The solubility of $CuSO_4.5H_2O$, for example, is about one fifth that of $CuCl_2.2H_2O$, and both nickel and ferrous chlorides are also more soluble than their sulphate counterparts.

High strength magnesium chloride brines offer significant advantages over other leaching media as follows:

- **High H^+ activity** – enables only a small amount of acid (HCl) to effect leaching of the value metals. This is because with high magnesium chloride concentration, the high hydrogen ion activity provides a considerably higher driving force than for equivalent acid strength in more dilute solutions, and in many instances, the amount of acid added is only marginally higher than the stoichiometric amount required for the nickel, cobalt, iron and magnesium.
- **High chloride concentration** – stabilises base metals in solution and leads to increased solubility. The formation of ferric anionic complexes such as $FeCl_4^-$ facilitates the recovery of iron by solvent extraction.

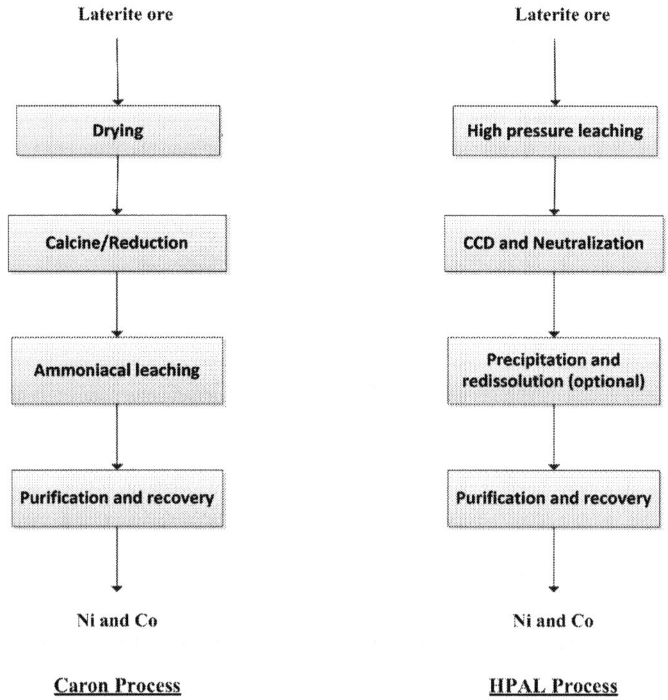

Figure 2: Hydrometallurgical Processing of Lateritic Ores

Based on the above considerations, an innovative process for the production of base metals (Ni, Co and Fe) from laterite ores has been developed by Process Research ORTECH Inc. (PRO), a world leader in the development of chloride metallurgy [8-22]. The PRO laterite process is environmentally friendly as the hydrochloric acid is regenerated using pyro hydrolysis and recycled to the leaching process and the tailings generated from leaching are inert. Furthermore the process involves pyrohydrolysis of iron-rich preg strip solution and bleed stream from the magnesium chloride containing raffinate from nickel solvent extraction. Iron oxide and magnesium oxide by-products generated from pyrohydrolysis can be sold in the market. Magnesium oxide can also be used for pH control and precipitation of Ni and Co compounds from respective preg stripping solutions. PRO is currently developing a hydrothermal hydrolysis process that can lower the energy cost for production of Fe_2O_3 and regeneration of HCl from brine.

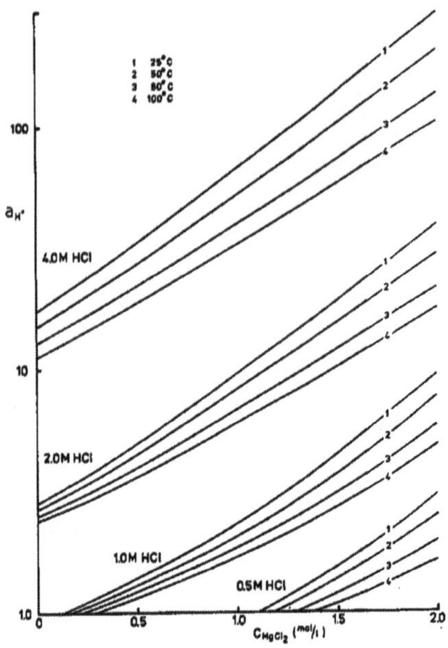

Figure 3: Effect of $MgCl_2$ Concentration on the Activity of H^+ in HCl Solutions [7]

PRO Laterite Process [23]

The process flowsheet developed by PRO is efficient and environmentally friendly. Novel leaching and solvent extraction separation steps have been developed to make this alternative more economically attractive in the current market conditions. The process flowsheet is shown below in Figure 4.

The PRO Laterite process employs the following innovative steps:

1. Leaching in mixed chloride media to bring Fe, Ni, Co and Mg in solution.
2. Innovative solvent extraction steps for successive extraction and stripping of Fe, Ni and Co ions to obtain high purity concentrated preg strip solutions of Fe, Ni and Co, respectively.

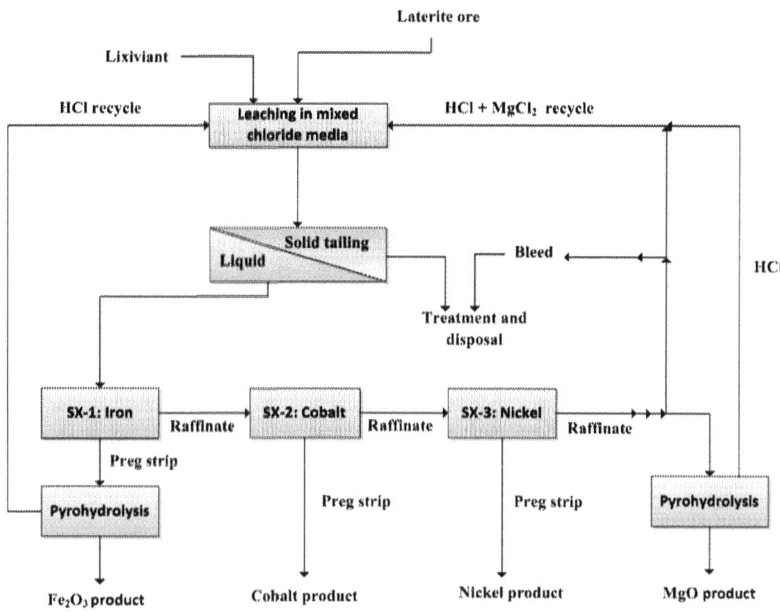

Figure 4: PRO Process for Treatment of Lateritic Ores [23]

3. Precipitation of high purity Ni and Co products from respective high purity concentrated preg strip solutions of Ni and Co via the addition of MgO generated from pyrohydrolysis.
4. Production of Fe_2O_3 and regeneration of HCl by pyrohydrolysis of the iron chloride rich preg strip solution.
5. Production of MgO and regeneration of HCl by pyrohydrolysis of bleed stream from the magnesium chloride containing raffinate from the nickel solvent extraction stage. The resulting MgO can be used for pH control as well as precipitation of Ni and Co compounds from their respective preg strip solutions. Excess MgO produced can be sold.
6. Recycling of HCl generated by pyrohydrolysis as described above.

Comparison of PRO Laterite Process with Commercial Processes

A comparison of PRO laterite process with commercial processes is shown in Table I.

Table I. A comparison of PRO laterite process with commercial processes

	Pyrometallurgical process	Caron process	HPAL process	PRO Laterite process
Ores that can be treated	Med-high Ni (1.8-2.5 %), low-med Fe (10-20 %) ores	Low Ni (1.2-1.5 %), high iron (>40 %) ores	Low Ni (1.2-1.5 %), high iron (>40 %) ores	Both low Fe and high Fe ores
Ni recovery	~ 95 %	~ 75 %	~ 95%	> 90 %
Co recovery	None	< 50%	~ 95%	> 90 %
Fe recovery	No	No	No	Yes
Mg recovery	No	No	No	Yes
Capital cost	High	Medium	High	Low
Operating cost	High	Medium	High	Medium
Energy requirement	High (due to water removal from ore and high temperature processing)	High (due to water removal from ore and high temperature processing)	Low (no water removal from ore and low temperature process)	Medium (no water removal from ore and low temperature process, however pyrohydrolysis needed for acid regeneration adds energy cost)
Reagents	Not recycled	Not recycled	Not recycled	Recycled
Residue amount	High volume	High volume	High volume	Low volume
Environmental	Slag disposal	Residue disposal containing ammonia	High volume residue disposal	Inert residue
Sulphur price	No effect	No effect	Very dependent	No effect
Fuel price	High effect	Medium effect	Lower effect	Medium effect
By-product credit	None	None	None	Fe_2O_3 and MgO credit

Conclusion

An innovative process consisting of leaching in mixed chloride media and selective solvent extraction steps has been developed for treating lateritic ores. This process offers significant advantages over existing commercial processes. Atmospheric leaching of the ore in mixed HCl-$MgCl_2$ lixiviant is highly efficient in metal recovery. Innovative solvent extraction steps are used to generate high purity preg strip solutions of Fe, Ni and Co individually. The pyrohydrolysis process is used to regenerate the acid and make the process sustainable and environmentally friendly. Magnesium from ore which is leached as magnesium chloride can be recovered as MgO by pyrohydrolysis and the regenerated acid (HCl) can be recycled to the leaching step. The process can treat low grade ores as well as rejects from higher grade ores. The process is also suitable for slags generated from pyrometallurgical operations during the production of Ni, Cu and Co. It has lower capital cost because it involves atmospheric leach as compared to autoclaves for a pressure leach process.

References

1. A.D. Dalvi, W.G. Bacon, and R.C. Osborne, "The past and future of nickel latertites", PDAC 2004 International Convention, March 7-10, (2004), 1-27. Available online: http://www.pdac.ca/pdac/publications/papers/2004/techprgm-dalvi-bacon.pdf.

2. J.P. Golightly, "Nickeliferrous Laterites: A General Description", International Laterite Symposium, D.J. Evans, R.S. Shoemaker, H. Veltman Eds., New Orleans, Louisiana, The Society of Mining Engineers of AIME, New York, (1979), pp. 3-23.

3. R.A. Alcock, "The Character and Resources Available to the Nickel Industry, Extractive Metallurgy of Nickel and Cobalt", Proceedings of a Symposium, The Metallurgical Society, C.P. Tyroler and C.A. Landolt Eds., 117th TMS Annual Meeting, Phoenix, Arizona, (1988), 67-89.

4. R. McCallum, "Nickel", Australian Commodities, 14 (1) (2007), 140-145.

5. M.G. King, "Nickel Laterite Technology – Finally a New Dawn", JOM, 57 (7) (2005), 35-39.

6. D. Marshall and M. Buarzaiga, "Effect of Magnesium Content on Sulfuric Acid Consumption During High Pressure Acid Leaching of Laterite Ores", International Laterite Nickel Symposium, W.P. Imrie and D.M. Lane, Eds., The Minerals, Metals and Materials Society, Charlotte, North Carolina, USA, March 14-18, (2004), 307-316.

7. J.J.C. Jansz, "Estimation of Ionic Activities in Chloride Systems at Ambient and Elevated Temperatures", Hydrometallurgy, 11 (1), (1983), 13-31.

8. V.I. Lakshmanan, R. Sridhar, G.B. Harris, and J. Magee, "The Jaguar Nickel Inc. Sechol Laterite Project Atmospheric Chloride Leach Process", International Laterite Nickel Symposium, W.P. Imrie and D.M. Lane, Eds., The Minerals, Metals and Materials Society, Charlotte, North Carolina, USA, March 14-18, (2004), 219-241.

9. P.G. Christies, V.I. Lakshmanan, and G.J. Lawson, The liquid-liquid extraction of copper (II) and iron (III) from chloride solutions using LIX 64N in kerosene, ISEC Proceedings, 1 (1974), 685-696.

10. V.I. Lakshmanan, G.J. Lawson, and P.S. Nyholm, "The extraction of copper(II) and iron(III) with Kelex 100 from aqueous media containing chloride ions", ISEC Proceedings, 2 (1974), 1169-1183.

11. G.B. Harris, V.I. Lakshmanan, R. Sridhar, and G. Puvvada, "A Process for the Recovery of Value Metals from Base Metal Sulfide Ores", (2007), Canadian Patent No. 2,478,516.

12. G.B. Harris, V.I. Lakshmanan, R. Sridhar, and G. Puvvada, "Process for the Recovery of Value Metals from Base Metal Sulfide Ores", (2010), US Patent No. 7,736,606 B2.

13. G.B. Harris, V.I. Lakshmanan, T.J. Magee, and R. Sridhar, "Atmospheric Chloride Leaching of Base Metal Sulfides", Hydro-Sulfides (2004), International Colloquium on Hydrometallurgical Processing of Copper Sulfides, Santiago, Chile, (2004), 384-398.

14. V.I. Lakshmanan, R. Sridhar, and R. Roy, "Application of Mixed Chloride technology for the recovery of TiO2, Base metals and PGM", Fray International Symposium, Cancun, Mexico, Nov. 27 - Dec. 1, (2011), 11 pages.

15. V.I. Lakshmanan, R. Sridhar, T. Sheikhzeinoddin, M.A. Halim, and R. Roy, 2012, "Extraction of Titanium and Vanadium by Chloride Leach Processes", T.T. Chen Symposium proceedings, The Minerals, Metals and Materials Society, Orlando, Florida, USA, March 11-15, (2012),8 pages.

16. V.I. Lakshmanan, R. Sridhar, and R. Roy, "Application of Mixed Chloride Technology for The Recovery of TiO_2 and Base Metals", ALTA 2012, Perth, Australia, May 26-June 2, (2012), 12 pages.

17. V.I. Lakshmanan, R. Sridhar, and R. Roy, "Synthesis of TiO_2 by an Innovative Atmospheric Mixed-Chloride Leach Process", Chloride 2011 proceedings, The Minerals, Metals and Materials Society, San Diego, USA, Feb. 28-March 3, (2011), 8 pages.

18. V.I. Lakshmanan, R. Sridhar, and R. Roy, "Recovery of Precious Metals from Base Metal Sulfide Ores by a Hydrometallurgical Process", Chloride 2011 proceedings, The Minerals, Metals and Materials Society, San Diego, USA, Feb. 28-March 3, 2011, (2011), 8 pages.

19. V.I. Lakshmanan, R. Sridhar, R. Roy, and V. Ramachandran, "Recovery of Value Metals from Ores by Mixed Chloride Extraction", Proceedings of Sixth International Symposium, Hydrometallurgy 2008, Phoenix, Arizona, Aug. 17-20, (2008), pp 895-902.

20. V.I. Lakshmanan, R. Sridhar, and R. Roy, "Selective Reduction Leach Process for Laterites", Conference of Metallurgists, Sudbury, ON, Aug 23-26, (2009), 11 pages.

21. M.C.E. Bell, R. Sridhar, and H. F. Bakker, "Gaseous Solubilization of Reduced Metal Values", (1977), Canadian Patent # 1013576.

22. G.B. Harris, V.I. Lakshmanan and R. Sridhar, "A Process for the Recovery of Value Metals from Material Containing Base Metal Oxides", (2006), Canadian Patent # 2467288.

23. V.I. Lakshmanan and R. Sridhar, "Separation of Iron from Value Metals in Leaching of Laterite Ores", (2011), US Patent Application No. 61/569314.

Ni-Co 2013
Edited by: Thomas Battle, Michael Moats, Violina Cocalia, Harald Oosterhof, Shafiq Alam, Antoine Allanore, Rodney Jones, Nathan Stubina, Corby Anderson, and Shijie Wang
TMS (The Minerals, Metals & Materials Society), 2013

REDUCTIVE LEACHING OF LIMONITIC LATERITES USING FERROUS SULPHATE

Mariela Zuniga, Edouard Asselin

The University of British Columbia, Vancouver, BC, Canada

Keywords: laterites, limonite, nickel, ferrous sulphate, leaching

Abstract

This study was carried out on a limonitic laterite from New Caledonia. The main objective was to evaluate the atmospheric leaching of the ore in ammonium sulphate media using ferrous sulphate as reductant. Different parameters were examined, such as total ammonia concentration, temperature and ferrous sulphate concentration. At 80°C, 130 g/l of total ammonia, 2.73 g $FeSO_4$/ g ore and pH 8.2, nickel extraction was 64 % while 20 % of cobalt remained in solution after 8 hours.

Introduction

Hydrometallurgical treatment for laterites has increased due to the importance these deposits have gained as sources of nickel and cobalt [1, 2]. In recent years many researchers have studied different ways to extract nickel and cobalt by ammoniacal or acidic solution leaching.

There are a large amount of studies involving acid leaching of laterites. These studies include high pressure and atmospheric leaching to release nickel and cobalt from laterites. In general, acid leaching presents good nickel and cobalt recoveries but the acid consumption, especially under atmospheric conditions, is high and leach liquors require neutralization prior to further treatment [1, 2].

One alternative to acid leaching, the Caron process, includes a reductive roasting stage followed by oxidative ammoniacal leaching. However, this process presents some limitations such as high operational cost, mainly from the drying and reductive roasting stages [3]. Also, low final cobalt recoveries (around 50%), which probably occur due to absorption of cobalt onto ferrihydrite and/or passivation of the Fe-Ni-Co alloys formed during roasting, are a significant economic downside [4,5,6].

In ammonia media, many reducing agents such as ferrous sulphate, glucose, thiosulfate, etc. have been tested to recover nickel and cobalt especially from deep sea manganese nodule ores [7]. The best results have been reported when ferrous sulphate was used with extractions of over 90% achieved for nickel and cobalt [8]. Also, there are some studies where synthetic goethite is directly reduced by ferrous sulphate in ammonia to produce magnetite, where the total transformation is achieved around 130°C [9]. Recent studies used metallic iron as a reducing agent to leach limonites, the results show a partial leach of the ore and the ferrous ion seems to play an important role on leaching kinetics [10].

The present work reports experimental results obtained during the leaching of a limonitic laterite in ammoniacal solutions using ferrous sulphate as a reducing agent under atmospheric conditions. This study includes the effect of temperature, ammonia concentration and ferrous sulphate concentration. Equation (1) shows the proposed overall chemical reaction for goethite. Based on the Pourbaix diagrams for the $Ni\text{-}NH_3\text{-}H_2O$ and $Co\text{-}NH_3\text{-}H_2O$ systems, mono- to hexa-ammines of nickel and cobalt (i.e. $1 \leq n \leq 6$) can form depending on pH and total ammonia concentration [11-16]. Equation (2) summarizes the reactions for nickel and cobalt complexed in ammoniacal media.

$$FeSO_{4(aq)} + 2FeO{*}OH + 2NH_{3(aq)} = Fe_3O_4 + (NH_4)_2SO_4 \quad (1)$$

$$M^{2+}_{(aq)} + nNH_{3(aq)} = M(NH_3)_n{}^{2+}_{(aq)} \quad (2)$$

Experimental Procedures

The samples were prepared as follows: 14 kg of limonitic laterite containing approximately 40 wt % water were dried. The P_{80} of the ore was 40 microns. The sample was dried and homogenized by using a coning and quartering procedure. One representative sample of approximately 3 kg was obtained. Further splitting by coning and quartering was performed to produce smaller laterite samples which were used in the leaching tests. The ore samples were characterized by chemical analysis (ICP), SEM and quantitative X-ray diffraction (QXRD). Selected elements present in the laterite are given in Table 2.1. Table 2.2 shows the crystalline phases found in the laterite as obtained through QXRD.

Table 2.1: Elemental content in the ore

Element	%
Fe	49
Ni	1.4
Co	0.36
Mn	1.4

Table 2.2: XRD quantitative phase analysis (wt.%)

Mineral	% laterite
Goethite	57
Maghemite	13.5
Hematite	5.7
Kaolinite	2.9
Gibbsite	1.3

Experiments were done batch-wise in a 1 L jacketed and baffled glass cell under atmospheric pressure. The chemical reagents were of analytical grade and were used as-received while deionized water was used to make the solutions. Temperature was controlled by circulation of hot water (thermostatically controlled water bath) through the reactor's water jacket. Leach solutions were prepared using ammonium hydroxide and ammonium sulphate. The pH was set above the buffer point of ammonia (which changed with temperature), to ensure the stability of aqueous ammonia and the nickel, cobalt and iron ammines. The solutions were sparged with nitrogen gas for 10 minutes prior to heating the solution, to decrease the presence of oxygen and favor the formation of iron ammines rather than hydroxides. The experiments were performed at 600 RPM stirring speed with an overhead mixer and a pitch-blade impeller at 10% solids. Samples were taken throughout the course of each experiment (approximately 10 ml of pulp). The samples were separated quickly using a centrifuge. The clear solution was then acidified with 2 M hydrochloric acid and analyzed by AAS for Ni, Fe, Co and Mn. The leach residues were dried at room temperature and then homogenized to obtain representative samples. The samples were analyzed by ICP and QXRD.

Results and Discussion

The best extraction result for nickel was 64% at 80°C using 2.73 g of $FeSO_4$ / g of ore and 130 g/l total ammonia over 8 hours. For both cobalt and manganese extractions, the best result obtained was 44% at 45°C using 0.37g $FeSO_4$/g of ore and 130 g/l of total ammonia.

Effect of temperature

Leaching experiments were carried out at 20, 45, 65 and 80 °C, keeping the other variables constant. Reducing agent concentration was 0.73g of $FeSO_4$/ g of ore. Figures 1 and 2 show nickel, cobalt, manganese and iron dissolution. Nickel extraction is very slow at low temperatures but it increases over 65°C. Ferrous is consumed during the reaction and its consumption is faster at higher temperatures. Thus, from Figure 1, it is possible to see that after 8 hours there is no ferrous available to further reduce goethite.

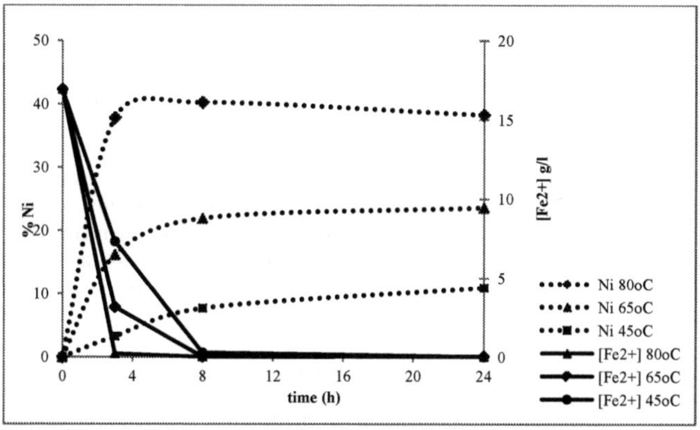

Figure 1: Effect of temperature on nickel extraction and iron behavior in ammonia solutions using ferrous sulphate as reducing agent at 600 RPM during 24 hours.

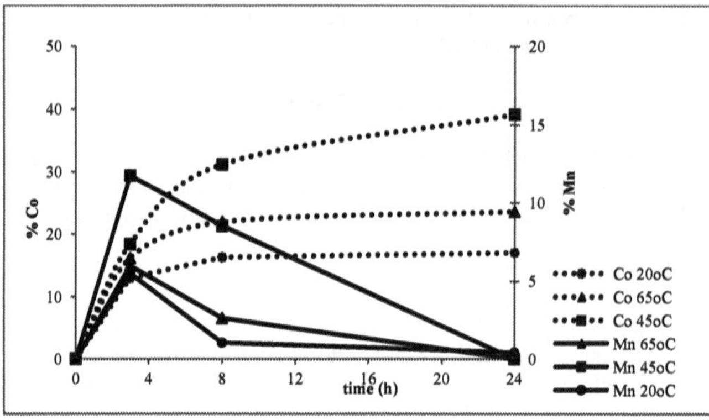

Figure 2: Effect of temperature on cobalt and manganese extraction in ammonia solutions using ferrous sulphate as reducing agent at 600 RPM during 24 hours.

Cobalt and manganese deportment is different than that of nickel. The best result was obtained at 45°C and increasing temperature is detrimental for cobalt and manganese recovery. These metals behave differently due to the fact that they are mainly associated with different mineral phases; cobalt and manganese to asbolane and nickel to goethite [17]. Thus it is not unexpected that these elements can present different kinetics under the same leaching conditions.

The solid phase analyses by QXRD show that the main phase in the residues is magnetite. Also, magnetite concentration increases with temperature. Thermodynamically, oxide formation is favored by temperature [14]. Magnetite formation could explain the low recoveries and losses of cobalt and manganese. There are numerous studies showing that cobalt can co-precipitate with iron and/or manganese oxides because it can replace these cations in the spinel, and that this phenomenon is favored by increased temperatures [7, 18].

Also, the cobalt and manganese may be adsorbed onto oxides and/or hydroxides during leaching. Research has shown that cobalt can be adsorbed onto manganese and iron oxy-hydroxides [19-23]. The capacity for adsorption of manganese oxy-hydroxides is reportedly greater than that of similar iron species [24].

Effect of ferrous sulphate concentration

Figures 3 and 4 show the effect of ferrous sulphate concentration, which included: 0.37, 0.73, 1.5 and 2.73 g of $FeSO_4$/ g of ore. Nickel extraction increased with ferrous sulphate concentration. At 1.5 g of $FeSO_4$/ g of ore, 64% Ni extraction was achieved after 24 hours of leaching. Very similar results are obtained using 2.73 g $FeSO_4$/ g of ore. These results suggest that, beyond 1.5 g of $FeSO_4$/ g of ore, added reducing agent does not improve extraction. This is probably due to the fact that magnetite precipitates around the goethite particles, thus the reaction between ferrous and goethite is very slow when the magnetite precipitate layer is thick. Figure 5 shows a particle leached partially during the process. Here, it was possible to see how the magnetite precipitates around of goethite particles.

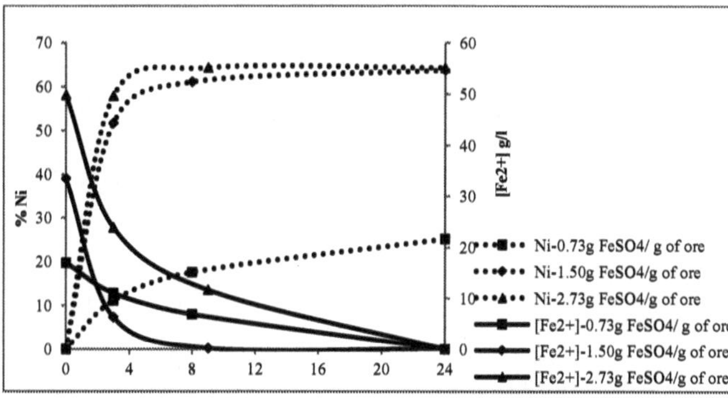

Figure 3: Effect of ferrous sulphate concentration on nickel extraction and iron behavior in ammonia solutions at 80°C, 600 RPM for 24 hours.

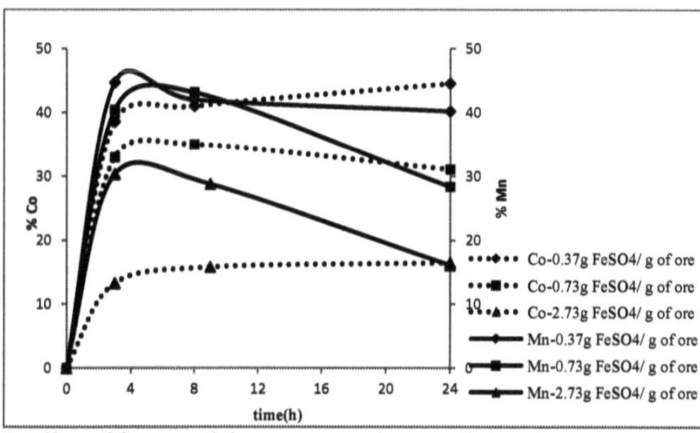

Figure 4: Effect of ferrous sulphate concentration on cobalt and manganese extraction in ammonia solutions at 80°C, 600 RPM for 24 hours.

Figure 5: SEM picture of the leached residue after 8 hours at 80°C.

The best results for cobalt and manganese extraction were obtained when the ferrous sulphate concentration was lowest. The maximum cobalt extraction of 44% was obtained after 24 hours of leaching whereas maximum manganese extraction of 44% occurred after 3 hours. High ferrous concentrations were detrimental for cobalt and manganese extraction, likely because they improve goethite reduction resulting in increased magnetite precipitation. Residue analyses show that the magnetite content increased with reductant concentration. These results are in agreement with the observed effect of temperature where precipitation during leaching was detrimental for cobalt and manganese extraction.

Effect of initial ammonia concentration

Figures 6 and 7 show the effect for initial total ammonia concentration. The experiments were conducted above the ammonium/ammonia buffer point to benefit nickel, cobalt and manganese ammine stability. The total ammonia concentration was varied between 65 and 130 g/l.

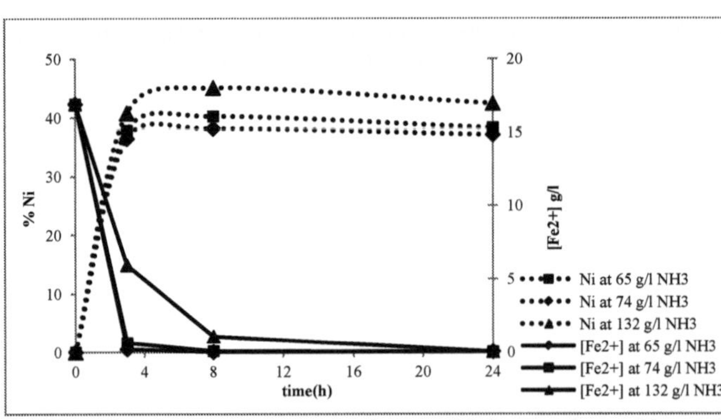

Figure 6: Effect of total ammonia concentration on nickel extraction and iron behavior in ammonia solutions at 80°C, 600 RPM for 24 hours.

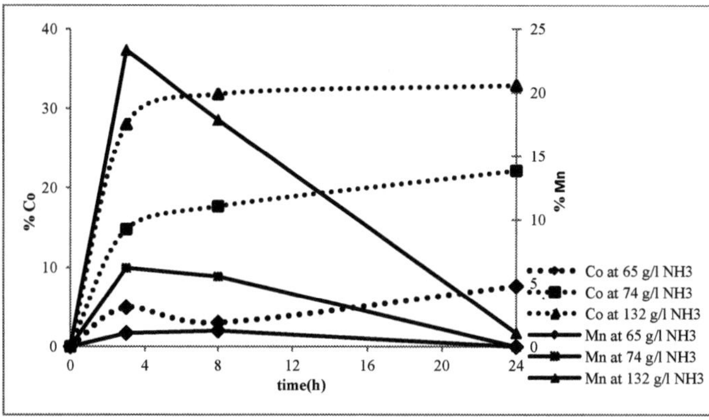

Figure 7: Effect of total ammonia concentration on nickel extraction and iron behavior in ammonia solutions at 80°C, 600RPM for 24 hours.

Increasing ammonia has a strong effect on cobalt and manganese extraction. For nickel, added ammonia concentration improves nickel extraction but the effect is not as significant as for cobalt and manganese. This positive effect is expected because ammonia is a ligand to keep the cations in solution. At higher ammonia concentrations, the thermodynamic stability zones for the ammines become larger. When the ammonia concentration is increased, ferrous is consumed more slowly, but the nickel extraction is higher (Figure 6). These results suggest that ferrous' capability to reduce goethite and asbolane concentration is higher when ammonia concentration is increased. The oxidation of some ferrous with oxygen during the leaching is expected, however increasing ammonia concentration seems to help decrease reductant losses.

Concluding Remarks

Results obtained from the present research suggest the limonitic laterite can be partially reduced by ferrous sulphate. Maximum nickel extraction was 64% at 80°C, 2.73 g $FeSO_4$/ g of ore, 130 g/l total ammonia in 10 hours. At the same conditions the cobalt and manganese extractions were only 16% and 43%, respectively. The different leaching behavior between nickel, cobalt and manganese is due to the fact that they are associated with different mineralogical phases.

Nickel extraction is improved by increased temperature, ammonia concentration and ferrous sulphate concentration; cobalt extraction requires low temperature and ferrous sulphate concentration as well as high ammonia concentration. High ferrous sulphates to ore ratios do not improve nickel extraction, likely due to magnetite precipitation around goethite particles, thus preventing contact between particle and reductant.

References

1. R.G. McDonald, B.I. Whittington, "Atmospheric acid leaching of nickel laterites review", Part I. Sulphuric acid technologies, Hydrometallurgy, 91 (2008), pp. 35–55.

2. R.G. McDonald, B.I. Whittington, "Atmospheric acid leaching of nickel laterites review", Part II. Chloride and Bio-Technologies, Hydrometallurgy, 91 (2008), pp. 56–69.

3. X. Meng, K. Han, "The principles and applications of ammonia leaching of metals", Mineral Processing and Extractive Metallurgy Review, 16 (1) (1996), pp. 23–61.

4. D. Kerfoot, D. Weir, "The hydro and electrometallurgy of nickel and cobalt", G. Tyroler, C. Landolt (Eds.), Extractive Metallurgy of Nickel and Cobalt (Proceedings of Intl. Symposium), TMS, Warrendale, Pa (1988), pp. 241–267.

5. A. Monhemius, "Treatment of laterite ores of nickel to produce ferronickel, matte or precipitated sulphide", Extractive Metallurgy of Nickel, A. R. Burkin (Editor), Soc. Chem. Ind., London, 1987, pp 51–75.

6. M. Nicol, A. Nikoloski, J. Fittock, "A fundamental study of the leaching reactions involved in the Caron process", W. Impie (Ed.) *et al.*, Proceedings International Laterite Nickel Symposium, Charlotte, NC, TMS, Warrendale, Pa (2004), pp. 369–384.

7. S. Acharya, S. Anand, S. Das, R. Das, P. Jena, "Ammonia leaching of ocean nodules using various reductants", Erzmetall, 42 (2) (1989), pp. 66–73.

8. R. Das, S. Anand, "Precipitation of iron oxides from ammonia-ammonium sulphate solutions", Hydrometallurgy, 38 (1995), pp. 161–173.

9. M. Mohapatra, S. Anand, R. Das, "Aqueous reduction of crystalline goethite under ammoniacal conditions", Hydrometallurgy, 65 (2–3) (2002), pp. 227–235.

10. M. Zuniga, E. Asselin, "Reduction of limonitic laterite in ammoniacal solutions using metallic iron", Hydrometallurgy, 104(2) (2010), pp. 260–267.

11. E. Asselin, ''Thermochemical aspects of the Fe, Co, Ni–NH_3–H_2O systems relevant to the Caron process'', Hydrometallurgy 2008 (Proceedings 6th International Symposium on Hydrometallurgy, Phoenix, Arizona), C. Young, P. Taylor, C. Anderson and Y. Choi (Editors), TMS, Warrendale, Pa., 2008, pp. 522–531.

12. Isaev et al., "The formation of iron(II) ammines in aqueous solution", Russian Journal of Inorganic Chemistry, 35 (8) (1990), pp. 1162–1164.

13. K. Osseo-Asare, "Application of activity-activity diagrams to ammonia hydrometallurgy: Fe–NH_3–H_2O, Fe–NH_3–H_2O–CO_3 and Fe–NH_3–H_2O–SO_4 systems at 25 °C", Transactions of Institute of Mining and Metallurgy, 90 (1981), pp. 159–164.

14. K. Osseo-Asare, "Application of activity-activity diagrams to ammonia hydrometallurgy II. The copper, nickel and cobalt ammonia water systems at elevated temperatures", Process Fundamental Considerations in Selected Hydrometallurgy Systems, M. Kuhn (Editor), AIME, New York, 1981, pp. 360–369.

15. K. Osseo-Asare, S. Asihene, "Heterogeneous equilibria in ammonia/lateriteleaching systems", D. Evans, R. Shoemaker, H. Veltman (Eds.), Proceedings International Laterite Symposium, TMS, Warrendale (1979), pp. 585–609.

16. S. Zhong, M. Hepworth, "A calculation method for determining equilibria in metal–ammonia–water systems", Hydrometallurgy, 38 (1) (1995), pp. 15–37.

17. D. Georgiou, V. Papangelakis, "Characterization of limonitic laterite and solids during sulfuric acid pressure leaching using transmission electron microscopy", Minerals Engineering, 17 (2004), pp. 461–463.

18. J. Wang, T. Deng, "Studies on the behavior of Co^{2+} and Zn^{2+} in the conversion from goethite to magnetite in NH_3(NaOH)–Fe^{2+}, Co^{2+}–H_2O medium under hydrothermal conditions", Hydrometallurgy, 92 (2008), pp. 107–114.

19. F. Esmadi, J. Simm, "Sorption of cobalt(II) by amorphous ferric hydroxide", Colloids and Surfaces A: Physicochemical and Engineering Aspects, 104 (2/3) (1995), pp. 265–270.

20. K. Osseo-Asare, D. Fuerstenau, "Adsorption phenomena in hydrometallurgy. 1. The uptake of copper, nickel and cobalt by oxide adsorbents in aqueous ammoniacal solutions", International Journal of Mineral Processing, 6 (2) (1979), pp. 85–104.

21. K. Osseo-Asare, D. Fuerstenau, "Adsorption phenomena in hydrometallurgy. 2. Electrophoretic behavior of oxides in aqueous ammoniacal solutions of copper, nickel and cobalt", International Journal of Mineral Processing, 7 (2) (1980), pp. 117–127.

22. K. Osseo-Asare, D. Fuerstenau, "Adsorption phenomena in hydrometallurgy. 3. Model for copper, nickel and cobalt uptake by oxides adsorbents in aqueous ammoniacal solutions", International Journal of Mineral Processing, 7 (3) (1980), pp. 219–234.

23. K. Osseo-Asare, D. Fuerstenau, "Adsorption of copper, nickel, and cobalt by oxide adsorbents from aqueous ammoniacal solutions", Journal of Colloid and Interface Science, 118 (2) (1987), pp. 524–542.

24. K. Osseo-Asare, and D. Fuerstenau, "Adsorption losses in ammonia leaching of copper, nickel and cobalt from deep-sea manganese nodules", Complex Metall. 78', (Proceedings Intl. Symposium, Bad Harzburg, Germany), M. Jones (Editor), IMM, London, 1978, pp. 42–47.

Ni-Co 2013
Edited by: Thomas Battle, Michael Moats, Violina Cocalia, Harald Oosterhof, Shafiq Alam, Antoine Allanore, Rodney Jones, Nathan Stubina, Corby Anderson, and Shijie Wang
TMS (The Minerals, Metals & Materials Society), 2013

SULFURIC ACID LEACHING CHARACTERISTICS OF Ni-DOPED GOETHITE

Guanghui Li, Wen Cai, Mingjun Rao[1], Qian Zhi, Tao Jiang

School of Minerals Processing & Bioengineering, Central South University, Changsha, Hunan 410083, China

Keywords: Nickeliferous Goethite, Ni-doped Goethite, Sulfuric Acid, Leaching, Kinetics

Abstract

In the atmospheric acid leaching process of low-grade nickelferous laterite, the leaching mechanisms still remain unclear when sulfuric acid is used as the leaching agent. It is therefore imperative to study the leaching characteristics of Ni-doped goethite. In this work samples of goethite and nickeliferous goethite were synthesized and characterized by chemical composition, surface area and XRD analyses. The results show that part of nickel is incorporated into the lattice of nickeliferous goethite. The synthesized nickeliferous goethite was pre-leached in diluted hydrochloric acid at 25°C to separate the adsorbed nickel. Subsequently, the Ni-doped goethite in which nickel is completely distributed within the lattice was leached in sulfuric acid. The results indicated that the extraction of Ni was incongruent with that of Fe. Sulfuric acid leaching kinetics of Ni and Fe are characteristic of chemical reaction and internal diffusion controlled processes, respectively.

Introduction

Due to the increasing demand for nickel around the world, the utilizization of nickeliferous laterite ores is imperitive as such ores account for approx. 72% of the world reserves of nickel. Classified from the main mineral constituents, limonitic laterite mainly contains goethite which itself contains low nickel concentrations and high iron concentrations, and account for about 28% of the world's Ni resource. Atmospheric acid leaching is a promising method to extract nickel from limonitic laterite. However, some problems still exist in this process such as low selective extraction of nickel relative to iron, large acid consumption, and long leaching time. Since mostnickel exists in goethite of limonitic laterite; it is necessary to study the leaching characteristics of nickeliferous goethite in order to reveal the acid leaching mechanisms of limonitic laterite.

Research studies conducted on the characterization and leaching behavior of pure and

[1] Corresponding Author: Dr. Mingjun Rao, Email: raomingjun2003@126.com, Tel: +86-731-88830542

nickeliferous goethites have been carried out by Rakesh et al. [1]who investigated the leaching behavior of nickeliferous goethites by adsorption and co-precipitation modes. The authors results indicated that high levels of extraction of nickel were achieved for the two types of nickeliferous goethites examined and that extraction of nickel of nickeliferous goethite from co-precipitation mode was superior to that observed from the adsorption mode. Work done by P. Lussiez [2] concerning the acid leaching behaviors of nickeliferous goethites, suggested that part of the Ni^{2+} ions were absorbed on the surface of synthesized goethite, and that the concentration of Ni^{2+} ions lead to an increased rate of dissolution Apparently, the reports mentioned above cannot fully interpret the leaching behaviors of nickeliferous goethite in sulfuric acid, due to the influence of Ni^{2+} ions absorbed on the surface of goethite. So far, it seems that there has been no such report of the leaching behaviors of goethite with Ni^{2+} ions completely incorporated into the crystal lattice when leached in sulfuric acid.

In this paper, the obtained Ni-doped goethite in which nickel was completely distributed within the lattice was leached in sulfuric acid, and the effect of leaching time and temperature on extraction of nickel and iron was studied. The controlling step of leaching process was revealed by evaluating the leaching kinetics of Ni and Fe. The results in this paper may be consulted for use in developing a more effective atmospheric pressure process for extracting nickel from laterite.

Materials and Methods

Reagents

Reagents involving ferric nitrate nonahydrate ($Fe(NO_3)_3 \cdot 9H_2O$), sodium hydroxide (NaOH), nickel nitrate hexahydrate ($Ni(NO_3)_2 \cdot 6H_2O$) and sodium thiosulfate pentahydrate ($Na_2S_2O_3 \cdot 5H_2O$) used in this paper are all of analytical grade.

Synthesis and Characterization of Pure and Nickeliferous Goethite

Pure goethite samples used in the present study were prepared by the co-precipitation method in the laboratory. Following the same procedure as described by Schwertman and Cornell [3, 4], 1 mol/L ferric nitrate and 10 mol/L sodium hydroxide solutions were mixed under the given conditions. preparation of nickeliferous goethite was achieved by simultaneously adding a 8.52 mmol/L $Ni(NO_3)_2$ solution with 10 mol/L sodium hydroxide solution drop wise to a 1 mol/L ferric nitrate solution The suspension was aged for 48 h at 70 °C, and then washed thoroughly with distilled water until a pH of 7-8 was obtained. The samples were subsequently dried and grinded. The total iron grade of both pure and nickeliferous samples was determined by titrimetry method, and the nickel content in the doped sample was determined by atomic absorption spectrophotometry. Samples of goethite and nickeliferous goethite were characterized by chemical composition, surface area and XRD analyses.

Chemical composition and surface area of the both samples of goethite are given in Table I, which shows that surface area has been increased from 47.44 to 57.46 m^2/g by doping Ni^{2+} ions in goethite matrix. The surface area increasing with Ni^{2+} ions doping has been also reported by M. Mohapatra [5].

Table I. Main chemical composition and surface area of pure and Ni-doped goethite

Samples	Surface area (m^2/g)	Fe (mass%)	Ni (mass%)
Pure goethite	47.77	61.81	—
nickeliferous goethite	57.46	62.94	0.84

XRD patterns of both samples of goethite shown in Fig.1, reveal that all of the diffractions of synthetic pure goethite could be indexed to a pure orthorhombic α-FeOOH phase, which was easily detected using data available in the Powder Diffraction File [6]. For nickeliferous goethite, only the intensity of the main diffraction peaks changed comparison to the pure goethite. Internal disorder always exists in synthetic goethite, the degree of which increases with increasing unit cell edge length, c [7]. c-values procured from the pure and nickeliferous goethite were 0.46149, 0.46175nm, respectively, indicating that the internal disorder of nickeliferous goethite was larger, thus demonstrating that nickel was incorporated into the lattice structure of goethite.

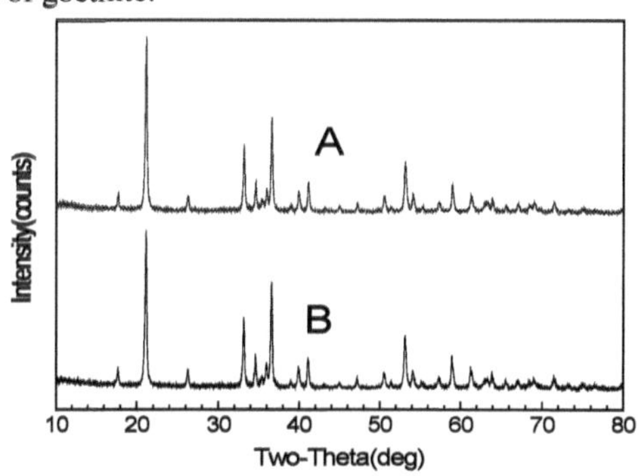

Figure 1. XRD patterns of pure (A) and Ni-doped goethite (B)

<u>Preparation of Ni-doped Goethite</u>

On the basis of evaluating the mechanisms of synthesizing α-FeOOH via the hydrothermal method [8] and characterization of nickeliferous goethite, part of nickel was incorporated into the crystal lattice of nickeliferous goethite, while the other part absorbed on the surface in the form of hydroxo complexes.

The research was focused on the acid leaching mechanism of goethite with nickel incorporated into goethite structure totally. Therefore, it was necessary to examine the synthesized nickeliferous goethite.

R. M. Cornell [9] determined that nickel hydroxo complexes could be removed by diluted hydrochloric acid liquor at room temperature, without effecting the internal structure of goethite. The procedure was as follows: A measured amount of synthesized nickeliferous goethite was leached in 0.4mol/L hydrochloric acid for 2h at 25 °C, and then filtered, washed, dried, and grinded.

Acid Leaching Tests

Leaching tests were conducted in a DY-8 autoclave equipped with eight 70 ml stainless steel vessels rotating end to end in glycerin bath. The bath was electrically heated and its temperature controlled by a thermostat. At the beginning of each trial, 5g of the Ni-doped goethite sample were combined with 50 ml of 8% (w/w) sulfuric acid inside the stainless steel autoclave vessels. The sealed vessels were subsequently soaked in the bath at a certaintemperature and rotated at 30 rpm. Filtration was performed immediately after leaching, followed by atomic absorption analysis of the filtrate to determine the nickel content, while the residue was dried and analyzed for total iron content by titration [10].

Results and Discussion

Sulfuric Acid Leaching

As shown in Fig.2 extraction of nickel increases with increasing temperature and time, however the variability of leaching differed at different temperatures, the higher temperature resulting in faster leaching of nickel with extended time. Rakesh et al. [1] also determined that extraction of nickel from synthesized nickeliferous goethite in sulfuric acid at lower temperatures increased considerably (>50%) within a short time, due to the absorption of Ni on the surface of the nickeliferous goethite as hydroxo complexes. For leaching of Ni-doped goethite in Fig.2, extraction of nickel decreased to approx. 12% within 15 minutes.

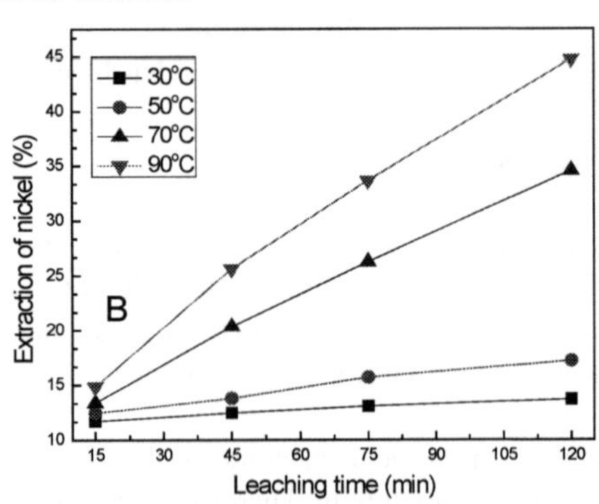

Figure 2. Effects of leaching time on extraction of nickel of Ni-doped goethite at

different temperatures

Fig.3 shows the effect of leaching time and temperature on the leaching of iron in sulfuric acid pre-leached Ni-doped goethite. Evident from the results was the low extraction of iron (< 9%), which increased with increasing temperature and time .

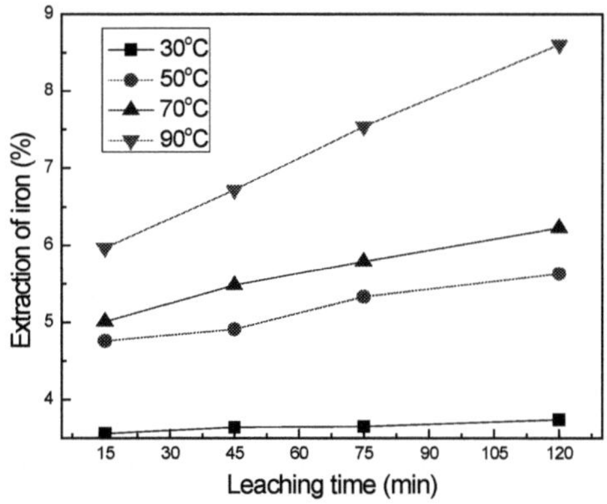

Figure 3. Effects of leaching time on extraction of iron at different temperatures (pre-leached Ni-doped goethite)

Fig.4 shows the relationship between the extraction of nickel and iron (leaching temperature 50 °C) in the acid leaching tests of Ni-doped goethite (in similar 20°C increments from 30 to 90°C). From the results it appears as though the extraction of nickel and iron is incongruent, which suggests that nickel is distributed non-uniformly within the lattice of goethite, possibly towards the core or periphery of the crystals. R. Lim-Nunez and R. J. Gilkes [11] proposed a similar belief that goethite doped with Co, Al, Mn all satisfy homogeneous leaching regularity, while goethite doped with Ni and Cr, did not due to that no statistically significant systematic displacement of unit-cell dimensions from those for unsubstituted goethites was observed for Cr- and Ni-goethites. A similar conclusion was made by R. M. Cornell [9] in that Ni distribution in α-FeOOH is non-uniform; a significant concentration of Ni observed concentrated near the surface of the crystals, furthermore this distribution is in accord with what would be expected from the crystallization kinetics of α-Ni, FeOOH. A. Manceau [12] indicated that the Ni site in synthetic goethite is strongly relaxed with a contraction of the goethite structure in the plane of edge-sharing double chains (bc plane), and an expansion in the direction of corner linkages (approximately the ab plane). This anisotropic relaxation of the Ni site locally reduces the distortion of the goethite structure.

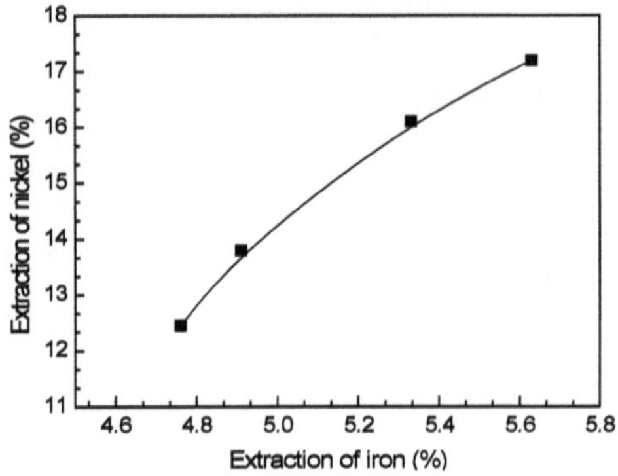

Figure 4. Variation of extraction of nickel with iron (50 °C)

<u>Leaching Kinetics of Ni and Fe</u>

To determine the kinetic parameters and rate-controlling step for the leaching of metals, the experimental data were analyzed according to the shrinking core model.

It was found that leaching of nickel fits for the equation $1\text{-}(1\text{-}R)^{1/3} = k * t$, while the equation $1\text{-}2R/3\text{-}(1\text{-}R)^{2/3} = (2D_2C_0/r_0^2\alpha\rho)t$ for iron, and both of them present straight lines for a good approximation (Fig.5). This suggests that leaching kinetics of Ni and Fe are characteristic for chemical reaction and internal diffusion controlled processes, respectively.

According to Arrhenius equation:

$$\ln k = -E/(RT) + \ln$$

The *lnk* vs *1/T* is plotted, and slope of the line is the activation energy of leaching reaction.

Arrhenius fitting curves of leaching of nickel and iron are shown in Fig.6, and the activation energies of nickel and iron are 46.1kJ/mol and 38.7kJ/mol, respectively.

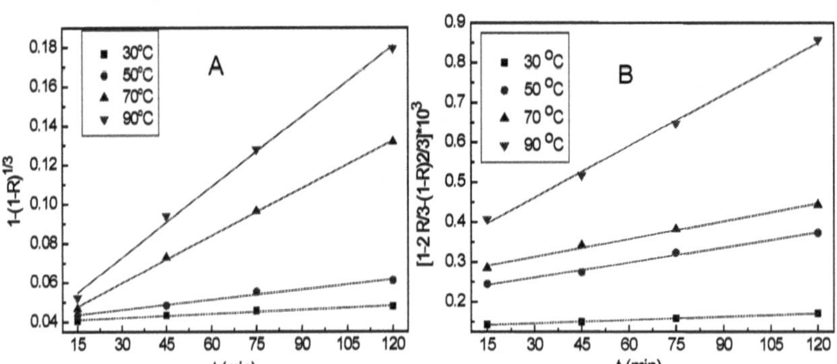

Figure 5. Curve of leaching kinetics of Ni (A), Fe (B)

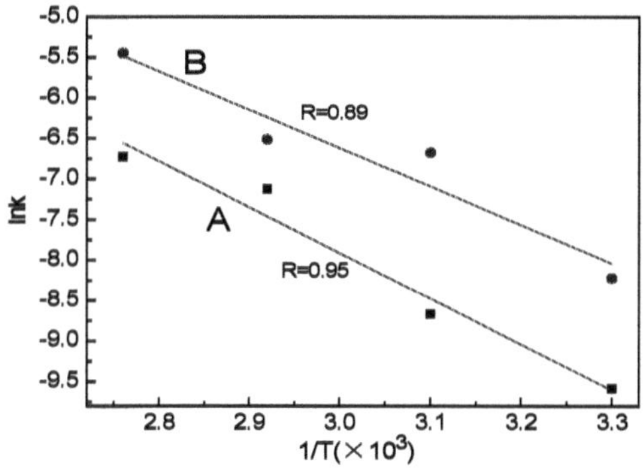

Figure 6. Arrhenius fitting curves of leaching of Ni (A) and Fe (B)

Conclusions

(1) Samples of goethite and nickeliferous goethite were synthesized and characterized by chemical composition, surface area and XRD analyses. The results show that the bulk of nickel was incorporated into the crystal lattice of nickeliferous goethite, and the rest of nickel was absorbed on the surface in the form of hydroxo complexes.

(2) The results obtained from the Ni-doped goethite in which nickel was completely distributed throughout the lattice was leached in sulfuric acid, indicated that extraction of nickel and iron increased with increasing temperature and time. As the leaching ratio of nickel was as low as 44.74% even at 90 °C for 120min, for that the internal structure of goethite is hardly destroyed. Meanwhile, the extraction of Ni was incongruent with that of Fe, which demonstrates that incorporated Ni is nonuniformly distributed in he lattice.

(3) Sulfuric acid leaching kinetics of Ni and Fe were characteristic of chemical reaction and internal diffusion controlled processes, respectively. The calculated activation energies for Ni and Fe were 46.1 and 38.7 kJ/mol, respectively.

Acknowledgments

The authors want to express their thanks to the National Science Fund (51174230/51234008) for The Ministry of education program for New Century Excellent Talents (NCET-11-0515) for financial support of this research.

References

[1] R. Kumar, R.K. Ray, and A.K. Biswas, "Physico-chemical nature and leaching behaviour of goethites containing Ni, Co and Cu in the sorption and coprecipitation mode," *Hydrometallurgy*, 25 (1990), 61-83.

[2] P. Lussiez and K. Osseo-Asare, "synthesis, characterization and dissolution

behavior of cupriferous and nickeliferous goethites," *Process Mineralogy: Extractive Metallurgy, Mineral Exploration; Energy Resources, Chicago*, 1981 , 291-304.

[3] S. Mustafa, S. Khan, and M. Iqbal Zaman, "Effect of Ni^{2+} ion doping on the physical characteristics and chromate adsorption behavior of goethite," *water research*, 44 (2010), 918-926.

[4] U. Schwertman and R.M. Cornell, *Iron Oxides in the Laboratory—Preparation and characterization* (New York, NY: VCH Publishers, Inc., 1991), 73-74.

[5] M. Mohapatra, S.K. Sahoo, and S. Anand, "Removal of As(V) by Cu(II)-, Ni(II)-, or Co(II)-doped goethite samples," *Journal of Colloid and Interface Science*, 298 (2006), 6-12.

[6] JCPDS PDF cards No. 29-713 and 81-0464 for α-FeOOH, cards No.86-2267, 74-2081 and 10-0325 for $NiFe_2O_4$; International Centre for Diffraction Data, Joint Committee on Powder Diffraction Standards, Powder Diffraction File, 1601 Park Lane, Swarthmore, PA 19081, USA.

[7] U. Schwertmann, P. Cambier, and E. Murad , "Properties of goethites of varying crystallinity," *Clays Clay Min*, 33 (1985), 369-378.

[8] D. Crerar, S. Wood, and S. Brantley, "Chemical controls on solubility of ore forming minerals in hydrothermal solutions," *The Canadian Mineralogist*, 23 (3) (1985), 333-352.

[9] R. M. Cornell, R. Giovanoli, and W. Schneider, "The Effect of Nickel on the Conversion of Amorphous Iron(Ⅲ) Hydroxide into more Crystalline Iron Oxides in Alkaline Media," *J. Chem. Tech. Biotechnol*, 53 (1992), 73-79.

[10] G.H. Li , M.J. Rao, and T. Jiang, "Leaching of limonitic laterite ore by acidic thiosulfate solution," *Minerals Engineering*, 24 (2011), 859–863.

[11] R. Lim-Nunez and R. J .Gilkes, "Acid dissolution of synthetic metal-containing goethites and hematites," *The Clay Minerals Society, Bloomington, Indiana*, 1987, 197-204.

[12] A. Manceau, M.L. Schlegel, and M. Musso, "Crystal chemistry of trace elements in natural and synthetic goethite," *Geochimica et Cosmochimica Acta*, 64 (2000), 3643–3661.

Ni-Co 2013
Edited by: Thomas Battle, Michael Moats, Violina Cocalia, Harald Oosterhof, Shafiq Alam, Antoine Allanore, Rodney Jones, Nathan Stubina, Corby Anderson, and Shijie Wang
TMS (The Minerals, Metals & Materials Society), 2013

CHARACTERISTICS OF NICKEL LATERITE CRUSHED ORE AGGLOMERATES

Adirek Janwong, Nikhil Dhawan, Thien Vethsodsakda and Michael S. Moats

Department of Metallurgical Engineering, College of Mines and Earth Sciences, University of Utah, 135 South 1460 East, Room 412

Keywords: Nickel Laterite, Crushed Ore Agglomeration, Agglomerate Size Distribution, Electrical Conductivity, Permeability

Abstract

Crushed ore agglomeration has been used and is considered as a pretreatment step to improve heap leaching performance of complex copper and gold ores. It is well known that proper agglomeration can prevent solution channeling and ponding, improve percolation within the heap and increase metal recovery rates. However, not much is published regarding nickel laterite agglomeration behavior. Agglomerate size distribution and bed permeability are important parameters to evaluate the quality of agglomerates. With proper considerations, electrical conductivity can be used to control the agglomeration conditions. In this study, a nickel laterite ore from Piaui, Brazil was agglomerated and subjected to different tests to evaluate agglomeration conditions. The study included determination of agglomerate size distribution, saturated hydraulic permeability and electrical conductivity response. It has been found that moisture content and acid concentration have opposite effects on the agglomerate size. Also, retention time and drum rotation speed have small effect on agglomerate size. As expected the permeability of the agglomerate bed is strongly correlated to agglomerate size. Electrical conductivity of the agglomerate bed is affected by agglomeration size, moisture content, acid concentration, and external compaction.

Introduction

Agglomeration has been used to improve the performance of heap leaching for many ores. Proper agglomeration improves permeability within the heap, prevents solution ponding and channeling, and increases the metal recovery. Suitable agglomerates for heap leaching can be considered from their quality. Agglomerate quality can be characterized using size distribution of the agglomerates, strength of agglomerates, and permeability of agglomerates and agglomerate bed.

In order to control the qualities of agglomerates, several techniques have been proposed such as electrical conductivity measurements [1], permeability of agglomerate bed measurements [2, 3], dip tests [4] and soak tests [5].

The agglomerate size distribution, permeability of agglomerate beds, and electrical conductivity of agglomerates were investigated in this study. It is believed that this is the first report of these parameters for a nickel laterite ore to be published in the open literature. Important agglomeration factors such as acid concentration, moisture content, drum rotation speed and mixing time were varied to obtain different agglomerates for characterization.

Experimental Procedures

A nickel laterite from Piaui, Brazil was used in this study. The top size of the ore was limited to 12.5 mm. The nickel laterite was composed mainly of quartz, silico-ferruginous plasma, and clay minerals. The nickel content is 0.9%. The main nickel–bearing minerals are chlorite, smectite, vermiculite, manganese oxide and silico-ferruginous plasma [6]. Since moisture content, acid concentration, retention time and drum rotation speed are operating factors in drum agglomeration, they were investigated in this study.

In nickel laterite ore agglomeration, the moisture contents reported lie within a wide range (10-30%) due to the nature of the ores [7; 8; 9; 10]. Thus, in this study the moisture content was varied between 10 to 25% on a dry ore basis. Moisture content includes water and acid in its calculation as shown in Equation 1.

$$Moisture\ content\ (\mu,\ \%) = \frac{Mass_{wet} - Mass_{dry}}{Mass_{dry}} x100 \qquad (1)$$

Sulfuric acid concentration in this study ranged from 102 to 972 g/l which overlaps the range indicated by Duyvesteyn et al. [11]. The amount of the sulfuric acid solution required for agglomeration was firstly calculated, and then the obtained acid solution amounts were used to calculate the amount of sulfuric acid and water applied to the drum via different hoses.

Retention time for an industrial drum agglomerator has been reported to be 1 to 4 minutes [12; 13]. As such, time was fixed at 3 minutes for the first set of experiments when moisture and acid strength were being investigated. Then to examine the effect of time, it was varied in the range of 3 to 6 minutes. Due to the experimental setup used, shorter mixing times were not obtainable in a safe and reproducible manner.

Speeds for commercial agglomeration drums have been reported to be 30-50% of the critical speed [12]. In this study, 31.5% of the critical speed was used in the first stage and then in the second stage, 15.7, 23.6, 39.3 and 47.1% of the critical speed were considered to observe the effect of drum speed.

Agglomeration was performed in batch mode. 4 kg of material were placed in a plastic drum used as the agglomerator. The drum had dimensions of 26.67 cm in diameter and 36.2 cm in length. The plastic drum had 4 lifters of 5.0 mm height, 20 mm width and 33 cm length spaced evenly around the circumference of the drum. To achieve homogeneity, the sample was mixed for 30 minutes in the drum at a pre-determined speed, corresponding to the speed used in the experiments. Water and acid were applied separately to imitate industrial practice. Peristaltic pumps were used to control the addition rate of each liquid. The solution applying time was 1/3 of the agglomeration time. Three batches were agglomerated in order to obtain enough sample (12 kg total) per condition.

Three tests - size distribution, electrical conductivity and permeability/bulk density - were performed to characterize the agglomerates. The size distribution of the agglomerates in this study was determined by hand screening. The hand sieving method were used to limit the breakage of agglomerates during the test. 350 g of agglomerates was collected by the cone and quartering technique. The agglomerates were then air dried for 24 hours to stabilize the agglomerates. The very fine particles which were smaller than 1.7 mm were difficult to determine by hand screening. Therefore, in this study the smallest sieve size was 1.7 mm. The agglomerate size distribution (ASD) curve, D_{10} and D_{50} of the agglomerates were obtained from the sieve data.

To measure electrical conductivity, 650 g of agglomerated sample were placed in a specially designed electrical conductivity measurement cell shown schematically in Figures 1. The cell was carefully dried prior to introduction of the sample to avoid changes in output voltage due to extra moisture. Five different direct current voltages were applied to the circuit with a maximum voltage of 3 V to avoid corrosion of the stainless steel electrodes.

After finishing the first conductivity tests with no mechanical load applied, a second conductivity test was performed at a load of 20.4 kg (this simulates the pressure that the agglomerates face at about 2 m height in the real heap). The amount of load was calculated from the bulk density of the ores. The experimental procedures were the same as when no load was applied.

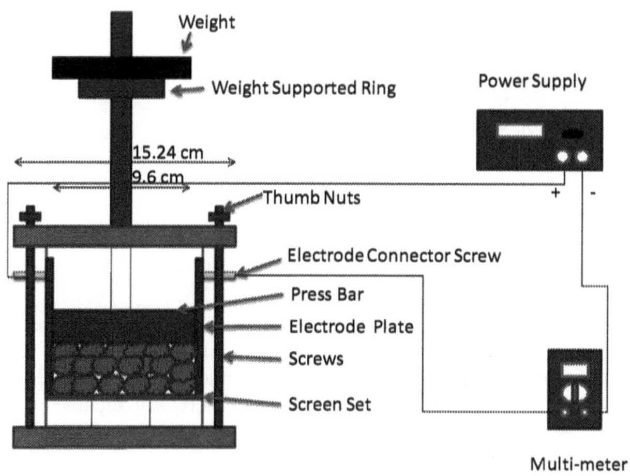

Figure 1. The schematic diagram of the electrical conductivity test set up.

After finishing the electrical conductivity tests, a hydraulic permeability test was performed based on ASTM D 2434-68 [14]. 10 kg of agglomerates were placed into a 19.05 cm inner diameter test column. The test set up is shown schematically in Figure 2. The agglomerated bed was flooded by introducing water from the top. The flow rate was adjusted to achieve a steady state value. A difference in pressure or head was then measured across the agglomerate bed from the attached manometer. Head as a function of various flow rates were obtained. The values of area and the height of the agglomerated ore bed were measured to allow the calculation of hydraulic permeability. Bulk density, before and after flooding, was also calculated from the collected data.

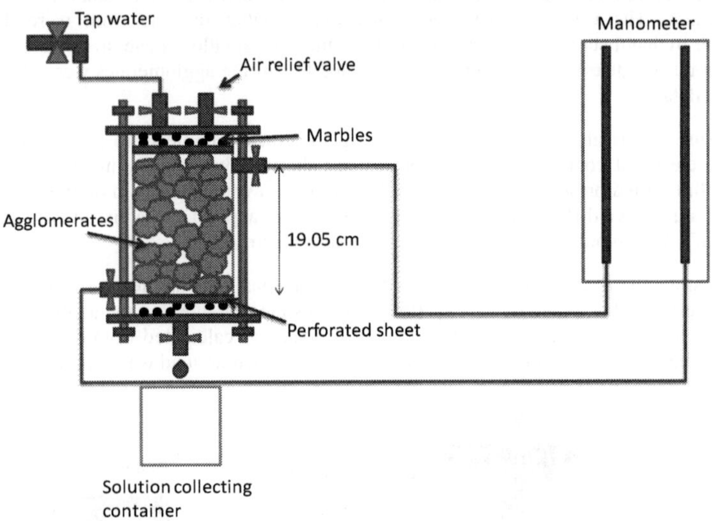

Figure 2. The schematic diagram of permeability test set up.

Results and Discussion

Agglomeration Size Distribution

The main objective of agglomeration is to reduce the number of fine particles in the ore body and produce suitable agglomerate size distributions (ASDs). The preferable size of agglomerates varies from ore to ore. The amount of fines left unattached to other fines or coarser particles affects the permeability of the agglomerate body. For this reason, unattached fine particles in the ore body should also be studied.

The effects of moisture content, acid concentration, drum rotation speed and mixing time on the agglomerate size distribution of the nickel laterite were studied. In the agglomerate size distribution figures, the legends have the form of A: B: C. A represents the moisture content in wt,, while B represents the acid solution concentration (g/L). Finally, C represents special conditions; for example NC is for drum critical speeds and min is for mixing times in minutes.

The effects of moisture content and acid concentration on the size of the nickel laterite agglomerates were studied first. Mixing time and drum rotation speed were keep constant at three minutes and 31.5% NC, respectively. Moisture contents were varied approximately from 10 to 25% on a dry basis. Acid concentration was varied from 102 to 972 g/L.

As expected the size distribution of agglomerates was coarser as moisture content increased as shown in Figure 3. As the moisture content increases obviously the volume of liquid added during agglomeration also increases. The liquid fills the void space between the adjacent particles creating capillary forces between those particles. The capillary forces provide enough force for nearby particles to attach to each other. If there is enough solution provided to the system, the liquid will go into the free surrounding surface of the nuclei agglomerates and attract free particles to attach to them and become coarser. The smaller particles and fines in the samples also play an important role in the agglomeration process. They can act as filling agents, filling the empty space between the coarser particles and bonding them together.

As the moisture content increases, the amount of smaller size agglomerates decreases and the amount of coarser size agglomerates increases. Bouffard [4] also observed the same effects of moisture content as she mentioned that higher moisture content produced a narrower size distribution of agglomerates. The amount of agglomerates smaller than 1.7 mm decreased as the liquid in the system increased. A significant decrease of less than one mm agglomerates can be observed when increasing the moisture content from 10.2 to 15.3 wt%. The agglomerates continue to grow as more liquid phase is introduced to the system. The coarse agglomerates which are larger than 13.33 mm can be observed as the moisture content increases from 15.3 to 20.4 wt%. The growth of coarser agglomerates as the moisture content increases is attributed to smaller agglomerates that attached to the larger agglomerates.

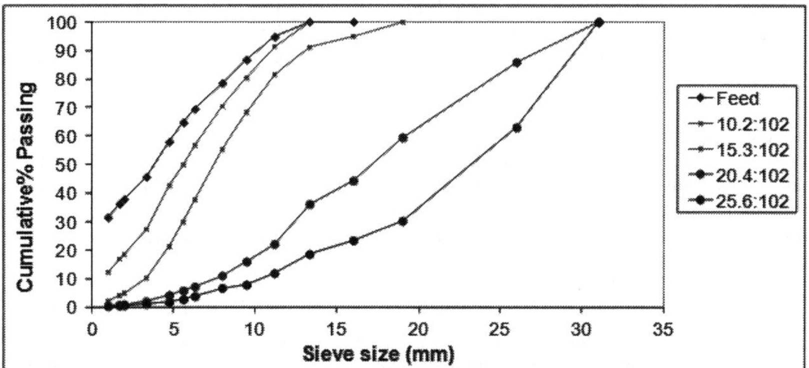

Figure 3. ASDs obtained with different moisture contents (102 g/L acid concentration) are compared to the feed particle size distribution.

Figure 4 illustrates ASDs for batches produced with relatively constant moisture content while acid concentration was varied. From the graph, it is clearly seen that an inverse relationship exists between agglomerated size and acid concentration. Agglomerates become smaller as the acid concentration increases. The liquid volume introduced to the agglomeration process decreases as the acid concentration increases while moisture content remains constant. Therefore the capillary force which helps to hold particles together decreases as less liquid is present in the empty spaces surrounding particles. This leads to smaller agglomerates observed.

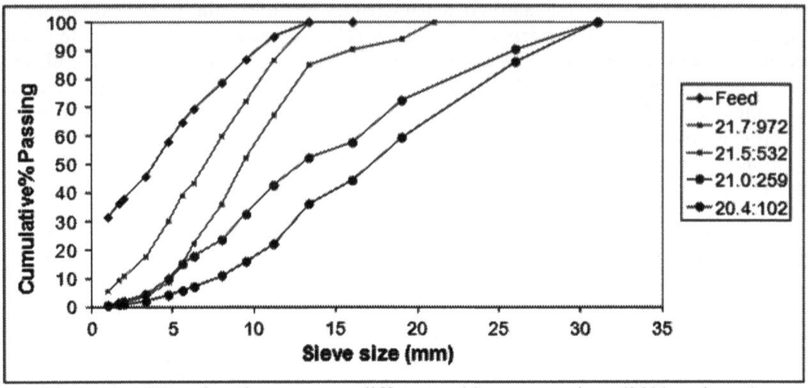

Figure 4. ASDs of agglomerates at different acid concentrations (20% moisture).

Chamberlin recommends that no -150 μm (-100 mesh) fines remain unattached after agglomeration [13]. To measure the amount of -150 μm particles wet screening is needed. As the agglomerates are held together by a liquid bridge, wet screening will destroy the agglomerates. Thus, agglomerates which are smaller than 1.7 mm (-10 Tyler mesh screen) are considered as fines in this study. Increasing moisture content generally leads to less fines and larger agglomerates.

While the agglomerate size increases with increasing moisture, the -1.7 mm material eventually reaches a minimum. Beyond this minimum, adding more liquid resulted in the formation of "mud". Considering the results in the first phase of experiments, about 20% moisture content is the maximum moisture content for agglomeration (unless very high acid strengths are used) as illustrated in Figure 5. Beyond this point the fine particles were not further decreased and the formation of mud was observed.

To illustrate that the volume of solution added is important, both D_{50} and D_{10} are plotted in Figure 6 versus the volume of solution applied during agglomeration. While there is some scatter in the data, a general trend shows that agglomerate size increases with increasing solution volume addition. At low solution addition, both D_{50} and D_{10} increase gradually. As the liquid volume increase beyond a point (~100 L/tonne), the D_{50} and D_{10} begin to increase. At this point, the solution addition volume is sufficient to provide a capillary force necessary for agglomeration.

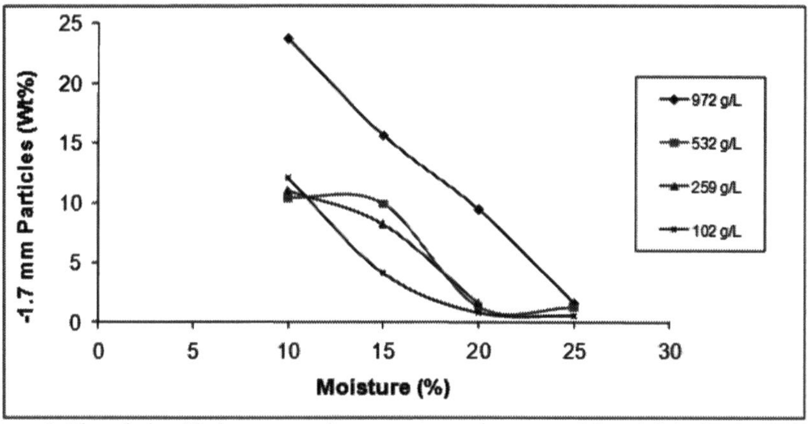

Figure 5. The relationship between moisture content and amount of fines left in the ore body at different acid concentrations as indicated in the legend.

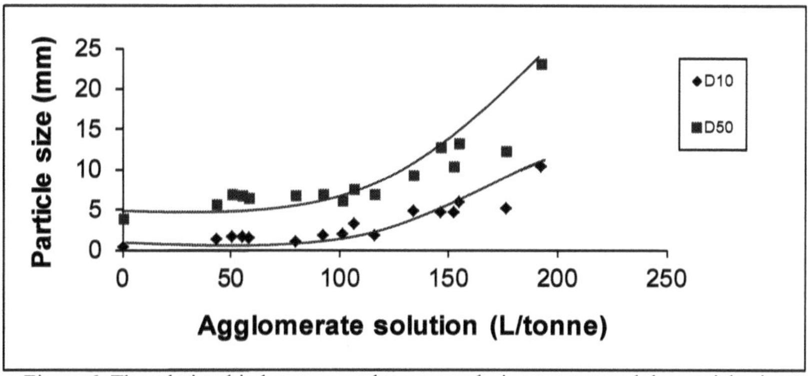

Figure 6. The relationship between agglomerate solution amount and the particle size distribution.

Drum rotation speed was examined while moisture content, acid concentration and mixing time were held constant at 20.4%, 102 g/L H_2SO_4 and three minutes, respectively. The drum rotation speeds were change from 15.7 to 47.1% percent of the critical speed. Figure 7 shows the ASD curves for agglomerates produced at different drum rotation speeds. The -1.7 mm weight percentage, D_{10} and D_{50} of these agglomerates are summarized in Table I. From the data, the agglomerate size distribution curves versus drum rotation speed are not significantly different.

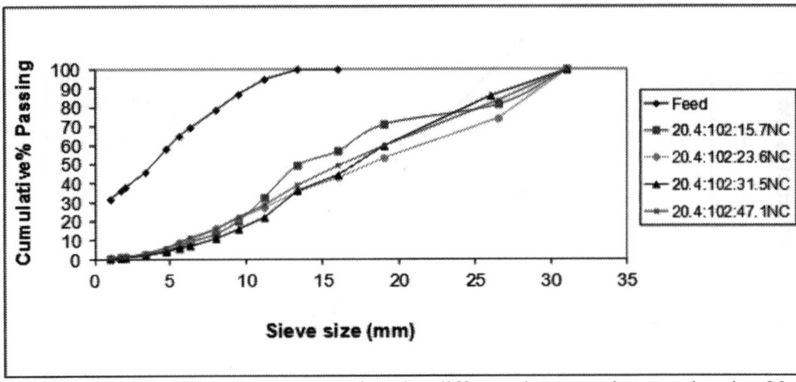

Figure 7. The ASDs of agglomerates produced at different drum rotation speeds using 20.4% moisture and 102 g/L H_2SO_4.

Table I. Fine particle, D_{10}, and D_{50} of agglomerates at different drum rotation speeds.

Drum rotation speed (%NC)	Fine particles less than 1.7 mm (%)	D_{10} (mm)	D_{50} (mm)	Percent of +19 mm particles (%)
15.7	1.03	6.63	13.53	28.90
23.6	1.39	6.21	18.05	46.80
31.5	0.77	7.55	17.12	40.54
47.1	1.03	5.86	16.18	40.40

Mixing (retention) time was studied while drum rotation speed, moisture content and acid concentration were fixed at 31.5% NC, 20.4% and 102 g/L. Mixing times were examined over the range of three to six minutes. Figure 8 shows the agglomerate size distribution curves of agglomerates at different mixing times. The -1.7 mm weight percentage, D_{10} and D_{50} of these agglomerates are summarized in Table II. In general, for the conditions examined, mixing time had some effect on the resulting agglomerate size distribution. The ASDs of the agglomerates increased slightly as the mixing time increased and a maximum was achieved at five minutes.

Considering the proportion of each agglomerate size within each agglomerate condition, increasing mixing time leads to coarser agglomerates. As time increases the smaller size agglomerate proportion decreases and the larger agglomerate proportion increases. The rolling action within the drum along with adequate liquid phase during agglomeration promotes the growth of agglomerates as the rolling time increases. However, the maximum point is reached at five minutes. At 6 minute mixing, the agglomerates start to decrease in size from those at five-minute mixing times as shown in Figure 8. The agglomerates continue to grow as long as adequate liquid, capillary force and rolling action is available during agglomeration. The agglomerates eventually reach their maximum size and then start to deteriorate as the rolling action introduces rubbing and impact force between the agglomerates causes them to break apart. These forces now are greater than the capillary force so the agglomerate size decreases.

Since agglomeration is often used to control the amount of fines, mixing time had minimal effect with the wt% -1.7 mm by <1% for all times. This indicates that shorter mixing times for this ore may be effective for agglomeration. A short mixing time would result in a shorter agglomeration drum in practice and reduce the necessary capital.

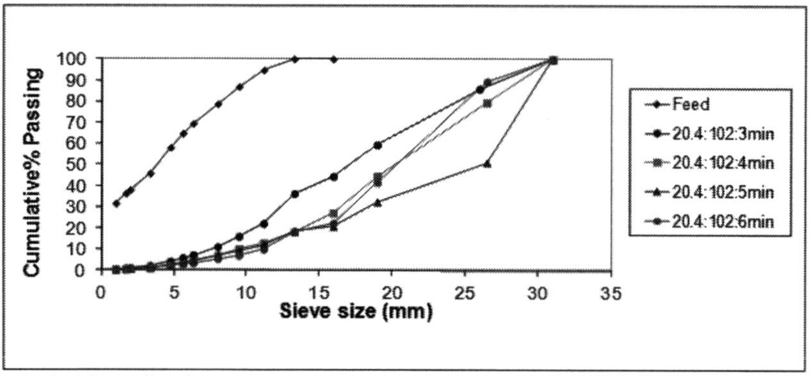

Figure 8. ASDs of agglomerates produced with different mixing times at 20.4% moisture, 102 g/L H_2SO_4 and 31.5% NC.

Table II. Fine particle, D_{10}, and D_{50} of agglomerates at different retention times.

Mixing Time (minute)	Fine particles less than 1.7 mm (%)	D_{10} (mm)	D_{50} (mm)
3	0.77	7.55	17.12
4	0.56	9.46	20.20
5	0.48	9.98	26.15
6	0.53	11.22	20.34

<u>Electrical Conductivity</u>

Electrical conductivity of agglomerates produced from copper ore can be affected by several factors such as the nature of samples, moisture content, solution volume, the compaction of the samples and acid concentration [1]. In this study, moisture content, solution volume, acid concentration and the compaction of sample effects on electrical conductivity of nickel laterite agglomerates were investigated.

Electrical conductivity was measured for agglomerates produced with various moisture contents and acid strength. Each condition was measured three times (e.g. one measurement per four kg batch, three batches per condition). The means, standard deviations and coefficient of variation for the measured electrical conductivity values are given in Table III along with agglomeration conditions. The coefficient of variation is the ratio of the standard deviation to the mean for a data set. These dimensionless numbers are useful to compare data sets with different units or means. Most of the coefficients of variation for the conductivity values at 20.4 kg applied load are smaller than those at no load applied. The median percent standard deviation for the conductivity measurements with no load and under a mechanical load were 11% and 6%, respectively. This indicates that electrical conductivity measured under a mechanical load provides a more precise measurement.

Table III. The mean, standard deviation and coefficient of variation of electrical conductivity of agglomerates at different moisture contents and acid concentrations.

Moisture content (%)	Acid concentration (g/L)	Mean conductivity (mS/m)		Standard deviation (mS/m)		Coefficient of variation	
		No load	20.4 kg load	No load	20.4 kg load	No load	20.4 kg load
15.3	102	12.5	32.8	4.5	3.2	0.036	0.096
15.7	259	22.0	66.2	6.5	15.8	0.294	0.238
16.0	532	22.4	57.3	1.8	2.5	0.082	0.044
16.1	972	11.8	41.4	2.2	2.8	0.188	0.066
20.4	102	24.4	103.5	5.1	26.2	0.210	0.253
21.0	259	35.6	113.4	2.6	5.8	0.072	0.051
21.5	532	52.7	152.4	2.0	1.3	0.038	0.009
21.7	972	22.3	74.3	3.1	3.0	0.139	0.040
25.6	102	22.4	57.3	1.8	2.5	0.082	0.044
26.3	259	265.2	818.4	223.1	382.1	0.841	0.468
27.0	532	110.3	346.0	3.1	19.6	0.028	0.057
27.2	972	33.7	97.0	5.7	12.5	0.170	0.129

A maximum in conductivity is observed for each % moisture near acid concentrations of 259 to 532 g/L (21.68-31.71 wt%). These results mirror the values of pure sulfuric acid conductivity values. The conductivity of sulfuric acid also varies with concentration and has its maximum value at about 30 wt% at $25\,^{\circ}C$.

Drum rotation speed and mixing time were varied while moisture and acid were fixed at 20.4% and 102 g/L H_2SO_4, respectively. Changing mixing time and drum rotation speed during the agglomeration process had no effect on electrical conductivity of agglomerates. The conductivity values of agglomerates were quite constant as the drum rotation speed and retention times changed due to the fact that the moisture content and the acid concentration of the solution were fixed. The agglomerate sizes were slightly different when varying mixing time and drum rotation speed, however, the size of agglomerates seems to have no effect on electrical conductivity of the agglomerate bed.

It appears from these tests, that electrical conductivity could be used as a moisture content control tool for laterite ores similar to copper ores [1] if the acid strength or conductivity of the agglomeration solution is known. As moisture content controls the formation of agglomeration, its measurement and control is very important to the agglomeration process. Measuring electrical conductivity when the agglomerate bed is subjected to a mechanical load provides a more consistent measurement.

Since various sulfuric acid concentrations were used in the study, a method to normalize its effect was developed. Conductivity ratio was calculated using equation 2 to mitigate the effect of sulfuric acid conductivity. The conductivity ratio is simply the ratio of the measured electrical conductivity of the agglomerate bed divided by the conductivity of the sulfuric acid solution used during agglomeration.

$$Conductivity\ Ratio = \frac{Measured\ Conductivity}{Conductivity\ of\ Sufuric\ Acid\ Solution} \quad (2)$$

The relationship between the conductivity ratio and the solution volume indicates an exponential relationship as shown in Figure 9. Although more solution volume leads to higher conductivities due to the viability of the electron pathways, other factors such as the conductivity of the solution, the natural conductivities of the minerals, and the area of void space in the packed bed have contributed to the conductivities. These factors affect the conductivities therefore the overall conductivities resulting from these effects produce some scatter in relationship with the solution volume. It should be noted that at the largest solution volume added, the conductivities of agglomerates with both no load applied and 20.4 kg of load applied drop significantly. This observation is not fully understood and needs to be investigated further.

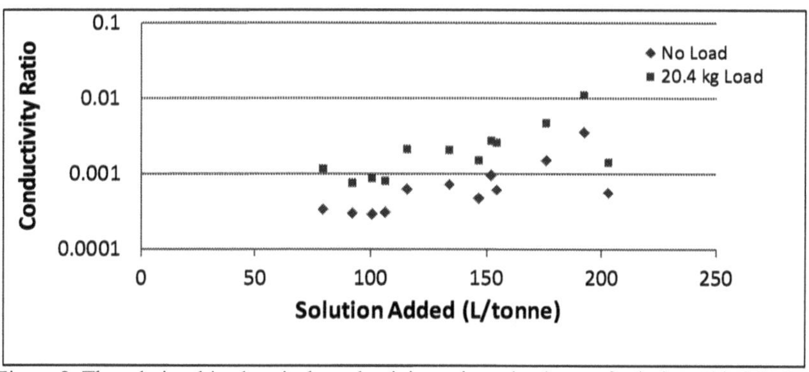

Figure 9. The relationship electrical conductivity ratio and volume of solution added during agglomeration.

Permeability

In general, permeability increases with increasing particle or agglomerate size in this study. As agglomerate size has already been shown to change as a function of moisture and acid strength, similar trends should and were observed with permeability. Permeability increased with increasing moisture at constant acid strength and decreased with increasing acid strength at constant moisture.

Eventually, the moisture content can become too high leading to "mud" formation. This will not cause larger agglomerates to form, thus decreasing the solution paths and decreasing permeability. This may have happened at 25.6% moisture with 102 g/L H_2SO_4.

Since permeability and particle size distributions appear to be correlated, the permeability data was plotted versus the volume of solution. Figure 10 illustrates that while scatter in the data exists, an obvious trend is observed that increasing the solution volume over the range examined increases permeability.

The agglomerates breaking down during permeability tests are another factor which influences the values of permeability of the agglomerate bed. From observations, higher acid concentrations used during agglomeration produce weaker agglomerates. These agglomerates tend to disintegrate during permeability tests resulting in lower permeability values. After the agglomerates broke down, the small particles or fines migrated down the column, filled the void spaces within the agglomerate body, and decreased the liquid flow paths.

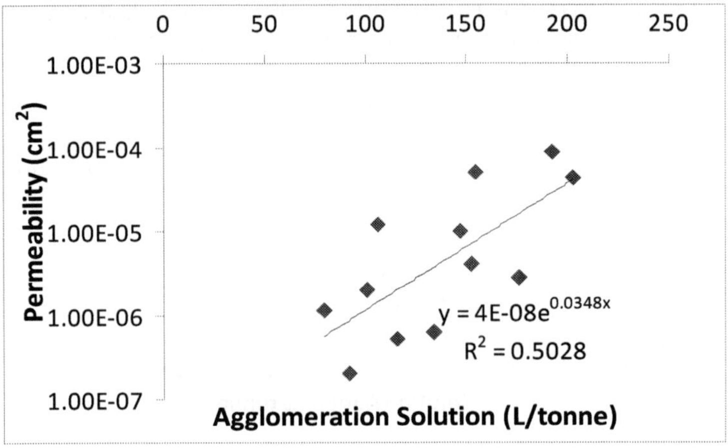

Figure 10. The relationship between agglomeration solution amount and permeability.

Table IV shows the bulk density and porosity values of the agglomerates before and after the permeability experiments. Bulk densities increased and porosities decreased during every permeability test. In general, this was caused by agglomerate degradation caused by the flowing water used in the test.

The bulk densities after flooding generally decreased as acid concentration was decreased with constant moisture. The final bulk densities were lowest (e.g. most porosity) with agglomerates produced with 102 g/L H_2SO_4 for every % moisture reported. Thus, examination of agglomerates by examining their bulk density during the permeability test indicated that agglomerates produced with 102 g/L H_2SO_4 provided better porosity and stability within the agglomerate bed resulting in reasonable permeability values.

Table IV. The bulk density and porosity of agglomerates before and after the permeability tests.

Moisture content (%)	H2SO 4 (g/L)	Bulk density (g/mL)		Δ Bulk density (%)	Porosity		Porosity change (%)
		Before Perm. Test	After Perm. Test		Before Perm. Test	After Perm. Test	
15.3	102	1.22	1.40	13	0.46	0.38	18
15.7	259	1.25	1.47	15	0.44	0.35	22
16.0	532	1.34	1.59	16	0.40	0.29	28
16.1	972	1.40	1.64	14	0.37	0.27	28
20.4	102	1.21	1.38	12	0.46	0.38	16
21.0	259	1.26	1.60	21	0.44	0.29	35
21.5	532	1.27	1.63	22	0.43	0.27	37
21.7	972	1.42	1.63	13	0.37	0.27	26
25.6	102	1.26	1.53	18	0.44	0.32	28
26.3	259	1.44	1.55	7	0.36	0.31	14
27.0	532	1.41	1.78	21	0.37	0.20	45
27.2	972	1.37	1.68	18	0.39	0.25	35

The drum rotation speed was varied from 15.7 to 47.1% NC. The permeability values of the agglomerates at different drum rotation speeds were not significantly different and in the range of 1.5 to $4x10^{-5}$ cm^2. Therefore, the drum rotation speed did not have a significantly effect on the nickel laterite agglomerate permeability.

The agglomeration times were changed from three to six minutes. The permeability values of agglomerates produced with different mixing times increased slightly with increasing mixing time. The permeability of agglomerates at different mixing times agrees well with the ASDs of the agglomerates. As the ASDs increase, the permeability also increases.

Conclusions

Agglomeration of a nickel laterite was evaluated in terms of size distribution, electrical conductivity and bed permeability. Agglomerate size distributions (ASDs) were affected by solution addition volume, acid concentration, and mixing time. Agglomerate size (D_{10} and D_{50}) increased as the amount of solution addition during agglomeration increased. More solution added during agglomeration provides more liquid phase to fill the void between and inside the particles creating capillary forces to attach the particles together. However, agglomerate sizes decreased as the acid concentration of the solution added during agglomeration increases since the liquid volume introduced to the agglomeration decreases as the acid concentration increases at constant moisture. The amount of fines (-1.7 mm) in the ores decreased as more solution was added during agglomeration. However, mud did form when excess liquid was added during agglomeration. Drum speed had minimal effect on size distribution. Increasing mixing time increased agglomerate sizes slightly.

Electrical conductivity of agglomerates was affected by the acid concentration, solution volume, compaction, and nature of the ores. The sulfuric acid concentration and moisture content had strong effects on electrical conductivity of the nickel laterite agglomerates. The electrical conductivity of nickel laterite agglomerates followed a similar trend as the electrical conductivity of sulfuric acid. The electrical conductivity increased as more solution volume was added during agglomeration. The electrical conductivities of agglomerates were less varied when 20.4 kg of load was applied.

Permeability of agglomerate beds was related to the agglomerate sizes. Coarser agglomerates create more void spaces within the agglomerate bed and allow more solution to flow through. Agglomerate strength is another factor that affected permeability. Agglomerates produced from high acid concentration tended to disintegrate during the permeability test and had low permeability.

References

1. G. Velarde, "Agglomeration Control for Heap Leaching processes," *Mineral Processing and Extractive Metallurgy Review,* 26 (2005), 219-231.

2. J.A. Garcia and M.K .Jorgensen, , "Agglomeration and Heap Testing Requirements for High Clay Ores," *Randol Gold Forum 97 Conference and Exhibition* (1997), 143-146.

3. D.T. Kinard and A.A. Schweizer,"Engineering Properties of Agglomerated Ore in a Heap Leach Pile," *Geotechnical Aspects of Heap Leach Design* (1987) 55-64.

4. S.C. Bouffard, "Agglomeration for Heap Leaching: Equipment Design, Agglomerate Quality Control, and Impact on the Heap Leach Process," *Mineral Engineering,* 21 (2008), 1115-1125.

5. K.A. Lewandowski and S.K. Kawatra, "Binders for Heap Leaching Agglomeration," *Minerals & Metallurgical Processing,* 26(1) (2009), 1-24.

6. F. Castro and G. Pereira "Characterization of the Nickel Laterite Sample and Guidelines for Column Leach Tests Project AMIRA P986," (Vale Mineral Development Center, Brazil, 2009).

7. S. Agatzini-Leonardou and D. Dimaki, "Heap Leaching of Poor Nickel Laterites by Sulphuric Acid at Ambient Temperatures," *Hydrometallurgy,* 94 (1994), 193-208.

8. S. Agatzini-Leonardou and I. G. Zafiratos, "Benefication of a Greek Serpentinic Nickeliferous Ore Part II. Sulfuric Acid Heap and Agitation Leaching," *Hydrometallurgy,* 74 (2004), 267-275.

9. M. Rodriguez and B.J. Wedderburn, "Hydrometallurgical Method for the Extraction of Nickel and Cobalt from Laterite Ores" (World Intellectual Property Organization, WO 2007/016737 A1, 2007).

10. N. Mora, A.S. Moroney, and S. Thomas, "Method of Agglomeration," (World Intellectual Property Organization, WO 2010/085857 A1, 2010).

11. W.P.C. Duyvesteyn, H. Liu, and M.J. Davis, "Heap Leaching of Nickel Containing Ore," (U.S. Patent 6,312,500, 2001).

12. S.C. Bouffard, "Review of Agglomeration Practice and Fundamentals in Heap Leaching," *Mineral Processing & Extractive Metallurgy Review,* 26 (2005), 233-294.

13. P.D. Chamberin, "Agglomeration: Cheap Insurance for Good Recovery when Heap Leaching Gold and Silver Ores," *Mining Engineering,* 38 (1986), 1105-1109.

14. "Standard Test Method for Permeability of Granular Soils (Constant Head)" (ASTM D 2434-68, ASTM International, 2006), 224-228.

Ni-Co 2013

Electrometallurgy

Session Chairs

Michael Moats
Nathan Stubina

Ni-Co 2013
Edited by: Thomas Battle, Michael Moats, Violina Cocalia, Harald Oosterhof, Shafiq Alam, Antoine Allanore, Rodney Jones, Nathan Stubina, Corby Anderson, and Shijie Wang
TMS (The Minerals, Metals & Materials Society), 2013

ACID MIST ABATEMENT IN BASE METAL ELECTROWINNING

Tim Robinson,[1] David White,[2] Ross Grassi,[3]

[1] Republic Alternative Technologies Pty Ltd, 11288 Alameda Drive, Strongsville, Ohio, 44149, USA

[2] Snowden Mining Industry Consultants, 45 Mudlark Way, Yangebup, Perth, WA 6164. Australia

[3] AMEC Mining & Metals Australia, Level 7 / 197 St Georges Terrace Perth, Western Australia, 6000, Australia

Keywords: Copper, Nickel, Zinc, Acid Mist, Electrowinning

Abstract

Acid mist abatement in base metal electrowinning cellhouses has advanced rapidly over the last 15 years as a result of both occupational health and economic pressures. Enablers include evolving designs and improved materials of construction. Cellhouse designs now typically include addition of surface tension modifying agents, cell hoods, and anode bags or anode skirts to reduce acid mist production. Automation is increasingly being considered to remove operators from the acid mist containing environment. In nickel electrowinning, chloride based anode bag technologies for chlorine recovery has been adapted to sulfate based systems. These improved technologies and design trends have resulted in cellhouse operating conditions giving greater operator comfort, while also achieving higher operational efficiency.

Introduction

As part of the reaction in base metal electrowinning (EW), the nascent oxygen or chlorine that is generated at the anode rises to the top of the electrolyte, and if not captured, creates acid mist in the tankhouse. Methods of acid mist suppression in the electrometallurgical industry over the years can be classified into four broad categories. These include:

- Electrolyte based, such as foams and balls (on top of electrolyte);
- Electrode design, such as anode bags/skirts, anode brushes and wiper systems;
- Cell design, such as cell hoods, cell freeboard, harvesting automation;
- Cellhouse design, such as cross flow ventilation systems, open wall cellhouses and overall automation.

Combinations of the above methods are typical in commercial EW operations.

Technology transfer between the base metal industries has made an important contribution to the continuous improvement of operating conditions in the EW cellhouse.

- Zinc EW was the first large-scale EW process to develop surface tension affecting foams and cross flow cellhouse ventilation designs, that were eventually used in copper;
- Large-scale copper EW applied the use of SX compatible foams, adopted cross flow ventilation and later ventilated cell hoods;
- Nickel EW led the development of anode bags and has recently adopted the use of cell hoods from copper.

Future acid mist reduction technology may eliminate anode gas evolution altogether with new and different reaction chemistries.

Electrolyte Based Mist Abatement

Electrolyte based acid mist abatement technology is focused on surfactants. Surfactants are added to electrolytes to produce a foam layer at the electrolyte surface and suppress acid mist. Foam is generated at the electrolyte surface as oxygen gas bubbles generated at the anode are released from the electrolyte to the environment Foams stabilize and prolong the life of bubbles at the electrolyte-air interface and thereby allow electrolyte to drain away from the bubble surface and hence reduce acid mist.

The first zinc and copper electrowinning plants operating at the beginning of the century did not use acid mist suppression methods. No foams were used and operators would typically have to wear clothes and cloth over the face to protect them from the mist [1].

The zinc industry was first to develop electrolyte–based surfactants on a large scale for electrowinning and has primarily focused on foams for mist suppression to the present day. Many different chemicals have been tested and used for zinc, including licorice extract, Dowfroth, Dowfax, soapwart, and metapara cresol (MPC). Licorice extract is a common mist suppressant in Zinc EW today [2].

Small plastic balls in layers on top on the cell were first introduced to suppress acid mist in copper. The first commercial Solvent Extraction - Electrowinning (SX-EW) plant, at Ranchers Bluebird in Arizona (1970) claimed the use of a thin layer of kerosene (Napoleum 470) as a mist suppressant on top of the electrolyte [3]. Besides the advantage of suppressing acid mist, Ranchers also claimed that this kerosene would also collect organic carryover from the SX plant so as not to have organic burn on the top sections of the copper cathode. Plastic beads (or BB's), that were much smaller than plastic balls (see Figure 1) were first used at Magma San Manuel in Arizona in the late 1980's and then later in other North American and Australian SX-EW plants.

Figure 1. Mist suppressant balls

Foams or surfactants such as DOWFAX and sulphonic acids (sometimes in combination with plastic balls) were used in the copper refinery liberators with success. These surfactants were tested with SX circuits and found to be detrimental to the organic. The surfactants would react with the SX organic and decompose the SX organic reagent and significantly affected phase separation. In 1980, 3M of the USA began developing with Inspiration (located in Miami, Arizona), a fluoro-based foam (called FC100) that was compatible with the SX reagent and operation. Operational problems observed with thick foam layers were fire hazards when cells contained SX organics, lower current efficiency and cell contact corrosion. In 1994, this chemical was further improved to minimize foam generation but still suppress acid mist and was called FC-1100. FC-1100 has become the most common mist suppressant surfactant around the world and is typically used with several layers of plastic balls [4].

The surfactant of choice for nickel EW has been sodium lauryl sulfate (SLS). Not only does SLS suppress acid mist but also makes a smoother cathode deposit. Small amounts of SLS decrease the surface tension on the nickel cathode surface and reduce adhesion of hydrogen bubbles on the cathode that can lead to pit formation. Not all nickel EW operations use SLS but Cawse Nickel (now shutdown) and Harjavalta have added SLS to the Ni EW electrolyte [5].

Recently research on electrolyte operating parameters concludes that electrolyte operating temperature in the cells has a significant effect on acid mist generation [6].

Electrode Based Design

The electrowinning reaction includes gas, either oxygen or chlorine evolving from the anode. Most electrode based technology for acid mist abatement is therfore centered around the anode.

Commercially pure nickel cathode is produced by electrowinning in either sulfate or chloride electrolyte, or a combination of both. Chloride based electrolyte systems generate chlorine gas at the anode The nickel industry to developed anode bag technology in order to capture chlorine gas at the source. Not only is the chlorine gas collected and removed from the operating environment in the cellhouse using anode bags, it is also returned upstream as an oxidant in the matte leaching process step.

The first operation to use anode bags was Nikkelverk, in Kristiansand Norway. In 1952, this operation converted from a sulfate electrolyte to a mixed chloride-sulfate electrolyte for nickel electrorefining. Cobalt was separated from the nickel stream as a cobaltic hydroxide, $Co(OH)_3$, using chlorine as the oxidizing agent. Chlorine gas was generated in a separate section of the nickel cellhouse using cells with graphite anodes [7]. In the late 1970's, the refining process was converted to a Chlorine Matte leach - chloride EW operation for both nickel and cobalt, using coated titanium anodes [8]. Nikkelverk, thus, has 60 years of experience with anodic chlorine generation in metal EW, using individual anode bags. Anode bags are also used at other chloride electrolyte nickel operations at Sumitomo in Niihama Japan, and Eramet's Le Havre, France operation [5].

Divided cell or diaphragm technology, using electrode bags, is also used in nickel or cobalt electrowinning from sulfate solution to separate the acidic anolyte from the cathode. This minimizes hydrogen gas generation at the cathode, thereby maintaining high electrowinning current efficiency and product quality, and reducing acid mist. In electrorefining operations, it is necessary to purify the electrolyte solution produced at the anodes, before it is re-introduced to the cellhouse at the cathodes.

In sulfate electrolyte systems for nickel and cobalt, either anode or cathode bags can be used. The advantage of the anode bags is that they suppress the acid mist caused by the oxygen at the anode. Newer nickel and cobalt EW cellhouse designs have made significant advances in reducing worker exposure to nickel and acid aerosol and mist exposure by using anode bag mist abatement technologies.

Various box and diaphragm bag frame designs have been used around the world, with polyethylene and polypropylene designs typical. INCO (now Vale) designed anode bags in 1983 for the new cobalt EW cellhouse at Port Colborne, Ontario [9] and Cawse Nickel in Western Australia (now shutdown, see Figure 2) implemented anode bags for commercial nickel cathode production in 1998 [10].

Figure 2. Anode bags for nickel EW [10]

A variation of the anode bag technology is the anode skirt, as shown in Figure 3 , which is a bag type that is open at the bottom and does not require a separate anolyte extraction system. The anode skirt does not collect gas but is useful for suppressing mist in cathode bag nickel EW systems. The skirt is made of hydrophilic polyester that collects the acid mist droplets and drains them back to the electrolyte while allowing air/oxygen mixture to escape [11]. Lead anode oxide corrosion product falls to the bottom of the cell to be cleaned and so the skirts do not have to be cleaned of lead sludge.

Zinc and copper commercial operations have not used anode bags, but some copper operations have installed wiper-type technology to the upper part of the anode to reduce acid mist. Plastic wipers attached to lead anodes initially had limited success at San Manuel (1986) and El Abra (1996). A modifed version using plastic brushes attached to the anode hanger bar has proved to be far more successful in reducing anode and cathode corrosion. In the late 1990's ASARCO Silver Bell EW plant near Tucson, Arizona, was the first EW tankhouse to convert totally to a 'Brush' mist suppression system. This technology was used in combination with FC 1100 surfactant, and has been operating successfully to date. Recently, Tecnocomposites of Chile have installed LateralFlow and LateralVex technology at the copper EW operations of Minera Sierra Miranda near Antofagasta, Northern Chile [13]. This technology includes flexible covers attached to the anodes and lateral overflow gutters (LateralFlow) and suction plenums (LateralVex) attached to the inner side of the cell.

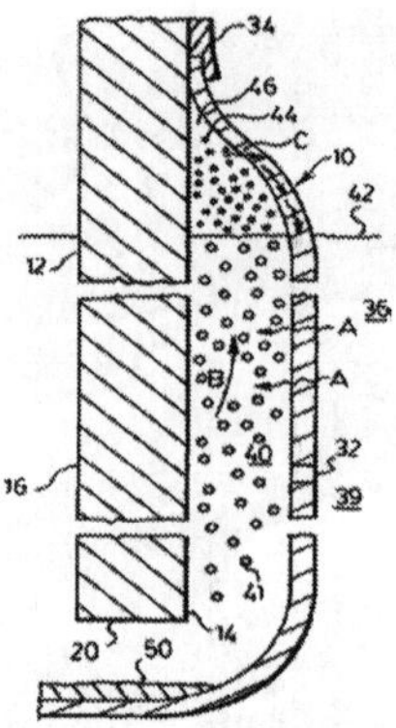

Figure 3. Schematic drawing of an anode skirt [11].

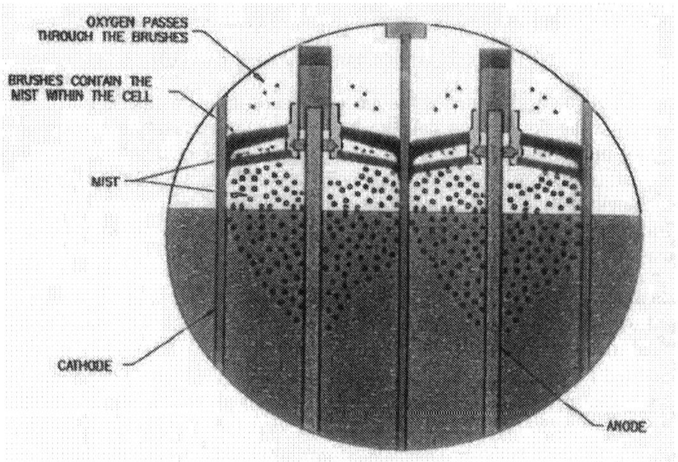

Figure 4. Schematic drawing of anode brushes

Cell Based Design

Cell hoods are the most common acid mist abatement technology that is designed around the cell structure [5, 14]. Fiberglass reinforced plastic (FRP) and plastic hoods on copper electrorefinery liberator cells over the years were standard practice but it was the Cerro Verde EW plant (now Freeport MacMoran) in Peru in 1977 that first applied cell hoods to a large commercial EW plant. These hoods were made of FRP and had lugs for crane lifting. They were later replaced by mist suppression balls to increase time efficiency in cathode harvesting. Copper SX-EW plants in colder climatic conditions at Los Bronces (first plant in 1996), Tintaya (high altitude Peru), OMG Kokkola (Finland), and Stillwater (small copper EW by-product circuit in Montana), were first to install cell top hoods. Today, ventilated cell hood technology has become standard in the copper industry and is commonplace, particularly in South America. The newer design of hood systems allow for the hood to be incorporated in the automated crane bale for efficient loading and unloading of electrodes. These hoods can be used in combination with surfactants and balls and are made of FRP. The nickel industry is now adopting the use of ventilated cell top hoods as shown in Figure 5 [15].

Cell freeboard is the distance from the electrolyte level to the top of the cell. The first designs of EW cells had large freeboards, typically of 300 mm or more, but in more recent cellhouse designs, this was reduced to around 150 mm to save costs. A potential opportunity for investigation exists to review freeboard impact on acid mist above the cells.

Figure 5. Cell hoods in a nickel electrowinning facility [15].

Cellhouse Based Design

Base metal cellhouse structure design can also assist with acid mist abatement. The simplest cellhouse designs have no side walls and are open to allow for natural ventilation. In Australian and African copper EW cellhouse designs, acid mist suppression methods have been characterised by the open tankhouse design (no tankhouse side walls), with the use of balls and foams. The mild African and Australian climate allows for this open cellhouse design. These include Australian operations such as Girilambone (1993), Nifty (1993), Cloncurry (1995) and Mt Cuthbert. Capital efficient cellhouse designs at relatively low altitude sites in Chile such as Andacollo (1995) and Lomas Bayas (1997) also used open tankhouse concepts.

Crossflow ventilation is a technology that has been designed into the cellhouse structure in the past to reduce acid mist. Amplats' Rustenberg plant in South Africa applied a cross flow principle first and this was applied on a larger scale at the zinc EW cellhouses in Canada (Cominco Trail in early 1980's) and Kidd Creek. From the mid 1990's until the more commonplace application of cell hoods, the large tankhouses in Chile typically used combinations of positive crossflow ventilation, balls and foams. The DESOM crossflow ventilation configuration, which became common practice in Chile, was first applied to the El Teniente SX-EW operation in 1985. Cerro Colorado, Quebrada Blanca, Chuqui SBL, Quebrada M and Escondida Coloso followed in 1994.

Most acid mist abatement technologies are focused on keeping the acid mist in the cell, or moving it efficiently away from the cellhouse operators. Permanent cathode technology removes operators from the cellhouse and hence acid mist exposure. Permanent cathode technology was first developed for the zinc EW industry at the beginning of the 20th century. Copper adopted the permanent cathode technology in the late 1970's in electrorefining and it was soon transferred to copper EW production [14]. Not only were productivity improvements seen in copper EW but also reduced operator exposure to acid mist with the more automated cellhouse. Nickel EW is also working with permanent cathode technology [5].

Combinations

In most base metal cellhouse operations around the world, a combination of the previously discussed technologies is implemented [4, 16]. Most copper EW plants around the world operate with a surfactant and mist suppression balls on top of the electrolyte. This can be combined with cell hoods or cross flow ventilation. Some zinc EW plants have combined cross flow ventilation systems in the cellhouse, with the use of surfactants in the cells, but typically most zinc plants use surfactants only [2]. Cawse Nickel combined anode bags with cross flow ventilation, as shown in Figure 6 [10]. The nickel industry is now also using cell hoods [5].

Figure 6. Cawse Tankhouse [10].

Future Technologies

Researchers from Freeport-McMoRan recently described work involving an alternative anode reaction technology (AART) using a non-lead anode [17, 18]. Whereas typical copper EW conditions produce oxygen gas and acid at the lead anode, AART oxidizes ferrous to ferric ion at the non-lead anode and thereby reduces cell voltage by approximately 50%. A further advantage is that no anode oxygen-generated acid mist is produced. Although promising, the technology has yet to be commercialized.

Conclusions

Modern cellhouse design aims to minimize operator and environmental exposure to cellhouse acid mist. The transfer of EW cellhouse technologies between zinc, copper, and nickel has led to improvements in acid mist capture technology. These technologies include electrode bags and skirts, more efficient cell ventilation and improved surfactants.

References

1. O.C. Ralston, *Electrolytic Deposition and Hydrometallurgy of Zinc*, (New York, NY: McGraw Hill Book Co Inc, 1921).

2. M. Moats, E. Guerra, and J. A. Gonzalez, "Zinc Electrowinning – Operating Data", *Zinc and Lead Metallurgy*, eds. L. Centomo, M. Collins, J. Harlamovs and J. Liu, (Montreal, Canada: Canadian Institute of Mining, Metallurgy and Petroleum, 2008), 307-314.

3. K.L. Power "Operation of the First Commercial Copper Liquid Ion Exchange and Electrowinning Plant," *Copper Metallurgy*, ed. R. P. Erlich (New York, NY: AIME, 1970).

4. T. Robinson et al., "Electrolytic Copper Electrowinning – 2007 World Tankhouse Operating Data. *Copper-Cobre 2007, Vol. V*, eds. G. E. Houlachi, J. D. Edwards, and T. G. Robinson, (Montreal, Canada: Canadian Institute of Mining, Metallurgy and Petroleum, 2007), 375-424.

5. F. K. Crundwell, M. S. Moats, V. Ramachandran, T. G. Robinson, and W. G. Davenport, *Extractive Metallurgy of Nickel, Cobalt and Platinum-Group Metals*, (Oxford, UK: Elsevier, 2011), 327-345, 575-578.

6. R. Al Shakarji, Y. He, and S. Gregory, "Performances of Mechanical Barriers in Reducing Acid Mist", *Hydrocopper 2011 - 6th International Seminar on Copper Hydrometallurgy*, eds. J.M. Casas, C. Lightfoot, and G. Tapia, 2011, Vina del mar, Chile.

7. J.R. Boldt and P.E. Queneau, *The Winning of Nickel*, (London, UK: Methuen, 1967), 360-361.

8. E.O. Stensholt, H. Zachariasen, and J.H. Lund," The Falconbridge Chlorine Leach Process," *Proceedings of the 25th Annual Conference of Metallurgists, Nickel Metallurgy, Vol. I Extraction and Refining of Nickel*, eds. E.Ozberk and S.W. Marcuson (Montreal, Canada: Canadian Institute of Mining, Metallurgy and Petroleum, 1986), 442-463.

9. R.G. Agnew, L.D. Kowal, and D. Young, "Process Optimisation at INCO's Cobalt Refinery," *International Symposium on Electrometallurgical Plant Practice,* eds. P.L. Claessens and G.B. Harris (Montreal, Canada: Canadian Institute of Mining, Metallurgy and Petroleum, 1990), 221-231.

10. A.D. Burvill and D.T. White, "Engineering Aspects of the Cawse Nickel/Cobalt Laterite Project," *ALTA 1999 Nickel/Cobalt Pressure Leaching & Hydrometallurgy Forum*, (Castlemaine, Victoria, Australia: ALTA Metallurgical Services, 1999).

11. G.M. Dunne and H.E. Holiday, *Electrode Cover for Preventing the Generation of Electrolyte Mist*, United States Patent 6,120,658, 19 September 2000.

12. Sunwest, "Anode brush brochure", (Tucson, AZ: Sunwest, 1999)

13. F. Penna, C. Villaseca, and R. Villarroel, "LateralFlow, a Concept to Control Acid Mist, Organic Removal, Height of Deposition and Other Benefits for Electrowinning Cells", *Hydrocopper 2011 - 6th International Seminar on Copper Hydrometallurgy*, eds. J.M. Casas, C. Lightfoot, and G. Tapia, 2011, Vina del mar, Chile.

14. T.G. Robinson et al., "Developments in Base Metal Electrowinning Cellhouse Design," *Electrometallurgy 2012*, eds. M. Free et al. (Hoboken, NJ: Wiley) 147-156.

15. T.G. Robinson et al., "Developments in Nickel Electrowinning Cellhouse Design," *ALTA 2012 Nickel/Cobalt/Copper Conference*, (Castlemaine, Victoria, Australia: ALTA Metallurgical Services, 2012).

16. M. E. Schlesinger et al., *Extractive Metallurgy of Copper, 5th edition*, (Oxford, England: Elsevier, 2011).

17. M. Free et al., "Electrometallurgy - Now and in the Future," *Electrometallurgy 2012*, eds. M. Free et al. (Hoboken, NJ: Wiley) 3-28.

18. S. Sandoval et al., "Demonstration of the Ferrous/Ferric Anode Reaction for Copper Electrowinning," *Proceedings of the 7th International Copper-Cobre Conference 2010: Volume 4 Electrowinning and -refining* (Clausthal-Zellerfeld, Germany: GDMB, 2010), 1617-1633.

Ni-Co 2013
Edited by: Thomas Battle, Michael Moats, Violina Cocalia, Harald Oosterhof, Shafiq Alam, Antoine Allanore, Rodney Jones, Nathan Stubina, Corby Anderson, and Shijie Wang
TMS (The Minerals, Metals & Materials Society), 2013

BOLEO COBALT ELECTROWINNING DEVELOPMENT

Jianming Lu[1], David Dreisinger[1, 2] and Thomas Glück[2]

[1]Department of Materials Engineering, University of British Columbia, 309-6350 Stores Road, Vancouver, BC, Canada, V6T 1Z4
[2]Baja Mining Corporation, Suite 500 200 Burrard St., Vancouver, Canada, V4R 2T9

Keywords: Cobalt, Electrowinning, Boleo

Abstract

The Boleo process of Minera y Metalúrgica del Boleo recovers cobalt from a complex oxide and sulfide ore. Following the recovery of copper and zinc products, cobalt is recovered by solvent extraction and electrowinning. The electrowinning process is conducted in a diaphragm cell in a sulfate medium. Cobalt is deposited on a coated stainless steel mandrel cathode with exposed circles to allow plating as disk shaped deposits over the surface of the cathode which can be easily stripped after 5-7 days of deposition. The bench-scale development work on cobalt electrowinning is presented, including selection of plating conditions and additives to control the cobalt disk morphology and pitting behaviour. The cobalt deposit internal stress and the conductivities of $CoSO_4$-Na_2SO_4-H_2SO_4 system were measured.

Introduction

The Boleo property is held by the Mexican company, Minera y Metalúrgica del Boleo S.A. de C.V. (MMB). MMB is owned 10% by Baja Mining Corp. and 90% by a consortium of Korean companies consisting of KORES – Korea Resources Corporation, LS Nikko Copper, Hyundai Hysco, SK Networks and Iljin Copper. The project is now under construction with an expected startup in 2013.

Average annual production, for the first six years is forecast as: 56,700 tonnes of copper cathode, 1,700 tonnes of cobalt cathode and 25,400 tonnes of zinc sulphate monohydrate. There is the potential to produce up to 220,000 tonnes per annum of manganese carbonate.

The project will consist of underground and surface mining operations, providing run-of mine ore to a fully integrated hydrometallurgical facility. The overall hydrometallurgical process for Boleo has been reviewed in earlier publications [1-3]. The process is a whole-ore leach in seawater using a combination of acid oxidation and acid reduction leaching to extract copper, cobalt, zinc and manganese.

The copper solvent extraction raffinate is split with a portion of the raffinate returning to leaching and the balance advancing to cobalt, zinc and manganese recovery. The raffinate for cobalt, zinc and manganese recovery is first neutralized to remove iron and aluminum from solution and then subjected to a novel direct solvent extraction process (DSX) developed by CSIRO (Australia) [4-6]. The direct solvent extraction process uses a mixture of Versatic® 10

(Shell) and LIX® 63 (BASF) extractants to separate cobalt and zinc away from manganese in the neutralized, iron-free raffinate. The raffinate from the DSX extraction may be treated to recover manganese by precipitation as manganese carbonate.

The cobalt and zinc are stripped from the DSX loaded organic solution and the strip solution advances to zinc recovery using a phosphinic acid extraction circuit. The zinc is stripped as a concentrated, high purity solution and will then be evaporated and granulated/crystallized to make a saleable zinc sulfate monohydrate product. The raffinate from zinc solvent extraction advances to cobalt solvent extraction and electrowinning to produce high purity cobalt cathode according to the following reactions:

Cobalt Extraction

$$CoSO_{4(aq)} + 2HR_{(org)} + Na_2CO_{3(s)} = CoR_{2(org)} + Na_2SO_{4(aq)} + CO_{2(g)} + H_2O_{(l)} \quad (1)$$

Cobalt Stripping

$$CoR_{2(org)} + H_2SO_{4(aq)} = CoSO_{4(aq)} + 2HR_{(org)} \quad (2)$$

Cobalt Electrowinning

$$CoSO_{4(aq)} + H_2O_{(l)} = Co_{(s)} + H_2SO_{4(aq)} + 0.5O_{2(g)} \quad (3)$$

In full-sheet cathode cobalt production, cobalt deposits become highly stressed and there is a tendency for them to peel off the blanks. Therefore, it was decided to produce round cobalt buttons in divided cells. A similar product is currently produced by Vale INCO, Canada [7]. Cobalt quality is affected by pitting of the electrodeposit. In the Vale INCO's cobalt recovery process, sodium laurel sulfate (SLS) used for foam stabilization also acts as an anti-pitting agent. [8]. In the Boleo cobalt recovery process, the anolyte is used to strip cobalt from the cobalt-loaded organic solution. Sodium laurel sulfate in the anolyte interferes with cobalt stripping and even extraction by causing problems with phase separation. Sodium laurel sulfate can therefore not be used in conjunction with a solvent extraction process.

The objective of this work is to eliminate or minimize pitting and find an optimum condition and/or appropriate anti-pitting agents for production of high quality of cobalt buttons.

Bench Scale Experimental Setup

Cobalt electrowinning was conducted in a divided cell which was placed in a water bath for temperature control (Figure 1). With the use of a peristaltic pump, the catholyte was circulated from the cell top to a mixing reservoir, then to a heating reservoir (in the water bath), and finally back to the cell at the bottom. The feed solution was pumped to the mixing reservoir using another peristaltic pump. The catholyte pH was controlled by adjusting the pH of the feed solution. The cathode and anode were separated by a polyester diaphragm. The cathode-anode distance was 100 mm. The catholyte flowed through the diaphragm to the anode compartment due to the hydraulic head.

The cathode blanks for cobalt button deposition were 304 stainless steel mandrels with a coating of dielectric ink supplied by Ionic Technology Limited. The diameter of the exposed plating circle was 24.7 mm. The anode was made of Pb-0.75% Ag alloy. To start a test, the starting catholyte was heated to a target temperature and used to fill the cell and the solution reservoirs. The catholyte composition was: 45 g/L Co as $CoSO_4$, 120 g/L Na_2SO_4, 0.07 g/L Mn as $MnSO_4$. The initial anolyte (36 g/L H_2SO_4, 45 g/L Co as $CoSO_4$, 120 g/L Na_2SO_4, 0.07 g/L Mn as $MnSO_4$) was heated to the target temperature and added to the anode compartment. A number of different cobalt salts, reagent grade sulfuric acid, manganese sulfate and sodium sulfate, and deionized water were used to prepare the solution. An activated carbon treatment of feed and catholyte startup solutions was conducted in a 6-L flask and the slurry was agitated using a magnetic stirrer for over 6 hours. The activated carbon was separated from the treated solution by vacuum filtration using Whatman No. 42 filer paper and a 0.2 μm membrane filter (for the final stage of activated carbon treatment).

The solution conductivity measurement was conducted using a Jenway 4320 conductivity meter and the solution temperature was controlled within ±0.1 °C using a water bath. The cobalt deposition stress was measured using equipment from Specialty Testing & Development Co.

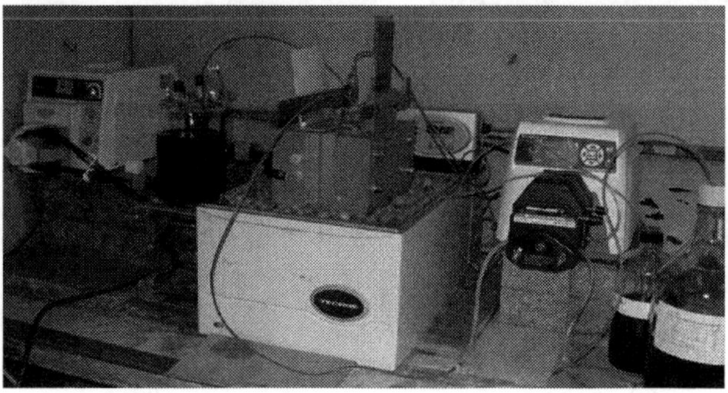

Figure 1. Cobalt electrowinning setup

Effect of Cobalt Raw Materials on Cobalt Button Quality

As shown in Figures 2 to 4, the cobalt button quality and morphology were dependent on whether the solution was treated with activated carbon and what raw cobalt material was used to prepare the solution since the impurities (especially organic compounds) varied for different cobalt raw materials and were removed or partially removed by activated carbon treatment. If the solution prepared using reagent grade cobalt sulfate from Alfa Asear was not treated with activated carbon, there were numerous pits in the cobalt button. Treatment with 4 g/L of Norit SX-Ultra activated carbon resulted in only a few pits in the button since some organic compounds was removed during the activated carbon treatment (Figure 2). For the solution prepared using cobalt carbonate from Alfa Asear, without activated carbon treatment, the cobalt button was brownish and dendritic while with activated carbon treatment, the cobalt button was bright, but there were several pits (Figure 3). The organic compounds in the cobalt carbonate significantly affected the cobalt deposition and facilitated the detachment of hydrogen bubbles.

For the solution prepared using cobalt hydroxide from Alfa Asear, without activated carbon treatment, there were no pits and the cobalt button had significant lateral growth (Figure 4). The cobalt hydroxide contained about1% of reducing organic compounds to prevent the oxidation of cobaltous hydroxide by oxygen. This organic compound significantly affected the cobalt deposition and suppressed pitting. The TOC level was so high that even after 4 stages of activated carbon treatment (20 g/L activated carbon for each stage), the measured TOC level was still 50-70 ppm. The cobalt button produced from such a treated solution was bright. There were, however, a few pits.

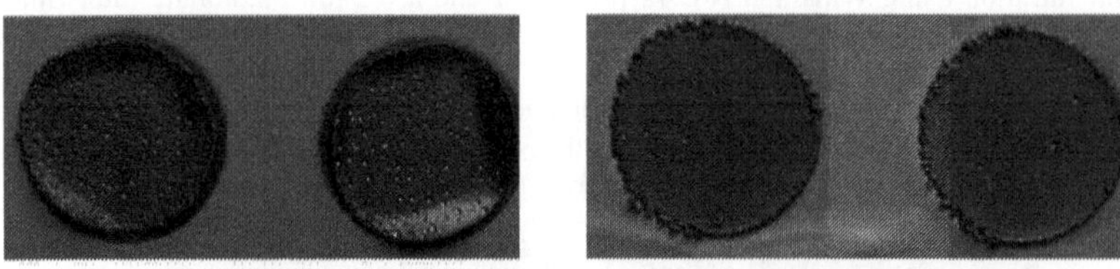

(a) (b)

Figure 2. Cobalt buttons produced from cobalt sulfate prepared using $CoSO_4$ from Alfa Asear (a) without activated carbon treatment (3 hours of deposition) and (b) treated with 4 g/L Norit SX-Ultra activated carbon (6 hours of deposition)

(a) (b)

Figure 3. Cobalt buttons produced from cobalt sulfate prepared using $CoCO_3$ from Alfa Asear (a) without activated carbon treatment (4 hours of deposition) and (b) treated with 4 g/L Norit SX-Ultra activated carbon (6 hours of deposition).

(a) (b)

Figure 4. Cobalt buttons produced from cobalt sulfate prepared using $Co(OH)_2$ from Alfa Asear (a) without activated carbon treatment (72 hours of deposition) and (b) treated with 4 stages of activated carbon (168 hours of deposition).

In order to remove the organic compounds to a sufficiently low level, the solution was treated with activated carbon in three stages (the usage of activated carbon: 20 g/L for the first and seconds stages and 10 g/L for the third stage). After three stages of activated carbon treatment, the solution surface tension was 77.1 mN/m and did not further increase with additional activated carbon treatment. The cobalt buttons produced from different solutions are shown in Figure 5. Pitting was significant for the cobalt buttons produced from all four solutions. The cobalt carbonate from Shepherd Chemical Company was made by dissolving cobalt metal in nitric acid and precipitating cobalt carbonate using bicarbonate. No organic compound was used in the production. Since it contained 0.24% of nitrate, it was purified by dissolution in sulfuric acid and re-precipitation using sodium carbonate. After the activated carbon treatment, the TOC level should be very low. However pitting was still significant (Figure 5d). It is expected that pitting is significant even in the absence of organic compounds since it is determined by interfacial properties (cobalt deposit-solution, gas-solution and cobalt deposit-gas). The current efficiency was 99.4 to 99.6% for all four tests.

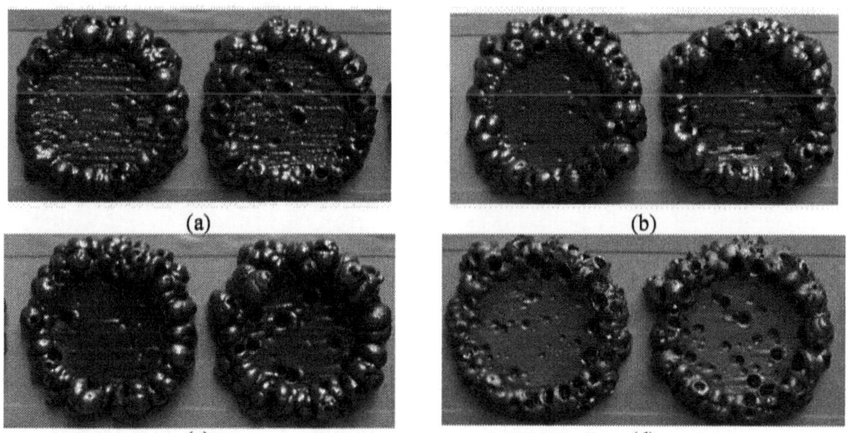

Figure 5 Cobalt buttons produced at 65 °C, 435 A/m^2, pH 3.5, and 5 days of deposition from the solutions prepared using (a) cobalt sulfate from Alfa Aesar, (b) cobalt carbonate from Alfa Aesar, (c) cobalt carbonate from Shepherd Chemical Company, (d) purified cobalt carbonate from Shepherd Chemical Company (3 days of deposition).

Effect of pH and Temperature

The effect of temperature and pH were investigated to improve cobalt button quality. With the use of purified cobalt carbonate from Shepherd Chemical Company, the number of pits decreased with decreasing temperature from 65 to 60 to 55 °C (Figures 5(a-d and 6). However, the deposit dendrites grew laterally significantly at a temperature below 60 °C.

With the use of reagent grade cobalt sulfate from Alfa Asear, there was no pitting in the non-edge area at 55 °C (Figure 7 b), and the deposit surface was rougher. The deposit dendrites grew upwards laterally like those with the use of purified cobalt carbonate from Shepherd Chemical Company. With increasing temperature to 60 °C, the dendrites were significantly suppressed (Figure 7a). However, pits appeared.

(a)

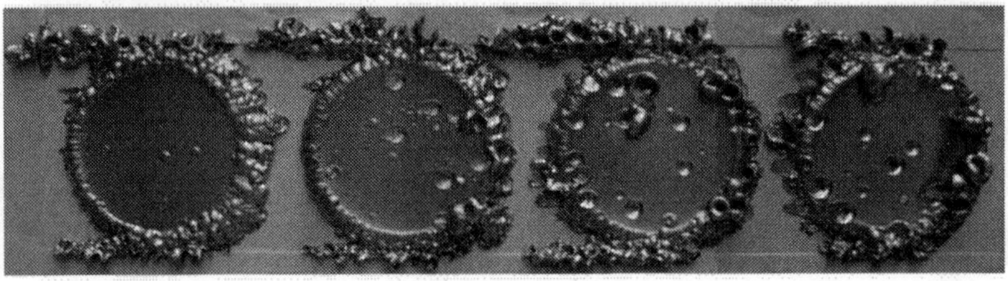

(b)

Figure 6. Cobalt buttons produced from the solution prepared using purified cobalt carbonate from Shepherd Chemical Co at (a) 60 °C and 2 days of deposition and (b) 55 °C and 1 day of deposition under the conditions of 435 A/m^2 and pH 3.5.

The reagent grade cobalt sulfate from Shepherd Chemical Company was very pure and the contents of nitrate and chloride were very low. With the use of this cobalt sulfate, at 65 °C, pitting was significant (Figure 8a) and the cobalt buttons were similar to those produced using reagent grade cobalt sulfate from Alfa Aesar. At 55 °C, there were practically no pits in the non-edge area (Figure 8b). The lateral dendrites did not grow like those with the use of cobalt sulfate from Alfa Aesar (Figure 7b) and cobalt carbonate from Shepherd Chemical Company (Figure 6), indicating some impurities affected cobalt deposition. The cobalt current efficiency decreased slightly from 99.9 to 99.4% with decreasing temperature from 65 to 55 °C.

At a pH above roughly 3, cobalt is deposited through a pH-dependent mechanism by the formation of a cobalt hydroxide complex ($Co(OH)_2$ or $CoOH^+$) [9, 10]. At a constant potential, the cobalt deposition current decreases with decreasing pH, indicating an increase in the inhibition of cobalt deposit. Therefore, dendrites should be suppressed at a lower pH. At a pH below roughly 3, cobalt electrodeposition occurs through the direct reduction of $Co^{2+}$and is practically independent of pH [10].

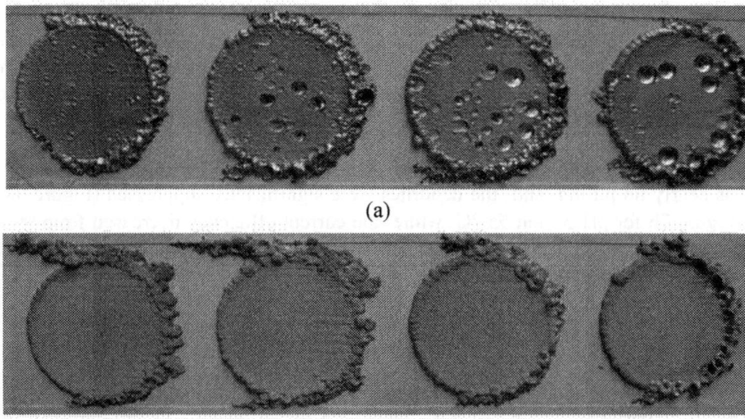

(a)

(b)

Figure 7. Cobalt buttons produced from the solution prepared using cobalt sulfate from Alfa Aesar at (a) 60 °C and (b) 55 °C under the conditions of 435 A/m^2, pH 3.5, and 1 day of deposition

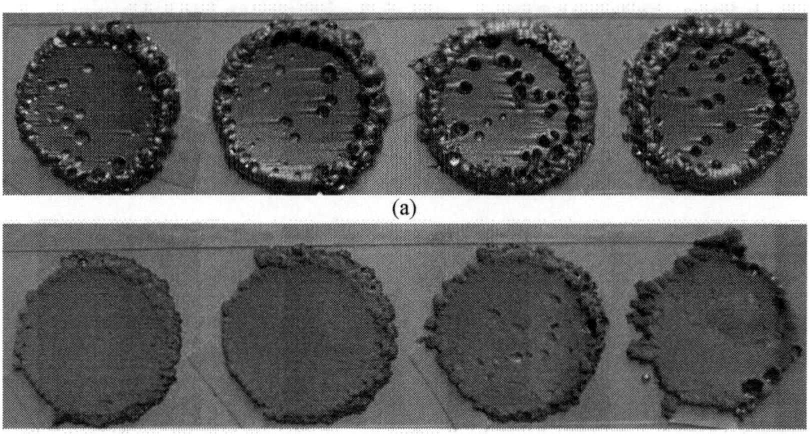

(a)

(b)

Figure 8. Cobalt buttons produced from the solution prepared using reagent grade cobalt sulfate from Shepherd Chemical Co at (a) 65 °C and 2 days of deposition and (b) 55 °C and 1 day of deposition under the conditions of 435 A/m^2 and pH 3.5.

At a lower pH, more hydrogen evolves and hydrogen bubbles quickly grow and detach, which in turn enhances the agitation of the catholyte near the cathode and facilitates the detachment of hydrogen bubbles. More atomic hydrogen is also adsorbed on the cobalt deposit surface, which affects the nucleation and crystal growth, and modifies the structure, morphology and shape of a cobalt round. Therefore, further investigation was conducted at a lower temperature and pH.

With the use of reagent grade cobalt sulfate from Alfa Aesar, as the pH decreased from 3.5 to 3.0, there was nearly no pitting, and the dendrites were significantly suppressed (Figure 9a for pH 3.0 and Figure 7b for pH 3.5) at 55 °C while the current efficiency decreased from 99.4 to 97.6%. With further decreasing pH to 2.5, the cobalt button did not change very much (Figure 9b) while the cobalt current efficiency decreased to 96.4%. The dendritic cobalt button edge was probably caused by the use of electroplating tape used for repairing the broken ink edges.

With the use of purified cobalt carbonate from Shepherd Chemical Company, the pitting and dendrites were also significantly suppressed as the pH decreased from 3.5 to 3.0 (Figure 6b for pH 3.5 and Figure 10 for pH 3.0). The cobalt current efficiency decreased to 97.7%.

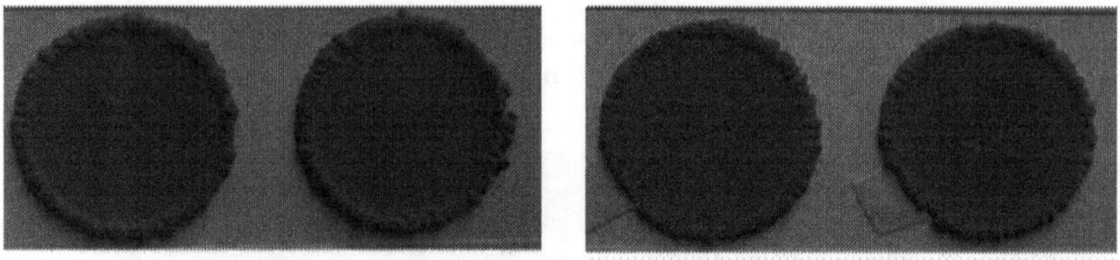

(a) (b)

Figure 9. Cobalt buttons produced from the solution prepared using cobalt sulfate from Alfa Aesar at (a) pH 3.0 and (b) pH 2.5 under the conditions of 55 °C, 435 A/m^2, and 1 day of deposition.

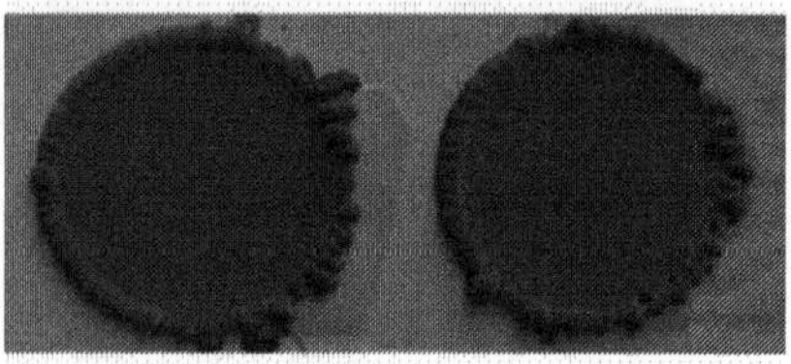

Figure 10. Cobalt buttons produced from the solution prepared using purified cobalt carbonate from Shepherd Chemical Company at 55 °C, 435 A/m^2, pH 3.0, and 1 day of deposition.

As discussed above, temperature and pH significantly affected the cobalt button quality, twelve 5-day tests were carried out to investigated the effects of temperature (50, 55, 60 and 65 °C) and pH (2.0, 2.5 and 3.0) and find an optimum condition for cobalt button production. The purified cobalt carbonate was used for all twelve tests. The summarized results are presented in Table I.

At pH 2.0, there were a few pits in the cobalt buttons produced at 50, 60 and 65 °C while there were slightly more pits in those at 55 °C (Figure 11). Smaller pits (less 1mm) were mainly developed after 3 and 4 days.

At a pH of 2.5, there were very few pits in the cobalt buttons produced at 50 °C while there were a few of pits at 55 °C and a lot of pits at 60 and 65 °C (Figure 12). Generally the current efficiency slightly increased with increasing temperature.

At pH 3.0, there was practically no pitting in the cobalt buttons produced at 50 °C while there were a few of pits at 55, 60 and 65 °C (Figure 13). The cobalt current efficiency was around 98% and almost independent of temperature. The deposit morphology and surface condition varied with changing temperature

Table I. Test conditions and results at different current densities at 435 A/m^2*

Temp. / °C	pH	Co CE /%	Average cell voltage / V	Energy consumption / kWh / kg Co
50	2.0	84.6	3.489	3.75
55	2.0	88.5	3.383	3.48
60	3.0	88.8	3.349	3.43
65	2.0	89.3	3.282	3.34
50	2.5	94.0	3.681	3.57
55	2.5	93.9	3.529	3.42
60	2.5	94.2	3.549	3.43
65	2.5	95.7	3.402	3.23
50	3.0	98.2	3.578	3.32
55	3.0	98.3	3.454	3.20
60	3.0	98.1	3.415	3.17
65	3.0	98.1	3.359	3.12

* pH was measured at room temperature.

As shown in Table I, generally the cobalt current efficiency increased from 88 to 98% with increasing pH from 2.0 to 3.0 while it was almost independent of temperature.

If cobalt electrowinning is operated at a pH range between 2.5 to 3.0, the temperature should be controlled around 50 °C to produce cobalt buttons with few pits. If it is operated at pH 2.0, a higher temperature (60-65 °C) can be used.

As shown in Figure 14, the maximum size and adhesion time of a growing hydrogen bubble is dependent on the wetting or contact angle (θ) which is determined by the surface (liquid-gas) tension ($\sigma_{l\text{-}g}$), solid-liquid interfacial tension ($\sigma_{s\text{-}l}$) and the solid surface free energy ($\sigma_{s\text{-}g}$) according to the following equation:

$$\mathrm{Cos}(\theta) = (\sigma_{s\text{-}g} - \sigma_{s\text{-}l}) / \sigma_{s\text{-}g} \qquad (4)$$

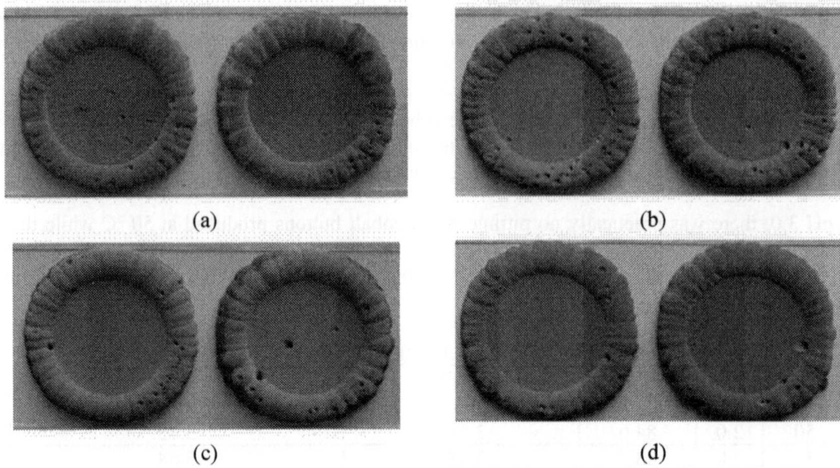

Figure 11. Cobalt buttons produced at (a) 50 °C, (b) 55 °C, (c) 60 °C and (d) 65 °C under the conditions of 435 A/m^2, pH 2.0, and 5 days of deposition.

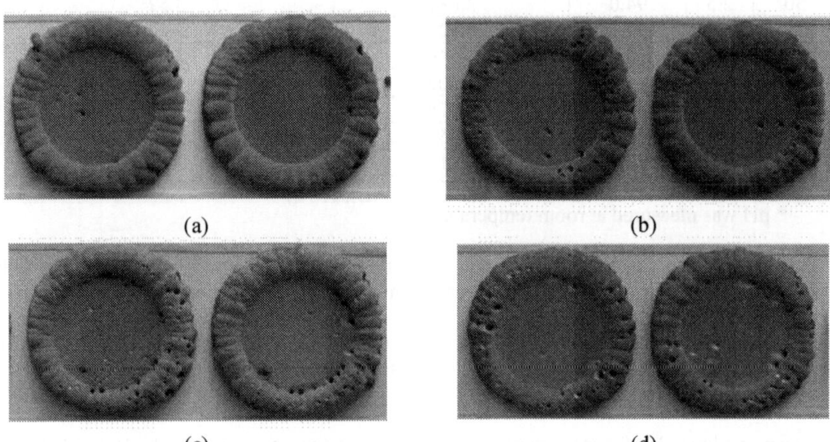

Figure 12. Cobalt buttons produced at (a) 50 °C, (b) 55 °C, (c) 60 °C and (d) 65 °C under the conditions of 435 A/m^2, pH 2.5, and 5 days of deposition.

Figure 13. Cobalt buttons produced at (a) 50 °C, (b) 55 °C, (c) 60 °C and (d)65 °C under the conditions of 435 A/m^2, pH 3.0, and 5 days of deposition.

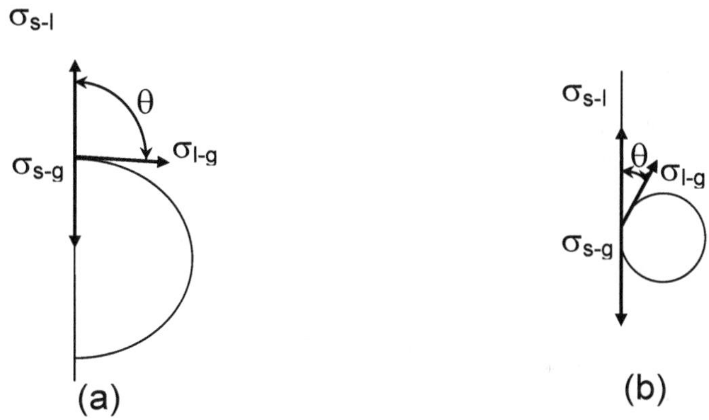

Figure 14. Adhesion of a gas bubble on metal deposit at (a) a large solid-liquid interfacial tension and surface tension and low solid surface free energy, and (b) a small solid-liquid interfacial tension and surface tension and high solid surface free energy.

A high solid surface free energy, low liquid-gas interfacial tension and low surface tension result in a small wetting angle, facilitating the detachment of a gas bubble and hence suppressing pitting. As the temperature and pH changed, the liquid-solid interfacial tension and cobalt deposit surface free energy varied. At 50 °C, the interfacial tension between the cobalt metal and catholyte decreased to such a low value and/or the deposit surface free energy increased to such a high value that hydrogen bubbles were easily detached and pitting was eliminated.

Surface morphology can also affect the detachment of a gas bubble. Equation 4 is only suitable for a clean, smooth surface. When a water droplet completely wets a rough surface, the effect of roughness on the contact angle is given by the Wenzel equation [11]:

$$\mathrm{Cos}(\theta_W) = r\,\mathrm{Cos}(\theta) \qquad (5)$$

where θ is the contact angle on a smooth surface, θ_W is the contact angle on a rough surface, r is the roughness ratio. Since the roughness ratio compares the true surface area of a rough surface with the surface of a comparably sized smooth surface, it is always larger than one. Therefore, surface roughness decreases the contact angle on a hydrophilic surface or increases the contact angle on a hydrophobic surface.

The water contact angle on cobalt metal is about 66° [12]. A rougher cobalt deposit should result in a smaller contact angle. As the temperature and pH varied, the surface morphology and roughness changed, affecting the contact angle and thus adhesion of a gas bubble on the deposit. Generally at 65 °C, the cobalt deposits were shining and smooth, and there were large pits. At 50 or 55 °C, the deposits were dark and rough, and there were either small pits or no pitting due to the decrease in the contact angle, which partially resulted from the increase in the deposit roughness.

In order to operate cobalt electrowinning in wider ranges of temperature and pH and produce high quality cobalt buttons, surfactants that do not interfere with solvent extraction should be tested. A proper surfactant does not necessarily need to decrease the catholyte surface tension significantly if it can reduce the interfacial tension between cobalt metal and the catholyte.

The addition of sodium sulfate improves the catholyte conductivity. However, it also increases the surface tension of the catholyte [13]. A lower concentration of sodium sulfate may be preferred.

Use of Anti-pitting Agents

Sodium lauryl sulfate (SLS) has been used as a foam stabilization and anti-pitting agent for cobalt electrowinning at Vale INCO's Port Colborne plant for a long time [8]. The typical SLS dosage used is 20-30 mg/L. At this dosage, SLS interferes with a solvent extraction process and make the organic/aqueous phase separation difficult. 3M FC1100 is a low foaming surfactant. It has been used as an acid mist suppressant in copper electrowinning and does not interfere with copper solvent extraction. 3M FC1100 may be used as an anti-pitting agent. Another three anti-pitting agents (AP1, AP2, AP3) were also selected for testing. The surface tension and interfacial tension between the aqueous and organic solution were measured using the ring method to evaluate these additives.

As shown in Figure 15, the surface tension was decreased rapidly from 77.1 to ~30 mN/m with the addition of SLS or AP2 while it decreased relatively slowly with increasing dosages of the other additives. FC1100 is the least effective. The decrease in the interface tension indicates that these additives have the potential to suppress pitting.

The organic/aqueous interfacial tension decreased with increasing concentrations of AP1, AP3, SLS and FC1100 (Figure 16). The interfacial tension did not decrease much by increasing the concentration of AP2 from 1 to 5 mg/L. Thermodynamically AP2 may potentially have the least effect on the disengagement of the aqueous and organic phases. However, the coalescence of droplets is affected by anti-pitting agents both thermodynamically and kinetically.

Drop-interface coalescence occurs through five consecutive stages: (1) approach of a drop towards the interface, resulting in both drop and interface deformation, (2) damped oscillation at the drop near the interface, (3) formation of a film of the continuous phase between the drop and its bulk (discontinuous) phase, (4) drainage of the film, its rupture and removal, with the initiation of the coalescence process, (5) transfer of the contents of the drop into its bulk phase [14]. Stages 1 and 2 are very fast, and Stages 3 to 5 constitute the time of coalescence. The time for Stage 5 is 0.06 to 0.08 seconds. The rate of coalescence is determined by Stages 3 and 4. Drop-drop coalescence is different from drop-interface coalescence. However, drop-drop coalescence still undergoes the steps similar to Stages 3 and 4.

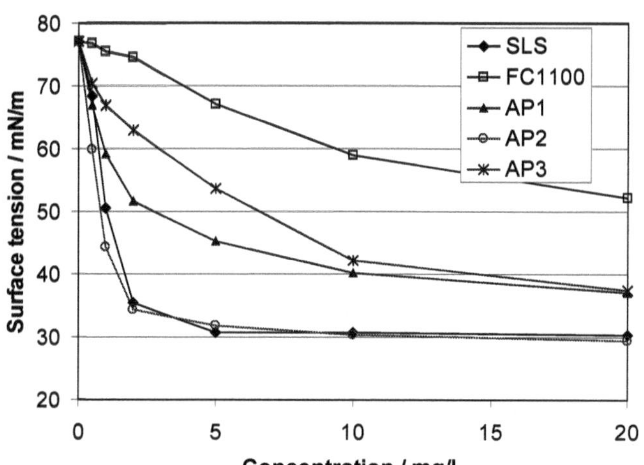

Figure 15. Feed solution surface tension as a function of additive concentration at room temperature and pH 2.5.

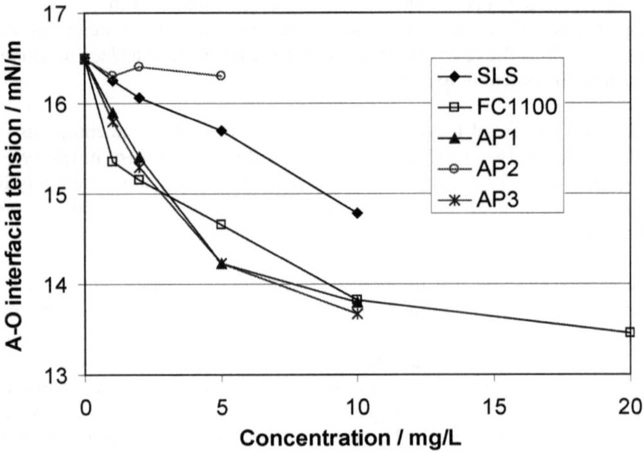

Figure 16. Interfacial tension between organic phase (30% Cyanex 272 in SX80) and feed solution (pH 2.5) at room temperature.

A high interfacial tension results in a high resistance to drop deformation by decreasing the drainage film area. Besides, flow of the film is inhibited by an increase in interfacial tension. Kinetically, coalescence rate normally decreases with increasing interfacial tension [14]. However, if the interfacial is too low, there is little driving force for coalescence, and then phase disengagement can be difficult. If the interfacial tension is below 1 mN/m, significant emulsification occurs, causing a huge problem in phase disengagement.

Surfactants accumulate at the interface and coalescence can be retarded. The effect of the surfactants on coalescence varies significantly since each surfactant is different. From the interfacial tension values shown in Figure 16, it is difficult to judge whether these additives will affect phase disengagement. The additives should be evaluated in both electrowinning and phase separation tests.

In the presence of 10 mg/L FC1100, pitting was not suppressed (Figure 17). Similar results were obtained at 20 mg/L FC1100. FC1100 did not function as an anti-pitting agent. With increasing SLS dosage from 0.5 to 5 mg/L, pitting was gradually suppressed (Figure 18). At 5 mg/L of SLS, there was no pitting except at the bottom edge. The deposit morphology changed as the SLS dosage varied.

Figure 17. Cobalt buttons produced at 65 °C, pH 3.0, 10 mg/L FC1100 and 1 day.

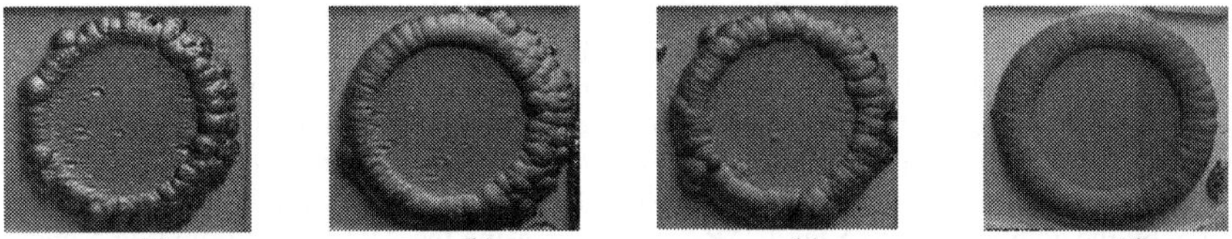

(a) (b) (c) (d)

Figure 18. Cobalt buttons produced at 65 °C, pH 3.0 and 5 days of deposition with the addition of (a) 0.5, (b)1.0, (c) 2.0 and (d) 5.0 mg/L SLS.

As shown in Figure 19, there were practically no pits in the three cobalt buttons produced with the addition of 2 mg/L AP1, 1 mg/L AP2 and 1 mg/L AP3, respectively. With the use of AP1, the catholyte surface tension gradually increased to a certain value after one day and then stabilized. In the successive test (as soon as the old cathode was taken out, a new cathode was immediately inserted the cathode compartment to start a new test), the cobalt round morphology was similar to that produced in the previous run. However, with the use of AP2 or AP3, in the successive run, there were some pits in the cobalt rounds. This indicates that AP1 degraded to some degree, but it still effectively suppressed pitting while AP2 and AP3 degraded to such a great degree that they were not able to effectively suppress pitting. The addition of 1 to 10 mg/L AP1, or 1-5 mg/L AP2 or 1-5 mg/L AP3 practically had no effect on cobalt current efficiency. The interfacial tension between the anolyte and organic solution was about 20 mN/m for all these three additives. When the anolyte was neutralized to pH.2.5 using 10 M NaOH solution, the interfacial tension between the neutralized anolyte and organic phase was 17.5 mN/m, which was slightly higher than between the feed and organic phase (16.5 mN/m). This indicates that these three additives degraded so significantly in the anode compartment that they did not function well as fresh surfactants.

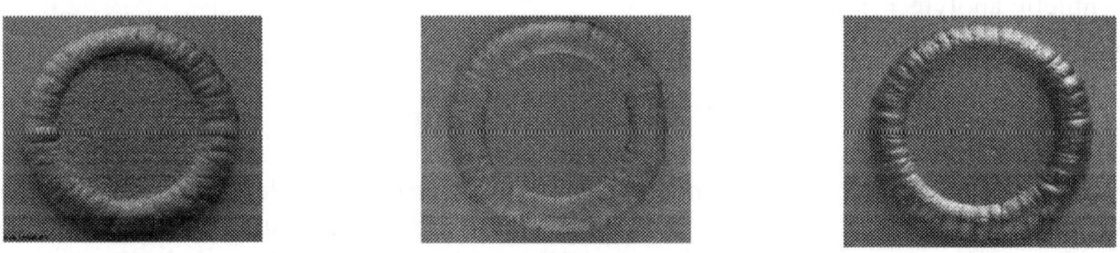

(a) (b) (c)

Figure 19. Cobalt buttons produced at 65 °C and pH 3.0 with the addition of (a) 2.0 mg/L AP1, (b)1.0 mg/L AP2 and (c) 1.0 mg/L AP3.

Organic-aqueous phase disengagement tests were conducted in a water-jacketed reactor with an inner diameter of 104 mm at 50 °C. The impeller (supplied by BASF) was driven by a Caframo high torque stirrer and its rotational speed was controlled at 1750± 17 rpm. The volumes of both the organic and aqueous phases were 375 mL. The organic phase (30% Cyanex 272 in SX80) was first loaded with 18 g/L Co before stripping (phase separation test). When the Co-loading was over 18 g/L, the viscosity of the loaded organic phase increased very quickly. The phase disengagement was tested at both organic phase continuity and aqueous phase continuity. If the aqueous phase needs to be continuous, the aqueous was added first and stirred, and then the organic phase was added. If the organic phase needs to be continuous, the organic was added first and stirred followed by the addition of the aqueous phase. The phase separation was considered complete when only a monolayer of bubbles remained at the interface. The results are summarized in Table II. For the continuous organic phase, the A-O phase disengagement was practically independent of the addition of 1 to 10 mg/L AP1 while it was affected very slightly by the addition of 1 to 5 mg/L AP2. For the continuous aqueous phase, the A-O phase separation was also independent of the addition of 1 to 10 mg/L AP1 while it was significantly affected by the addition of 1 to 5 mg/L AP2. With increasing the AP2 dosage from 0 to 5 mg/L, the separation time increased from 36 to 150 seconds. AP2 interfered with the phase separation while AP1 and AP3 did not interfere with the phase separation. AP2 was adsorbed on the organic droplet, producing energy barriers surrounding the droplet and thus retarding coalescence.

Table II. Phase separation time (seconds) under the different conditions

Aqueous phase	Continuous aqueous phase	Continuous organic phase
Synthetic anolyte	37	27
Anolyte produced from feed with 2 mg/L AP1	34	27
Synthetic anolyte + 1 mg/L AP1	34	26
Synthetic anolyte + 2 mg/L AP1	36	26
Synthetic anolyte + 5 mg/L AP1	36	25
Synthetic anolyte + 10 mg/L AP1	35	25
Anolyte produced from feed with 2 mg/L AP1	42	28
Synthetic anolyte + 1 mg/L AP2	85	30
Synthetic anolyte + 2 mg/L AP2	137	30
Synthetic anolyte + 5 mg/L AP2	150	30
Anolyte produced from feed with 1 mg/L AP3	35	27
Synthetic anolyte + 1 mg/L AP3	34	25
Synthetic anolyte + 2 mg/L AP3	34	24
Synthetic anolyte + 5 mg/L AP3	34	23
Synthetic anolyte + 10 mg/L AP3	33	22

Cobalt Deposition Stress

Cobalt deposition is associated with a high internal stress in the deposit. As more cobalt is deposited, the cobalt deposit tends to peel off from the cathode. It is important to know how the cobalt deposition stress changes as the operation conditions vary. The cobalt deposition stress was calculated according to the method given by Specialty Testing & Development Co. and discussed by Leaman [15]. After the cobalt deposition, the cathode was bent toward the anode, indicating a tensile stress. The deposition stress as a function of temperature and current density is shown in Figures 20 and 21, respectively. At 450 A/m^2, the stress first increased from 111.9 to 112.3 M Pa with increasing the temperature from 50 to 55 °C, and then decreased to 92.3 MPa with increasing the temperature to 65 °C. This trend is generally in agreement with that obtained from chloride media [16]. At 65 °C, the stress decreased from 97.6 to 84.9 MPa with increasing the current density from 400 to 550 A/m^2. At a higher temperature and current density, the cobalt deposition stress is lower and hence a cobalt button adheres to its cathode better. After 5 days of deposition, the adhesion force of a cobalt button to the cathode was around 100 kg from push testing. The cobalt button was easily stripped by vibration. The deposition stress as a function of AP1 concentration is shown in Figure 22. The deposition stress increased with increasing AP1 concentration.

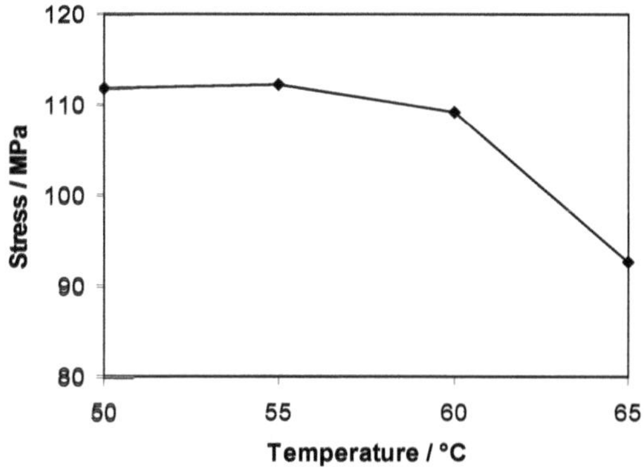

Figure 20. Cobalt deposition stress as a function of temperature at 450 A/m^2.

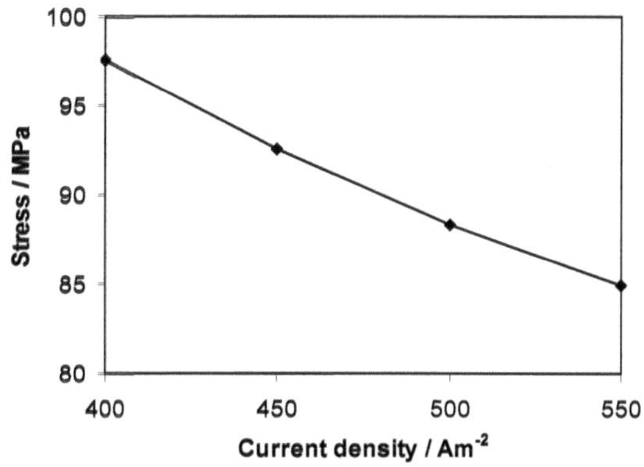

Figure 21. Cobalt deposition stress as a function of current density at 65 °C.

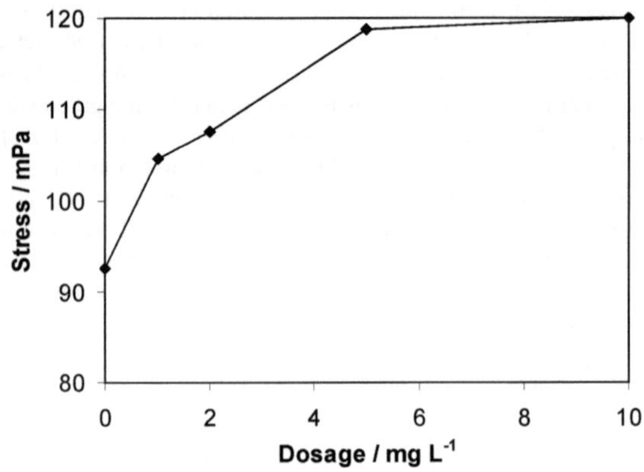

Figure 22. Cobalt deposition stress as a function of AP1 dosage at 65 °C and 450 A/m^2.

Conductivity of $CoSO_4$-Na_2SO_4-H_2SO_4 Solutions

Electrowinning can be significantly affected by the electrical conductivity of the electrolyte. A reasonably high electrical conductivity can greatly save energy consumption and improve cobalt deposit quality. However, in the literature, there is very little information available that can be directly used for Minera y Metalúrgica Del Boleo cobalt electrowinning process design. The objective of this measurement is to measure the electrical conductivity of the $CoSO_4$-Na_2SO_4-H_2SO_4 system and support the design of a cobalt electrowinning cell.

The conductivity as a function of sodium sulfate concentration depends on sulfuric acid concentration and temperature (Figures 23 to 26). At 25 °C (Figure 23), when the sulfuric acid concentration is below 25 g/L, the conductivity increases with increasing sodium sulfate concentration, while it decreases with increasing sodium sulfate concentration at a sulfuric acid concentration above 25 g/L. At 65 °C (Figure 24), when the sulfuric acid concentration is below 50 g/L, the conductivity increases with increasing sodium sulfate concentration while at 75 g/L of H_2SO_4, it practically does not change with increasing sodium sulfate concentration.

The above phenomena are related to the distributions of the species of Co^{2+}, Na^+, $NaSO_4^-$, SO_4^{2-}, $HSO_4^-$and H^+, the conductivities of these species, and the interaction between these species. There are no direct ionic conductivity data in the corresponding solution. However, there are data for equivalent conductivities at infinite dilution for comparison. The equivalent conductivities of Co^{2+}, Na^+, H^+ and HSO_4^- at infinite dilution are 5.5, 5.0, 35 and 5.2 mS m^2 mol^{-1} [17]. The conductivity of the proton is the highest and the solution conductivity is affected most greatly by the dissociation of sulfuric acid. At a higher temperature and lower sulfuric acid concentration, $CoSO_4$ and H_2SO_4 are dissociated to a higher degree and the dissociated species interact less, and the addition of Na_2SO_4 reduces their dissociation to a lesser degree and then overall increases the conductivity. At a lower temperature and higher sulfuric acid concentration, $CoSO_4$ and H_2SO_4 are dissociated to a less degree and the dissociated species also interact more. The addition of sodium sulfate reduces their dissociation (especially H_2SO_4) to a higher degree and then generally decreases the conductivity.

The conductivity as a function of temperature is shown in Figures 25 and 26. The conductivity increases with increasing temperature.

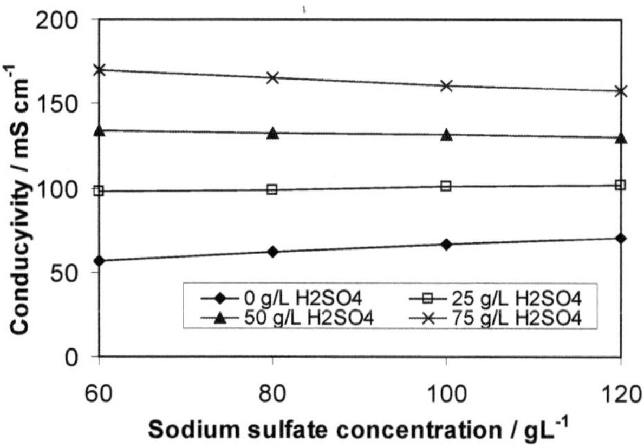

Figure 23. Conductivity as a function of sodium sulfate concentration at 25 °C.

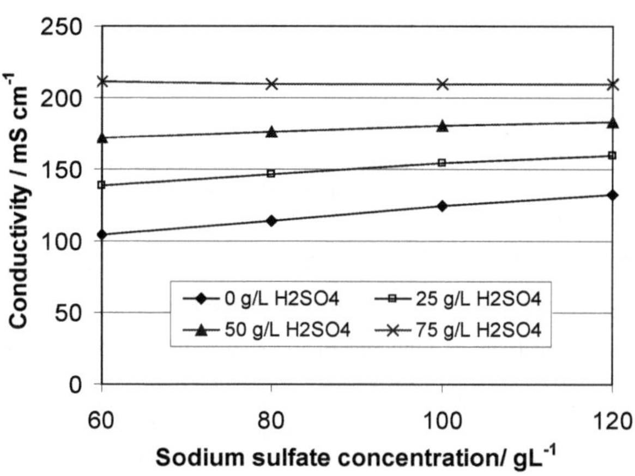

Figure 24. Conductivity as a function of sodium sulfate concentration at 65 °C.

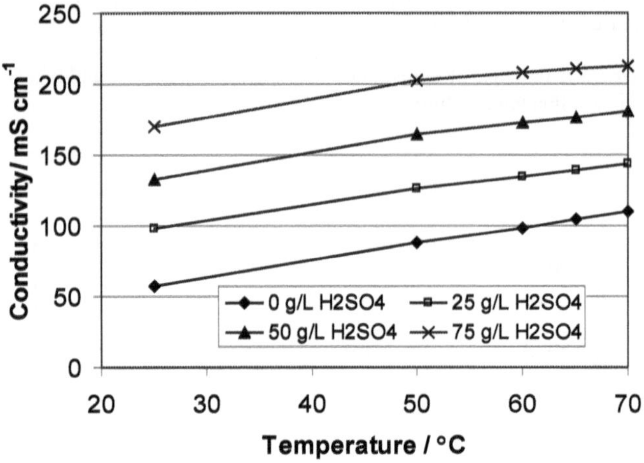

Figure 25. Conductivity as a function of temperature at 60 g/L of Na_2SO_4.

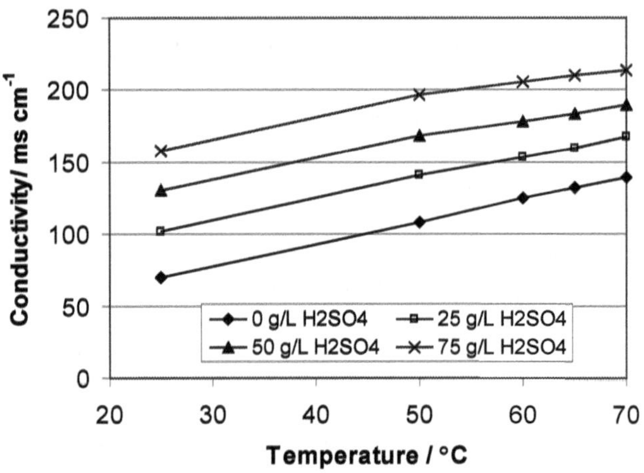

Figure 26. Conductivity as a function of temperature at 120 g/L of Na_2SO_4.

The design concentrations of cobalt in the catholyte and feed are 45 and 60 g/L, respectively. The cobalt bite is 15 g/L, indicating 25 g/L of H_2SO_4 in the anolyte. With increasing cobalt bite, the sulfuric acid concentration in the anolyte increases, resulting in the increase in the conductivity and then energy saving.

Conclusions

The cobalt round quality varied in the presence of different impurities (especially organic compounds). Pitting is the main problem for production of high quality cobalt buttons. Some organic compounds cause pitting while others suppress pitting. Pitting can still occur in the absence of organic compounds because pitting is caused by the adhesion of hydrogen bubbles on the cobalt deposition. The adhesion of hydrogen bubbles is in turn determined by the solution surface tension, the cobalt deposit surface free energy and the interfacial tension between the cobalt deposition and the solution. Temperature affects surface properties, such as the surface free energy and the interfacial tension, and hence influences pitting.

With decreasing temperature, pitting was suppressed while dendrites grew laterally, indicating that temperature significantly affected cobalt deposition. It can be expected that at a lower temperature, the interfacial tension between cobalt deposit and catholyte was lower, and/or the cobalt deposit surface free energy was higher, facilitating the detachment of hydrogen bubbles and suppressing pitting. With decreasing pH, the dendrite growth was suppressed due to the increased inhibition of cobalt deposition. At a pH below roughly 3, the lateral growth of dendrites was suppressed to an acceptable degree.

In the temperature range from 50 to 65 °C and the pH range from 2.0 to 3.0, the cobalt current efficiency increased from 88 to 98% while it was slightly affected by temperature. At 50 °C and a pH of 3.0, there was almost no pitting, indicating the interfacial tension between the catholyte and cobalt deposit decreased below a critical value and /or the cobalt deposit surface free energy exceeded a critical value such that hydrogen bubbles easily detached. The change in the deposit surface morphology was another factor.

With the use of appropriate additives (anti-pitting agents), pitting in the cobalt round can be suppressed without interfering with phase disengagement in cobalt solvent extraction. Cobalt deposit stress decreased with increasing temperature and current density.

Acknowledgements

The authors would like to thank Minera y Metalúrgica del Boleo for permission to publish this work.

References

1. D. Dreisinger, E. Norton, R. Molnar, K. Baxter, M. Holmes, and J. Riordan, "The Recovery Of Zinc From Baja Mining Corp's El Boleo Orebody," *Zinc and Lead Metallurgy*, ed. L. Centomo, M. Collins, J. Harlamovs and J. Liu, (Montreal, Ontario: CIM), 343-357.
2. D. Dreisinger, W. Murray, K. Baxter, M. Holmes, H. Jacobs and R. Molnar. "The metallurgical development of the El Boleo copper-cobalt-zinc project" ALTA Copper 2005, Perth, ALTA Metallurgical Services, Melbourne, Australia.

3. D. Dreisinger, K. Baxter and R. Molnar. "New development in the Boleo copper-cobalt-zinc-manganese project". ALTA Copper 2006, Perth, ALTA Metallurgical Services, Melbourne, Australia.
4. C. Y. Cheng and M. D. Urbani. "Solvent extraction process for separation cobalt and/or manganese from impurities in leach solutions". Patent Publication No. WO 2005/073415 A1.
5. C. Y. Cheng and M. D. Urbani. "Solvent extraction process for separation cobalt and/or nickel from impurities in leach solutions". Patent Publication No. WO 2005/073416 A1.
6. C. Y. Cheng, "Solvent extraction of nickel and cobalt with synergistic systems consisting of carboxylic acid and aliphatic hydroxyoxime", *Hydrometallurgy*, 84 (2006), 109-117.
7. R. Agnew, P. Garritsen and D. MacVicar, "Production of cobalt rounds at INCO's cobalt refinery", *Extractive Metallurgy of Nickel and Cobalt* ed. G. P. Tyroler and C. A. Landolt (Warrendale PA: The Metalurgical Society of AIME, 1988), 555-561.
8. V. Ettel, E. Devuyst, and J. Babjak, "Process for recovery cobalt electrolytically", US Patent, No. 4274930, 1981.
9. O. E. Kongstein, G. M. Haarberg and J. Thonstad, "Current efficiency and kinetics of cobalt electrodeposition in acid chloride solutions. Part I: The influence of current density, pH and temperature", *Journal of Applied Electrochemistray*, 37 (2007), 669-674.
10. J. T. Matsushima, F. Trivinho-Strixino and E. C. Pereira, "Investigation of cobalt deposition using the electrochemical quartz crystal microbalance", *Electrochimica Acta*, 51 (2006), 1960-1966.
11. R. Wenzel, "Resistance of solid surfaces to wetting by water", *Industrial and Engineering Chemistry*, 29 (1936), 988-994.
12. D. A Winesett, H.Ade, J. Sokolov, M. Rafailovich and S. Zhu, "Substrate dependence of morphology in thin film polymer blends of polystyrene and poly(methyl methacrylate)", *Polymer International*, 49 (2000), 458-462.
13. N. F. Mamontov and A.I. Chernov, "Connection between surface tension of nickel solution with the formation of pitting", *The Soviet Journal of Non-ferrous Metals*, 5 (1974), 37-38.
14. G. M. Ritcey and A. W. Ashbrook, *Solvent extraction – Principles and Applications to Process Metallurgy – Part I*, (Amsterdam: Elsevier Publishing Company, 1984), 210-214.
15. F. Leaman, "A new frontier for stress measurement", (Paper presented at the SurFin Conference, Cleveland, Ohio, USA, 15 August 2004).
16. M. A. Zhamagortsyants, A. A. Yavich and Z. N. Pilikyan, "Influence of solution pH on the electrodeposition of iron-group metals at temperature between and 25 and 175 °C", *Soviet Electrochemistry*, 14 (1978), 25-29.
17. P. Vansek, "Ionic conductivity and diffusion at infinite dilution", *CRC Handbook of Chemistry and Physics, 92th edition*, ed. W. Haynes, (Boca Raton, Florida: CRC Press, 2011-2012) 5-76.

Ni-Co 2013
Edited by: Thomas Battle, Michael Moats, Violina Cocalia, Harald Oosterhof, Shafiq Alam, Antoine Allanore, Rodney Jones, Nathan Stubina, Corby Anderson, and Shijie Wang
TMS (The Minerals, Metals & Materials Society), 2013

COMPARISON OF INTERCELL CONTACT BARS FOR ELECTROWINNING PLANTS

Chris Boon, Rob Fraser, Tim Johnston, Douglas Robinson

Hatch Electrometallurgy Group
Hatch Associates; 61 Petrie Terrace Brisbane, Australia 4000

Keywords: Intercell contact bar, Contacts, Resistance

Abstract

In normal practice, the design and sizing of intercell contact bars (ICCB) for electrowinning is based on previously used designs, empirical testwork, or rules of thumb, rather than from first principles. A major operating cost for electrowinning is power consumption. The ICCB is responsible for a portion of the power costs due to its electrical resistance, which comprises electrode contact resistances and the bulk resistance of the ICCB.

This paper presents the methods used to calculate the electrical resistances and the results for a number of different widely accepted ICCB designs currently used in electrowinning plants. The paper also makes comparisons between the systems on other key parameters for ICCB performance in nickel and cobalt tankhouses.

Introduction

Hatch has completed a study to compare a number of intercell contact bar designs currently available in the market for their use in electrowinning and electrorefining plants. The different designs have been compared on a number of bases to determine their ability to produce consistent product and minimize operating costs.

Operating costs are reduced when the power consumption per tonne of metal produced is reduced. It will be shown that a significant proportion of the power consumption is due to contact resistances and the ICCB resistive drop.

Better quality deposits and higher current efficiency result when each cathode in a cell has equal current meaning that current distribution is even. This is a key goal in an EW/ER plant design and specifically in ICCB design since the current distribution will be influenced by any variation in the contact system direct resistance values. In practice there will also be other factors that influence current distribution. These include:

- Dirty contacts or formation of oxide films.
- Damaged contact bars.
- Uneven electrode spacing.
- Different internal resistances of electrodes.

This paper presents the theory that governs calculation of resistive drop and power consumption for intercell contact bars and the electrode header bars that rest on them. A comparison is then made of the different contact systems to assess their relative efficiency with particular reference to nickel and cobalt plants. The systems are then also qualitatively compared on the basis of other practical considerations for ICCB design.

Background and Theory

Intercell Contact Bar Designs

The intercell contact systems considered in this paper include:

- Standard Walker intercell contact bar.
  - Triangular intercell contact bar.
  - Dog-bone intercell contact bar.
- Outotec's DoubleContact™ intercell bar.
- Cominco's spool contact bar.
- Notched contact bar.
- Freeport McMoRan double contact bar.

It is important that the different types of contacts used in all of these systems can be classified as either line contacts or point contacts according to the shape of the interface between the electrode header bar and the intercell contact bar.

A line contact occurs when the electrode header bar and intercell contact bar form a contact interface on an edge. This occurs when the flat surface of one member contacts the edge of the other. A point contact occurs when the two contacting items contact at a point. This can only occur when an edge from one member contacts the edge of another member or when an edge from one member contacts a convex surface of another member.

Intercell Bar – Standard Walker System. The Walker System uses continuous intercell contact bars which electrically connect adjacent cell electrodes in a parallel fashion. This system was patented in 1901 [1] and appears to be the basis of the more refined contact bar arrangements developed over time. The typical Walker system bars are generally of a dog-bone or triangular profile [2].

Triangular Intercell Bar. One of the most common profiles used when adopting the Walker System is the triangular profile [2]. It produces a line contact for both the cathode and anode.

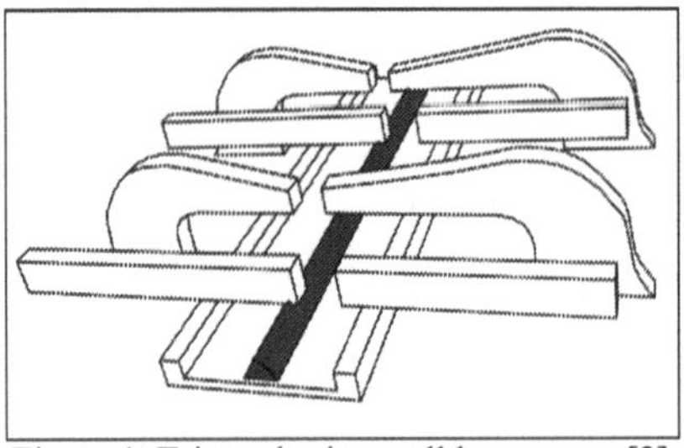

Figure 1. Triangular intercell bar system [3]

Dog Bone Intercell Bar. The other common profile used is the dog bone [2]. Similar to the triangular intercell bar, the cathode and anode each have separate line contacts but often on an edge formed by a radius, as opposed to a sharper edge formed by a triangular bar.

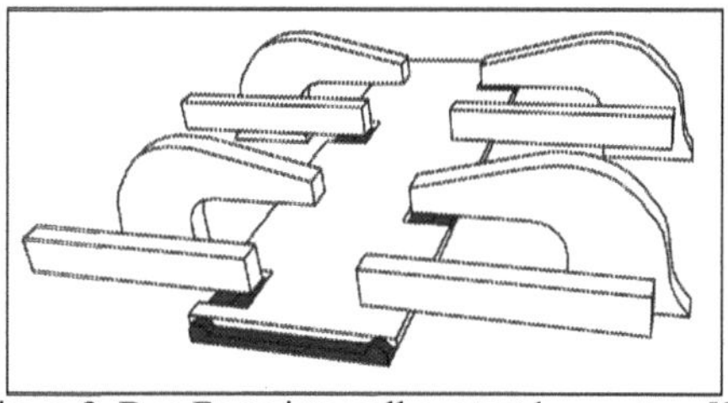

Figure 2. Dog Bone intercell contact bar system [3]

DoubleContact™ – Outotec. Outotec's DoubleContact™ intercell contact bar system is designed so that all electrodes have an electrical contact on each end. These electrical contacts are all line contacts. Outotec has claimed that the principal benefit of using the double contact system is a more even current distribution [4]. The basis of this design is that if there is a poor contact between an electrode and the contact bar, the electrode will still receive current from the contact on the other end of the electrode from the equalizer bar. Outotec claims that a more even current distribution is likely to reduce energy consumption and lead to better quality cathodes [4]. In this paper, the impact of the equalizer bars is not considered. This results in this system being considered as a simple dog-bone system.

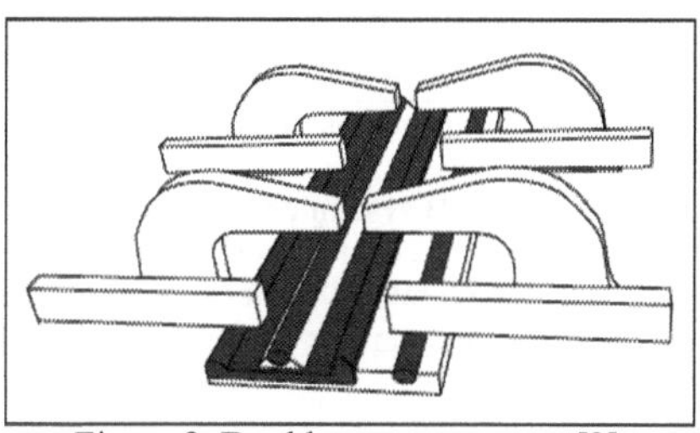

Figure 3. Double contact system [3]

Spool system – Cominco. The spool or sloping-tangent contact bar was developed in the 1970s by Cominco Ltd and utilizes a contact bar similar to that shown in Figure 4. This bar is claimed to provide excellent electrical contact via the notched portion of the header bars [5]. In theory this arrangement results in each electrode having four points of contact with the intercell contact bar.

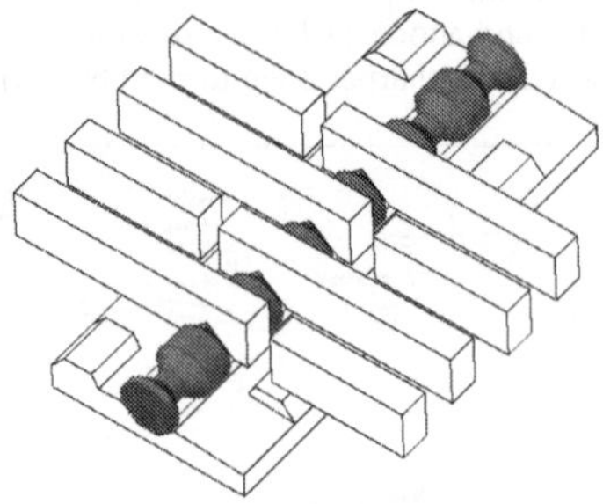

Figure 4. Illustration of a Cominco spool (developed from drawings in [5])

Notched Intercell Contact Bar. The notched intercell contact bar provides a double line contact for the cathode at the primary current transfer end as detailed in Figure 5.

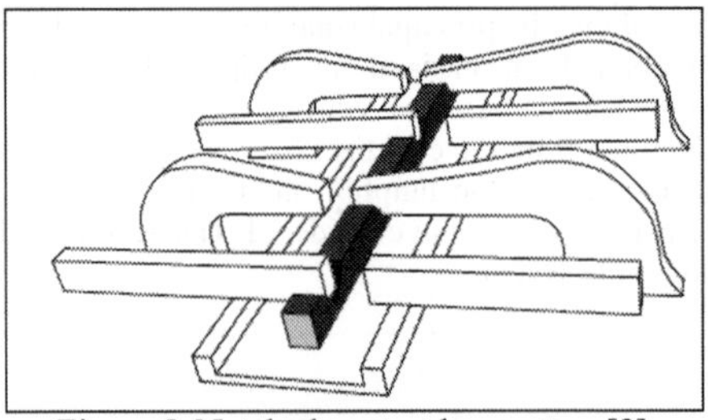

Figure 5. Notched contact bar system [3]

Freeport ICCB System. The Freeport ICCB system combines a double contact bar system with a notched contact bar generating a double line contact as shown in Figure 6, rather than the Outotec single line contact. As with the Outotec DoubleContact™ system, the equalizer bars are not considered in the calculations completed in this paper. This results in it being considered as a simple notched bar system.

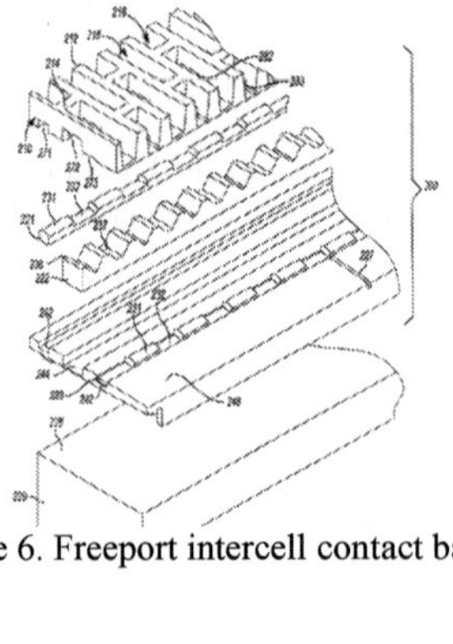

Figure 6. Freeport intercell contact bar [6]

Table I provides a summary of the intercell contact bars described above and their typical applications.

Table I. ICCB System Uses

ICCB	Type of Contact	Operation
Triangular	Line	Copper EW
Dog-Bone	Line	Copper ER, Cobalt, Nickel, Lead, Manganese
Outotec DoubleContact™	Line	Copper, Cobalt, Nickel, Lead, Manganese
Notched	Line	Zinc*, Manganese
Freeport McMoRan	Line	Copper
Spool	Point	Zinc

* may require watercooling

Methodology for Calculating Resistance

The total resistance of the intercell contact bar network is the sum of the 'contact' resistances between the electrodes and the intercell contact bar, and the bulk resistance, as shown in Figure 7.

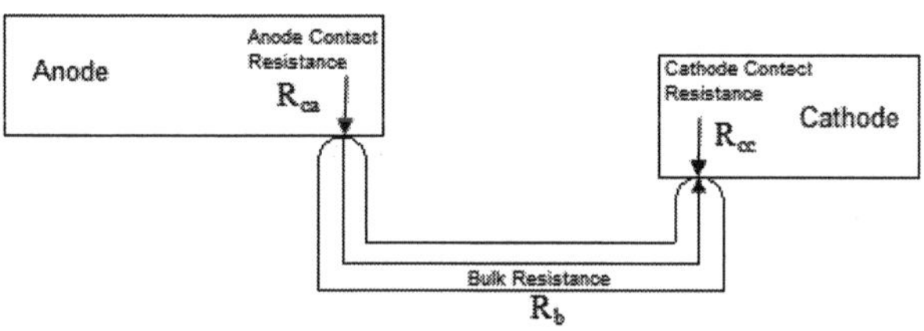

$$R_t = R_{ca} + R_b + R_{cc}$$

Figure 7. Resistances an in intercell bar network (dogbone configuration indicated)

Contact Resistance. To enable calculation of the cathode and anode contact resistances for each of the intercell contact bar systems, the equations presented in the literature have been assessed. For the case of two items being brought into contact, the area in apparent contact is much smaller than would be expected due to the microscopic high points (asperities) on each face. The resistance due to this smaller than expected area, is called the constriction resistance [7]. Additional resistance can occur due to thin films such as oxide deposits occurring on the contact surfaces. The sum of the constriction resistance and film resistance makes up the contact resistance for a particular contact.

According to Sawada et al. [8], the constriction resistance is dominant over film resistance when the loads exceed 10N (1kg) and when the film is thin. Due to the mass of electrowinning electrodes (>>1kg), it can be assumed that the film resistance is negligible, and hence, the contact resistance is equal to the constriction resistance. On this basis, the contact resistance is given by the following equation [9]:

$$R_c = \sqrt{\frac{\rho^2 \pi H}{4F}} \quad (1)$$

Where ρ = Electrical resistivity of material (ohm.m)
H = hardness of the material (Pa)
F = load normal to the contact bar surface (N)

It is important to note that based on Equation 1 the contact resistance is only dependent on the applied force and the hardness of the softer contact material, not on the apparent area of the contact. Due to this, whether the contact is a point or line, or the number of contacts per electrode, makes no difference to the electrode contact resistance by this theory.

However, Equation 1 assumes that the load is normal to the contact surface and for certain intercell contact bar designs this is not the case. Equation 1 can be re-arranged to take into consideration varying angles, and the mass of the electrode being split evenly on both sides of the cell.

$$R_c = \sqrt{\frac{\rho^2 \pi H}{2mg\cos\theta}} \tag{2}$$

where m = mass of electrode (kg)
g = acceleration due to gravity (9.81 m/s^2)
θ = the contact angle between weight vector and normal force vector as shown in Figure 8.

Buoyancy effects result from electrolyte displacement by the electrodes and other submerged components attached to the electrodes. These effects are taken into account by modifying equation (2) as follows:

$$R_c = \sqrt{\frac{\rho^2 \pi H}{2(mg - F_b)\cos\theta}} \tag{3}$$

where F_b = buoyancy force (N)

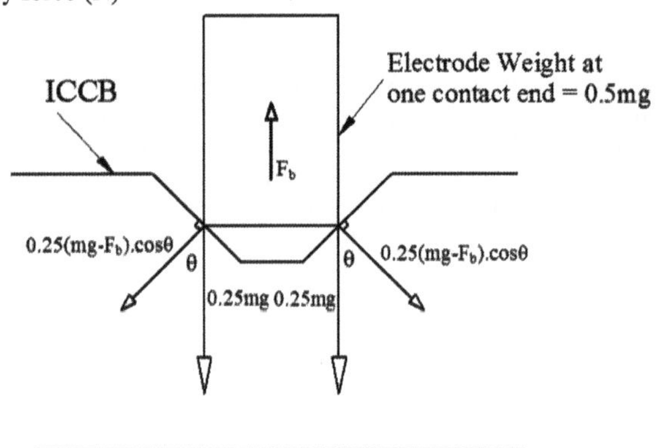

$F = 0.5(mg-F_b).\cos\theta$
Figure 8. Contact force derivation

Buoyancy Effect of Bagged Anodes. Bagged anodes are utilized in nickel and often in cobalt electrowinning to keep acid generated at the anodes away from the cathodes where it can decrease current efficiency. The bags are generally supported and sealed on the anodes by a frame. The anode gas is collected in the bag and this gas displaces liquid in the bag volume and thus increases the buoyancy of the anode/bag/frame system.

Bulk Resistance. For a material of length L, with electrical resistivity ρ and constant cross-sectional area A, the resistance along the length of the material is given by the following equation:

$$R = \frac{\rho L}{A} \quad (4)$$

Oxide Films on Copper. Oxide films are worthy of particular mention since thin oxide films are always formed on copper surfaces. When they are thin, these films do not lead to any significant increase in contact resistance [8]. However, the thickness of oxide films increases with time [10]. As the thickness increases, there is an increase in contact resistance [11] in the form of film resistance; hence, there is a need for regular contact bar cleaning in every plant. If sufficient cleaning is not carried out, the contact temperatures will increase due to this higher resistance and the rate of oxide formation will also increase since the rate of formation of oxide films is also known to increase with temperature [12]. The rate of oxide formation can become rapid if the temperature is allowed to continue to escalate [11].

In the extreme case, if the heat generated by current flow cannot escape an ICCB system fast enough then the system will overheat and will not sustain the required current flow. This has been seen to happen where the ICCB is undersized or where contact resistances are too high.

Some work has been done to find literature data on the impact of electrode weight or contact pressure on film resistance values but little information has been found. From the observations above and the personal experiences of one of the authors it appears that contact pressure may be an important factor when currents are high enough to make the contacts hot.

Heat removal from contact systems. The main method of heat removal from contact systems is conduction into the electrode header bars and into the electrolyte heat sink through the contact points. Other contributions are made through:

- Convection of heat from electrode header bars.
- Convection of heat from the exposed surfaces of the ICCB.
- Water cooling of ICCBs where this is employed.

The amount of heat rejected through convection is directly proportional to the amount of surface area exposed.

Results and Discussion

The resistance of the contacts and through the bulk of the ICCB is important because:

- Variations can lead to uneven current distribution.
- Power consumption can be affected.
- The amount of heat that needs to escape the ICCB system can be impacted.

The major components of the resistance and differences between their values for the various intercell contact bar types are reviewed below.

Bulk Resistance

To calculate the bulk resistance certain assumptions were made. These include:

- The current through each cathode was equal to 400A.
- The 'rule-of-thumb' $1A/mm^2$ was used to determine an initial cross-sectional area of the intercell contact bar – cross-sectional area was $400mm^2$.
- The distance between anode-cathode was 50mm.
- Electrical resistivity of ICCB material (copper) – $1.7x10^{-8}$ Ωm [7].
- No current was passing through the equalizer bar in the Outotec DoubleContact system or the Freeport McMoRan system.

With the above assumptions in place the bulk resistance is calculated by Equation 3 to be 2μΩ for all systems.

Contact Resistance

The contact resistances for each of the ICCB systems were calculated using the following assumptions and basis:

- Hardness of ICCB material (copper) – 500MPa [7].
- The contact angle, θ, for the spool contact bar is 45°.
- The contact angle, θ, for the notched contact bar is 30°.
- Thermal effects – temperature impacts on resistivity have not been considered.
- Film resistances were assumed to be negligible and contacts were assumed to be clean.
- Cathode mass at start and end of deposition cycle are 25kg and 85kg, respectively.
- Anode masses do not change during the deposition cycle. Two options were considered, a lead anode and a lightweight titanium catalytic anode. Their masses were 110kg and 20kg, respectively.
- Titanium anode dimensions and materials are taken from the conference paper by Sandoval et al. [13] which details the development of alternative anode technology.
- Bag contained volume is $0.07m^3$ and gas displacement of liquid in the bag is a maximum of 10% when cell is operating. This data has been taken from Hatch internal designs and calculations.
- Anode bags and frame materials have similar density to the electrolyte and thus assumed to have no impact on buoyancy.
- A typical nickel cell potential of 3.6V [14] is used.

Table II compares the different intercell contact bars mentioned, and displays the contact resistances that have been calculated for each ICCB system on the basis of no anode bags being used.

Table II. Calculated Contact Resistances and Total Maximum Resistance with No Anode Bags

ICCB	θ	Contact Resistances (μΩ)				R_t Pb Max (μΩ)	R_t Ti Max(μΩ)
		R_{ca} Pb	R_{ca} Ti	R_{cc} Start	R_{cc} End		
Triangular	0	15	35	33	18	50	70
Dog-bone	0	15	35	33	18	50	70
Notched	30°	16	38	35	19	53	75
Spool	45°	18	42	39	22	59	83

When the calculations are revised for the presence of anode bags the effective lead anode weight is decreased by 18% to 890N and titanium anode weight by 52% to 95N, at the maximum 10% gas hold up case. This would increase contact resistances by a maximum of 6% for bagged lead anodes and 50% for bagged titanium anodes.

There are several general observations from these results:

- The contact resistances are dominant over the bulk resistance of the intercell contact bar at between 95 and 98% of the total resistance.
- For unbagged lead anodes the total ICCB system resistance accounts for approximately 0.6% of the cell power consumption and for unbagged titanium catalytic anodes for approximately 0.9% of the cell power consumption.
- Cathode resistance changes from start to end of a deposition cycle as the weight of the electrode increases.
- There is a clear difference between the lead anodes and the catalytic titanium anodes – the titanium anodes have significantly higher contact resistance by approximately 20μΩ or 33% which equates to a power loss difference of 0.2% of the overall power consumption.
- It is interesting that unbagged titanium anode contact resistances are similar to the cathode resistances at the start of a cycle.
- Buoyancy effects are not substantial for the cathodes and both lead and titanium anodes at a maximum of 5% impact, when anode frames are not used.
- When anode bags/frames are used, the lead resistances are still low whereas titanium anode contact resistance increases dramatically. Unless bolted connections for bagged titanium anodes are used, resistive power loss will increase to 1.1% of the overall power consumption and heat generation will also increase significantly.

Comparative comments for the ICCB systems include:

- The triangular and dog-bone systems have lower contact resistance than the notched and spool bar systems. This difference is purely due to the difference in contact angle for the different systems. The steeper the contact angle the greater the contact resistance. The magnitude of the difference between the angled contacts and flat (contact angle of 0°) contacts is approximately 17%, which equates to a power loss difference of 0.1% of the overall power consumption which is minor from a power consumption perspective but generates more heat to be convected or conducted away from the ICCB.
- The difference between the Notched and Spool system results from the assumption made about the different contact angles and it should not be concluded that one is better than the other in practice.

Other means of comparison

Even electrode spacing. In Table III , the contact bar system methods of obtaining even electrode spacing are compared. It is expected that insulator blocks, notches and spool shape can provide similarly regular electrode spacing. Where the crane is relied upon to position the electrodes, the spacing reliability will depend upon the degree of automation of the crane and positive placement of the crane bale.

Table III. Electrode Spacing Mechanism

ICCB	Electrode spacing mechanism
Triangular	Insulator block or crane
Dog-Bone	Insulator block or crane
Outotec DoubleContact™	Insulator block
Notched	By notch
Freeport McMoRan	By notch and insulator block for equalizer bar
Spool	By spool shape

Heat removal. In Table IV, the systems are compared with respect to their ability to convect heat away from their surfaces. The approximate surface areas of one electrode pair ICCB section for the different geometries are calculated and presented.

Table IV. Surface Area for Heat Rejection

ICCB	Comparative surface area (no insulator block) [mm^2]	Comparative surface area (with insulator block) [mm^2]
Triangular	2000	800
Dog-Bone	5000	1500
Outotec DoubleContact™	5000	1500
Notched	3000	3000
Freeport McMoRan	3000	800
Spool	1000-3000	1000-3000

The following comments can be made:

- For the no insulator block case the dog-bone and Outotec DoubleContact™ arrangements allow for the largest amount of surface area.
- The notched, Freeport McMoran and Spool systems have the next highest amount of exposed surface area when no insulator blocks are used.
- The Spool system surface area is unclear because the exact dimensions have not been defined and significant reduction in spool bar diameter may be possible for this low current density system which could decrease surface area.
- When positioning insulator blocks are installed, the triangle bar and Freeport McMoRan systems have the smallest areas and the dog-bone and Outotec systems have the next smallest areas.
- The notched contact bar does not require insulator blocks for electrode spacing so would be expected to maintain the highest surface area.
- Where a crane is used for electrode positioning the dog-bone system may have highest surface area of all but this may introduce positioning errors and other issues.
- The importance of exposed surface area depends on the amount of convective heat removal from the ICCB compared to conductive heat removal through the header bars. This requires more work to assess.

Dirty contacts

In Table V, the type of contacts formed, and comments regarding redundancy, contact shear forces and contact pressure are made.

Table V. ICCB Design Considerations

ICCB	Type of Contact	Comments on contacts
Triangular	1 Line	No redundancy, no shear, low pressure
Dog-Bone	1 Line	No redundancy, no shear, low pressure
Outotec DoubleContact™	1 primary line, 1 equalizer bar line	Redundancy on equalizer bar, no shear, low pressure
Notched	2 Line	Redundancy, some shear, low pressure
Freeport McMoRan	2 primary lines, 1 equalizer bar line	Redundancy, some shear, low pressure
Spool	4 Point	Redundancy, some shear, high pressure

The following points can be made:

- Where there is a single line contact such as with triangular or dog-bone ICCB a dirty contact can significantly increase contact resistance.
- Multiple contacts provide some redundancy in the system so that if one contact is dirty the other(s) may still make good contact. It should be noted that the weight is distributed across the contacts so that if one contact is dirty the others will have higher voltage drop since each contact's resistance is higher than for a single contact system.
- The Freeport McMoran and Spool systems have redundancy at the main contact end whereas the Outotec system redundancy is provided on the equalizer bar. This equalizer bar contact will have low contact resistance since it carries full electrode weight but there will be an additional contact resistance and equalizer bar resistance needed to carry the current to that contact.
- Some shear between the header bar and contact system when the header bar is placed may provide benefits when contacts are dirty by self-cleaning the surfaces in contact. Regular contact cleaning is important for all contact systems but systems with shear may reduce the frequency required for this, or the criticality.
- The importance of film resistance is uncertain. Shear may cut through oxidation on the ICCB surface and reduce film resistance. In addition, contact pressure may be important and this will be highest for the point contacts on the Spool system. More work is required to investigate this factor.

Conclusion

This paper described the most common intercell contact bars currently available. These bars have been analyzed to determine how much electrical resistance can be expected due to the intercell contact bar system and have also been compared on the basis of other practical plant considerations. It was found that:

- The power consumption due to the ICCB system is between 0.5-1.1% of total power consumption.
- Increasing the load normal to the surface decreases the resistance of the ICCB such that a standard Walker bar results in a reduction in contact resistance, compared with the notched and spool bar alternatives for an impact of approximately 0.1-0.2% of power consumption.
- Lightweight anodes that are not bolted or otherwise positively fixed to the contact bar have higher electrical resistance than their lead based alternatives.
- The addition of anode bags increases the contact resistance due to buoyancy. This increase is minimal for lead anodes, but very significant for titanium anodes.

- Insulator blocks on ICCBs reduce the amount of surface area of ICCBs available for convecting heat away.
- A Notched ICCB appears the best configuration to reject heat through convection, whilst maintaining electrode positioning, where an accurate crane positioning system is not used.
- A subjective comparison has only been possible with respect to redundancy, contact shear and contact pressure but these warrant further investigation.

Further Work

This paper introduces the concept of determining the contact resistance of an intercell contact bar interface, and to use this information along with other parameters to develop the most efficient solution in the future. In future work the following additional aspects will be considered:

- Thermal effects need to be investigated in a more quantitative way to further improve understanding.
- In completing a thermal analysis, methods of bar cooling need to be examined to determine their cost effectiveness.
- More literature and practical testwork on film resistance and oxide film impacts is required particularly with respect to the importance of contact shear and contact weight/pressure.
- A detailed analysis of the equalizer bar concept to gain a better understanding as to whether they impact the overall resistance and hence power consumption.
- An assessment of cost impacts of the ICCB systems incorporating the differences in capital cost of the various systems as well as operating costs.
- Methods of reducing the contact resistance for titanium anodes.
- The presentation of the Hatch ICCB system being developed.

References

1. I.S. Laitenen, "Modelling, Simulation and Optimization of Copper Electrolysis Cell Group" (Thesis, Tampere University of Technology, 2009), 10-13.

2. N. Aslin, D. Stone , and W. Webb, "Current distribution in modern copper refining" (Xstrata ISAKidd Technical Papers, 2005) 12-13.

3. G.A. Vidal et al., "Performance of intercell bars for electrolytic applications: A critical evaluation," *Hydrometallurgy 2003*, 2 (2003), 1381-1393.

4. Outotec, Copper SX-EW technology, (Finland, 2007).

5. R. Deane et al., "Contact bar for electrolytic cells" (United States Patent No 4,035,280, July 12, 1977).

6. B. Ashford et al., "Double contact bar insulator assembly for electrowinning of a metal," (United States Patent No 2011024369, November 24, 2011).

7. Ragner Holm, *Electrical Contacts: Theory and Application* (New York: Springer-Verlag, 1967), 2, 44.

8. S. Sawada et al., "Prediction of Electrical Contact Resistance of Tin-Plated and Silver-Plated Terminals," *SEI Technical Review*, 71 (2010) 37-43

9. R.S. Timsit, "Electrical Contact Resistance: Properties of Stationary Interfaces," IEEE Transactions on Components and Packaging Technology, 22 (1) (1999), 85-97.

10. William D. Callister, *Materials Science and Engineering: An Introduction* (New York:John Wiley & Sons, 2003), 592-594.

11. Copper Development Association, Copper in Electrical Contacts: Technical Note 23 (Hempstead, United Kingdom, 1980).

12. P. Adegbuyi et al., "Effects of Temperature on the Oxidation Kinetic of Copper Alloys" *The Pacific Journal of Science and Technology*, 10 (2) (2009), 163-171.

13. S. Sandoval et al., "Development and Commercialization of an Alternative Anode for Copper Electrowinning," *Proceedings of Cu 2010*, (2010), 6-7.

14. F. Crundwell et al., *Extractive Metallurgy of Nickel, Cobalt and Platinum-Group Metals* (Amsterdam: Elsevier, 2011), 328-331.

Ni-Co 2013
Edited by: Thomas Battle, Michael Moats, Violina Cocalia, Harald Oosterhof, Shafiq Alam, Antoine Allanore, Rodney Jones, Nathan Stubina, Corby Anderson, and Shijie Wang
TMS (The Minerals, Metals & Materials Society), 2013

HIGH CURRENT DENSITY ELECTROWINNING OF NICKEL IN EMEW® CELLS

Jeremy Robinson[1], Ian Ewart[1], Michael Moats[2], and Shijie Wang[3]

[1] Electrometals USA, LLC, O'Fallon, Missouri, USA

[2] Materials Research Center, Department of Materials Science and Engineering, Missouri University of Science and Technology, Rolla, Missouri, USA

[3] Rio Tinto Kennecott Utah Copper, Magna, Utah, USA

Keywords: Nickel, Electrowinning, EMEW, Current Density

Abstract

Modern nickel electrowinning from sulfate electrolytes is beset by several processing challenges. This includes the need to operate a covered divided cell to minimize nickel mist emissions and ensure nickel plating. Electrometals' electrowinning (EMEW®) technology overcomes these challenges through the modified geometry of its revolutionary cell design. The EMEW® system exploits a higher solution flow rate in a sealed tubular cell. In this paper, data from commercial scale cells will be presented demonstrating the plating of high purity nickel from lower nickel tenor solutions at higher current densities while eliminating the need for diaphragms. Additionally, the EMEW® design consists of enclosed, round cells, which has the dual benefit of maintaining a healthier, cleaner work environment, as well as producing coherent, tubular cathodes. The cylindrical shape of the cathodes makes them easily harvestable, and also mitigates the effect of the well-known internal stresses inherent to nickel electrowon from sulfate solution.

Introduction

The production of high purity nickel occurs by several methods. For the recovery of nickel from solution, the most common methods are electrowinning and hydrogen reduction. Electrowinning can occur from either sulfate or chloride based electrolytes. In sulfate electrolytes, nickel electrowinning cells are designed to (1) be robust and economical, (2) allow cathodes to be easily harvested, (3) separate the anode from the cathode by a diaphragm, (4) maintain pH at 3.0-3.5 near the cathode, (5) provide a continuous flow of electrolyte and electricity, and (6) minimize and/or vent off gases [1]. These requirements are driven by thermodynamics, kinetics, and environmental considerations.

The desired reduction reaction in a nickel electrowinning cell is the plating of metallic nickel at the cathode:

$$Ni^{2+}_{(aq)} + 2\,e^- \rightarrow Ni_{(s)} \qquad E^\circ = -0.257 \text{ V vs. NHE} \qquad (1)$$

Competing against the plating of nickel at the cathode is the evolution of hydrogen gas:

$$2\,H^+_{(aq)} + 2\,e^- \rightarrow H_{2(g)} \qquad E^\circ = 0.0 \text{ V vs. NHE} \qquad (2)$$

Since the standard reduction potential for hydrogen gas evolution is more positive than that of nickel deposition, it is thermodynamically favored. To reduce the likelihood of hydrogen evolution, the concentration of hydrogen ion is lowered. Thus, an electrolyte pH of 3.0-3.5 is needed in sulfate electrolytes to effectively deposit nickel.

To achieve this pH, a diaphragm is needed to separate the electrolyte near the cathode (catholyte) and the electrolyte near the anode (anolyte). The anolyte is acidified during operation as water decomposes to oxygen and protons as the anodic reaction [2]. The need for a diaphragm adds to the operational complexity and difficulty of conventional electrowinning cells.

The oxygen gas generated at the anode creates an electrolyte mist as bubbles burst at the air/electrolyte interface. This mist is unhealthy for workers. As such, anodes are either encased in an anode bag for gas capture [3] or the cell is covered by a hood for ventilation [2].

Finally, nickel electrodeposits from sulfate electrolytes are notorious for the stress generated in them [4]. This stress has resulted in the use of starter sheets or the plating of crowns [5]. Recently, full-size, seven-day plates have been produced on mother blanks [2] by plating nickel through holes in titanium blanks. However, the longevity of the blanks and a stripping machine with continual shearing of nickel pins is still being demonstrated.

The production of stressed deposits, the need to control pH in the catholyte and the generation of mist limits nickel electrowinning to current densities less than those used in other base metal electrowinning (e.g. copper and zinc). Thus, conventional nickel sulfate electrowinning is beset by processing challenges. To overcome these challenges, high performance EMEW® cell technology has been employed to recover nickel from sulfate electrolytes at high current density without the use of diaphragms.

EMEW® Cell Description and Advantages for Nickel Electrowinning

The EMEW® cell design has been reported previously [6-8]. The basic design features include a pair of concentric tubular, rather than planar, electrodes. The ends of the assembly are fitted with plastic end caps, thus forming a closed chamber through which the target solution can be pumped at a high rate. The high flow and efficient mixing result in 'forced' and continual supply of metal ions to the surface of the cathode by improving the hydrodynamics of the cell.

For the electrowinning of nickel, a tubular anode constructed from titanium with a mixed metal oxide coating is employed. In a standard design EMEW® cell, the anode is approximately 50mm in diameter. The cathode is a stainless steel tube approximately 151 mm in diameter. Both the anode and cathode are fixed in the cell. A thin stainless steel starter sheet is rolled and inserted in the cell to deposit nickel. The deposited nickel cathode is removed from the cell as a tube. The stainless steel starter sheet then springs off the deposit to release the nickel cathode.

These inherent design features offer significant advantages for electrowinning nickel using EMEW® technology which are outlined below and discussed in more detail in the paper.

1. The high flow rate ensures high mass transfer in the cell enabling the cell to operate at a high current density over a wide range of nickel concentrations.

2. The key benefit offered by the high flow rate in the cell, however, is the very short residence time in the cell which enables the cell to be operated without a diaphragm for acid control. Acid control is performed outside the cell.
3. The fixed anode and cathode in the cell ensure a consistent anode to cathode gap and thereby an even current distribution in the cell.
4. The enclosed cell eliminates acid or nickel mist in the work environment.

Modeling of the EMEW® cell by Roux [9] has shown that the cell has very high mass transfer due to the high flow rate and the spiral flow in the cell. This enables the cell to operate at high current density over a wide range of nickel concentration at a fraction of the limiting cathode current density. This results in a high purity nickel product from a given nickel electrolyte.

The flow rate in the cell is approximately 6-10 m^3/hr per cell which is equivalent to a residence time of approximately 6-15 seconds in the active zone of the cell. Normal operating cathode current in the cell is 500 A/m^2 which is equivalent to 250 A in each cell. On this basis the expected sulfuric acid generation at the anode (assuming 100% anodic current efficiency for generation of oxygen and acid) is approximately 0.045-0.076 g/L per pass. The cells are typically operated in a parallel-series configuration with up to 6 cells in series. The maximum change in acid concentration between feed and tank return is then 6 times the per pass change for a total of 0.27-0.45 g/L. This small change in acid concentration does not affect the overall cathode current efficiency for nickel plating at a pH of around 3.5. Typical cathode current efficiency for nickel electrowinning in EMEW® cells under the conditions described is 90-95%.

Examples of Purification Processes Prior to EMEW Nickel Electrowinning

Electrolyte purification is essential for conventional cells and EMEW® cells to ensure a consistently good quality nickel cathode product. A significant amount of development work has been done on various nickel feed solutions in order to provide a custom fit purification circuit to suit various nickel feed streams. A few examples of electrolytes and key purification steps are provided for the reader.

<u>Purification of Industrial Waste Solutions Containing Nickel</u>

Industrial waste solutions by their very nature are typically mixed metals with different counter anions such as nitrate, chlorides and sulfates and may contain various chelating agents for specific applications such as electroless plating. Purification for industrial nickel waste solutions typically involves two stages, see Figure 1. The first stage is pH adjustment to precipitate iron and other impurities. The second stage is ion exchange to recover the nickel selectively and produce a clean nickel carbonate feed for the EMEW® circuit. Obviously, if other valuable metals are present in large quantities, they can be removed by EMEW or ion exchange to provide by-product credits.

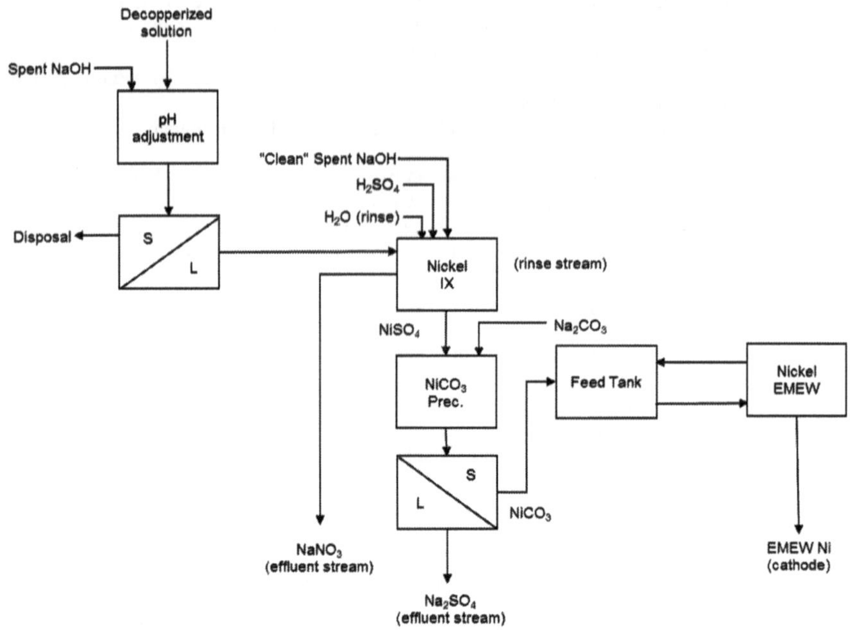

Figure 1. A flowsheet for recovering nickel from industrial waste streams by solution purification and electrowinning.

Purification of Nickel Sulfate from Copper Refinery Bleed Streams

Copper refinery bleed streams contain various amounts of nickel mixed with copper, iron, arsenic and other impurities. Separation of the impurities is achieved by pH adjustment to precipitate the impurities. A nickel carbonate intermediate product can be produced to feed the EMEW® circuit and control pH during electrowinning. A schematic diagram showing a basic flowsheet for processing crude nickel sulfate from copper refinery bleed is provided in Figure 2. In cases where the bleed from the copper refinery is processed directly by an acid purification unit [10, 11], the return of acid to the refinery is accomplished and the cost of neutralization is significantly reduced.

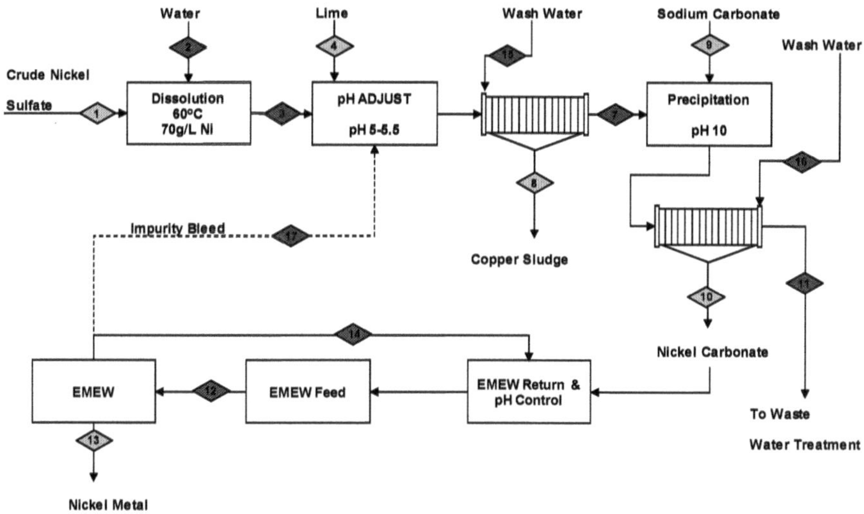

Figure 2. Schematic diagram of flow sheet to produce electrowon nickel from crude nickel sulfate produced from a copper electrorefining bleed.

The overall goal of the complete process is to produce cathodes from the EMEW® cells that are 99.5% Ni + Co, resulting in a high value and easily saleable nickel product. The purification process produces nickel carbonate slurry which is used to control the pH of the stock solution and replenish electrowon nickel during plating. Additionally, it is expected that the precipitated impurities can be removed in one filter cake, consisting primarily of copper and arsenic, as a suitable copper smelter feed.

Electrowinning of Nickel in Full Sized EMEW® cells

The EMEW nickel recovery technology has been demonstrated on a wide variety of byproduct and waste streams containing nickel including:

- Crude nickel sulfate from copper refinery by-products
- Copper refinery solutions
- Waste plating solutions
- Electroless nickel solutions
- Waste nickel filter cake
- Nickel sulfamate solutions

Typical conditions employed during the electrowinning of nickel in EMEW® cells are:

- Cathode current density between 400 - 500 A/m^2
- Electrolyte flow rate between 6-10 m^3/hr

- Temperature (T) = 50-60°C
- pH between 3.5-4.0
- Buffer with boric acid at 5-10g/L

Examples of operating data and product quality produced in some of these systems are presented in the following sections.

Crude Nickel Sulfate from Copper Refinery Bleed

Three crude nickel sulfate (CNS) materials (see Table I for compositions) were processed in a similar manner to the flow sheet given in Figure 2. The resulting purified electrolyte composition is presented in Table II indicates excellent separation of nickel and cobalt from other metals during purification. This electrolyte was used to produce a high purity nickel cathode as shown in Table III. The production of nickel occurred at high current density and with high current efficiency as shown by operating data presented in Table IV. The energy consumption during the trials was 5.0-6.4 kWhr/kg. This value is higher than a conventional nickel electrowinning cell (3.4-3.8 kWhr/kg based on data presented by Crundwell et al. [2]) because of the low electrolyte conductivity. A mature electrolyte is expected to have higher conductivity and the energy consumption for the EMEW cell is expected to approach 4.0 kWhr/kg.

Table I. Crude Nickel Sulfate Feed Compositions

Element	Analysis (wt %)		
	CNS1	CNS2	CNS3
Cu	0.97	0.83	0.43
Fe	0.31	0.23	0.37
Ni	25.16	24.55	28.77
H2SO4	5.15	5.76	4.4
Zn	0.23	0.2	0.13
Pb	0.01	0.009	ND
Co	0.98	0.94	0.54

Table II. Purified Nickel Sulfate Solution Feed to EMEW®

Element	Analysis (ppm)	Element	Analysis (ppm)
Ni	92400	Pb	5.2
Co	2600	Sb	3.9
Ca	336	Cu	1.8
Mg	187	Bi	1.1
Al	26	Cd	0.6
Zn	13.8	Fe	<0.5
Mn	6.8	Cr	<0.5
As	5.4		

Table III. Typical Nickel Cathode Composition Produced from Crude Nickel Sulfate using EMEW Technology

Element	Cathode 1	Cathode 2
Ni+Co	99.83%	99.80%
Cu	0.119	0.0611
Ca	0.025	0.0178
Bi	<0.0001	<0.0001
Sb	0.0042	0.00006
As	0.003	0.0001
Fe	0.015	0.1164
Zn	<0.0001	0.0001
Pb	0.0023	0.0005

Table IV. Summary of Selected Operating Data during the Recovery of Nickel by Electrowinning in an EMEW Cell

Time (Hr)	Current Density (A/m^2)	Ave Current Efficiency (%)	
		Cell 1	Cell 2
60	500	97.4	97.4
55	500	88.0	79.7
96	400	90.4	90.4
38	400	90.1	90.1
144	400	88.8	87.2
68	500	90.0	86.0

Copper Refinery Bleed Solution

A copper refinery bleed solution after acid purification was processed to produce nickel cathode. The composition of the feed solution is given in Table V. The resulting EMEW® nickel cathode composition is summarized in Table VI. Examples of nickel cathode deposited from EMEW® cells are shown in Figure 3.

The cylindrical shape of the cell helps utilize the stress of the nickel deposit. With planar electrodes, stress within the deposit causes the plate to prematurely peel from the mother blank to alleviate this stress. Since an EMEW® cell deposits a tubular product, the deposit stress does not cause premature peeling from the blank.

Table V. Feed Solution Composition from Copper Refinery Bleed Prior to Purification

Element	g/L	Element	g/L
Ag	<0.001	Mg	0.178
Al	0.143	Mn	0.009
As	3.192	Ni	15.3
Ba	<0.001	Pb	0.006
Be	<0.001	Se	<0.001
Ca	0.441	Sn	0.002
Cd	0.002	Te	<0.001
Cr	0.004	Tl	<0.001
Cu	2.87	V	0.001
Fe	1.063	Zn	0.108

Table VI. Cathode Composition Produced by Electrowinning after Solution Purification

Element	Ni	Co	Zn	Cu	Mn	S
Assay (wt %)	99.5*	0.298	0.075	0.039	0.028	0.027
Element	As	Fe	Pb	Cd	Na	
Assay (wt %)	0.007	0.006	0.005	0.002	0.001	

*By difference

Figure 3. EMEW® nickel cathode products produced in commercial cells.

Conclusions

The ability to electrowin nickel from sulfate electrolytes in a closed cell without a diaphragm and at high current density has been demonstrated in EMEW® cells. High purity nickel has been produced from various feed sources including industrial waste solutions, crude nickel sulfate and copper refinery bleed. The closed cell removes mist generation from the workplace. The elimination of the diaphragm removes an operational difficulty encountered in conventional electrowinning operations. The ability to operate at twice the conventional current density helps reduce capital costs for an EMEW® installation. Flow sheets incorporating solution purification and electrowinning have been demonstrated and in some cases commercialized.

References

1. T. Robinson and E. Tuuppa, "Developments in nickel electrowinning cellhouse design," *ALTA 2002 Nickel/Cobalt* (Melbourne, Australia: ALTA Metallurgical Services, 2002).
2. F. Crundwell et al., *Extractive Metallurgy of Nickel, Cobalt and Platinum Group Metals*, London, UK: Elsevier, 2011).
3. A. G. Pavlides, "Developments in cobalt and nickel electrowinning technology," Bateman Engineering. Retrieved from November 7, 2012 from http://www.bateman.com/papers/Ni%20%20Co%20EW%20Developments.pdf
4. D. Kittely and M. J. Nicol. "The electrocrystallization of nickel and its relationship to the physical characteristics of the metal," *Electrometallurgy 2001*, eds. J. E. Dutrizac et al. (Montreal, Quebec: CIM, 2001), 361-373.
5. E. O. Stensholt, "Recent improvements in the Falconbridge nickel refinery", *Extractive Metallurgy of Nickel and Cobalt*, eds. G. P. Tyroler and C. A. Landolt (Warrendale, PA: TMS, 1988), 403-413.
6. P.A. Treasure, "Performance characteristics of the EMEW cell," *Electrochemistry in Minerals and Metal Processing VI*, eds. F.M. Doyle, G.H. Kelsall and R. Woods (Princeton, NJ: Electrochemical Society, 2002), 367-378.
7. V. Escobar, T. Treasure and R.E. Dixon, "High current density EMEW copper electrowinning," *Hydrometallurgy 2003*, eds. C. Young, et al. (Warrendale, PA: TMS, 2003), 1369-1380
8. E. Roux, J. Gnoinski, I. Ewart and D. Dreisinger, "Cu-removal from the Skorpion circuit using EMEW technology", *Base Metals 2007*, (SAIMM, 2007), Retrieved from Nov. 7, 2012 from http://www.saimm.co.za/Conferences/BM2007/ on.
9. E. Roux and M. J. Nicol, "The effects of tangential or perpendicular inlet flow on mass transfer in an annular electrochemical cell", *Hydrometallurgy 2008*, eds. C. A. Young, et al. (Littleton, CO: SME, 2008), 625-631.
10. S. Wang, "Impurity Control and Removal in Copper Tankhouse Operations," JOM, 56 (7) (2004), pp. 34-37.
11. M. Sheedy, P. Pajunen and B. Wesstrom, "Control of copper electrolyte impurities – overview of the short bed ion exchange technique and Phelps Dodge El Paso case study," *Proceedings of the Sixth International Copper-Cobre Conference: Volume V Copper Electrorefining and Electrowinning*, eds. G. Houlachi, J. Edwards and T. Robinson (Montreal, Quebec: CIM, 2007), 345-357.

Ni-Co 2013
Edited by: Thomas Battle, Michael Moats, Violina Cocalia, Harald Oosterhof, Shafiq Alam, Antoine Allanore, Rodney Jones, Nathan Stubina, Corby Anderson, and Shijie Wang
TMS (The Minerals, Metals & Materials Society), 2013

HELM TRACKER™ CATHODE CURRENT SENSING TECHNOLOGY

"Rob Fraser, Tim Johnston, John Yesberg, Sebastien Nolet, Chris Boon"

"Hatch Associates; 61 Petrie Terrace, Brisbane, QLD, 4000, Australia"

Keywords: "electrowinning, electrorefining, short circuit, poor contact, current measuring"

Abstract

In electrowinning and electrorefining operations, the detection of operational issues associated with electrode currents has long been a highly labour intensive reactive operations task. The Hatch *HELM tracker*™ system is set to change the way tankhouses are operated with its ability to measure and record the current flow of every cathode and cell voltage in real time. This information is then presented to provide the ability to identify shorts, poor contacts, and any uneven current distribution. Benefits expected include increases in current efficiency, and thus reduction in operating costs, or increases in production. The software also facilitates simple management of cathode inspection/straightening, stripping and cell cleaning cycles.

Hatch has been demonstrating the robustness and performance of the *HELM tracker* system at commercial copper electrowinning plants. This paper describes the development of this exciting new technology and the expected benefits to the electrowinning and electrorefining industry with specific coverage of its application to the electrowinning of nickel and cobalt.

Introduction

In electrolytic metal plating processes, short circuits can occur due to dendrites or nodules growing out from the cathode surface until they reach the anode. This allows current to pass between anode and cathode without any cathodic metal deposition, thus reducing current efficiency. Short circuits can also occur due to misalignment of cathodes and anodes to the extent that they contact each other, as well as from delaminating or self stripping deposited metal bridging the gap between anode and cathode.

In nickel electrowinning and refining and some cobalt plants, filter cloth style separators between anodes and cathodes are used to allow catholyte pH control and separate capture of anode gas. In those cases, short circuits due to nodules and dendrites also damage the separators which may ultimately lead to poor catholyte pH control and a loss of current efficiency.

In chloride electrolyte systems, titanium anodes with iridium or ruthenium compound catalytic coatings are used to allow evolution of chlorine at the anodes. For these anodes, each short circuit from nodules and dendrites not only destroys the local active coating and its catalytic effect but also completely destroys any ability for that area to oxidize chloride to chlorine. Shorts thereby reduce the life of these anodes which is important since these anodes are relatively expensive compared to lead alloy anodes normally utilized in sulphate based systems. The use of such catalytic anodes is expanding with some sulphate electrolyte copper electrowinning plants now moving to a form of these titanium catalytic anodes; a nickel plant has these anodes for mixed chloride/sulphate duty; and there appears scope for further expansion into nickel sulphate electrowinning.

Short circuits can be caused by a number of factors including the presence of impurities, additive dosing problems, particulates in electrolyte, temperature control problems, a low or high concentration level of the depositing metal, poor pH control, high current density, uneven current distribution and poor electrode alignment. In plants, preemptive actions such as manual adjustments to electrode alignment, analytical concentration determinations and process control of flow, temperature and current and design advancements such as double contact intercell contact systems are used to combat these effects in order to reduce the impact on current efficiency and deposit quality.

Short circuits always lead to an increase in current flow through the affected cathodes. In contrast, dirty intercell contact bar or header bar contacts can lead to high resistivity connection between electrode header bars and the intercell contact bar which reduces current flow through specific cathodes. In this case the current is redistributed amongst the remaining cathodes leading to more uneven current distribution within the cell. Uneven current distribution can also occur due to the influence of the current distribution in neighbouring cells. If sufficiently uneven, certain cathodes start to generate poor morphology such as nodules and dendrites which in turn can lead to shorts or poor deposit morphology.

The *HELM tracker* system (patent pending) measures the current flow through every cathode in a tankhouse which is provided to operators in real time through human machine interfaces (HMI). It is capable of directing the operators to the exact location of any uneven current distribution including shorts and bad contacts allowing rapid actions to be taken to increase current efficiency, improve deposit quality and either decrease operating costs or increase revenue. This paper describes the basis for the technology, its technological features, its stage of development and its particular applicability to nickel and cobalt plants.

Historical Methods of Current Distribution Monitoring

There have been a number of systems developed to detect current distribution problems and a few have been applied across electrowinning and electrorefining operations for a variety of metals. The most common methods include infra-red imaging, cell voltage monitoring, and hand held gauss and individual cathode current measurement devices.

Infra-red imaging relies on temperature differences which occur as a result of the difference in resistive heat generated from poor electrode contacts, short circuits and normal cathodes. Poor contacts result in a cold spot in the array of header bars due to a lack of current flowing through that point. Short circuits result in a highly localised current flow through the specific cathode. IR cameras may be hand held or crane mounted. The disadvantage of this system is that sufficient time has to pass before the thermal signatures of shorts can be detected. Hand held devices require operators to be constantly scanning the tankhouse to find the thermal anomalies.

Figure 1. Use of handheld thermal device to scan electrowinning cell contacts for signs of current distribution problems[1].

Cell voltage monitoring devices such as the Outotec Cell Sense[1] and MIPAC Cell View[2] devices provide wireless monitoring of individual cell voltages. The cell voltage is the voltage difference from one side of a cell to the other. Current distribution problems such as short circuiting or poor contacts may result in a change in cell voltage, however, the measurement is coarse and the problem needs to be significant before being able to be detected. Once a problem is detected it is not immediately clear which electrode in a cell has the problem and therefore other methods are needed to locate the problematic electrodes. A similar system has also been developed for the Optibar busbar configuration[3].

Handheld Gauss or individual electrode current measuring tools require operators to be continuously roaming the tankhouse to take measurements. They often walk on top of the electrode header bars, which imposes both safety issues and in some cases can increase the likelihood of moving the electrodes out of alignment and actually causing electrode short circuits.

Description of *HELM tracker*

To overcome the disadvantages of these previously described systems, the *HELM tracker* system allows continuous measurement of the current flow in every cathode in an electrolytic tankhouse. The system allows for instantaneous detection of the magnitude and specific location of electrode current distribution anomalies such as short circuits and poor contacts, which is displayed to the operators through the systems HMI. The system requires no electrical or physical contact with the cathodes, and can be easily retrofitted into an existing plant, without interfering with the operation.

The *HELM tracker* works by measuring the magnetic field that is generated by the current passing through the cathode header bar. Ampere's circuital law, a static version of one of Maxwell's equations, relates the magnetic field B around contour C to the current I_{enc} flowing through any conductors enclosed by C.

$$\oint_C B.dl = \mu_0 I_{enc}$$

If we idealise the header bar as a thin conductor that is very long (compared to the distance at which we measure the field), then this reduces to the Biot-Savart law which, when integrated expresses the magnetic flux density B at a distance r, due to the current flowing at all infinitesimal lengths of the conductor. (The direction of the B field is given by the right-hand rule.)

$$|B| = \frac{\mu_0 I}{2\pi r}$$

Thus the magnitude of the magnetic flux density B at a point near the header bar is proportional to the current, and inversely proportional to the distance. The *HELM tracker* "bar" contains an array of sensors, which detects the tangential flux density at a point under each cathode header bar.

The magnetic flux density at a given sensor will not only be due to the local cathode current, there will be an effect due to the current in neighbouring cathodes. As shown in Figure 2, this neighbouring field is not oriented in the same direction. If the sensor-cathode distance is small compared to the cathode spacing, then only the small tangential component of the neighbouring cathode's field will be detected by the sensor. While this is fortuitous, a technique has been developed by which these effects may be calculated and removed by mathematical manipulation.

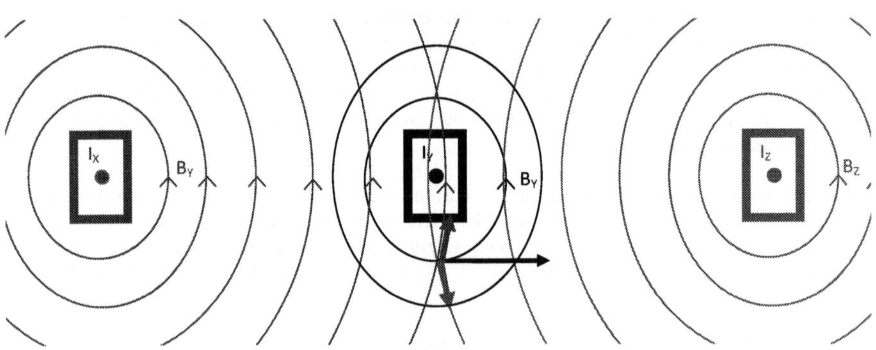

Figure 2. Magnetic fields B_X, B_Y, B_Z surrounding three cathode header bars X, Y, Z, with currents I_X, I_Y, I_Z, showing their contributions to the net field at a sensor under cathode Y.

The magnetic field signal under each cathode header bar is detected, amplified, and converted to digital form by a small microcontroller. Some calibratory transformations are also performed before the current data is communicated wirelessly to a central tankhouse server, where additional transformations can be applied.

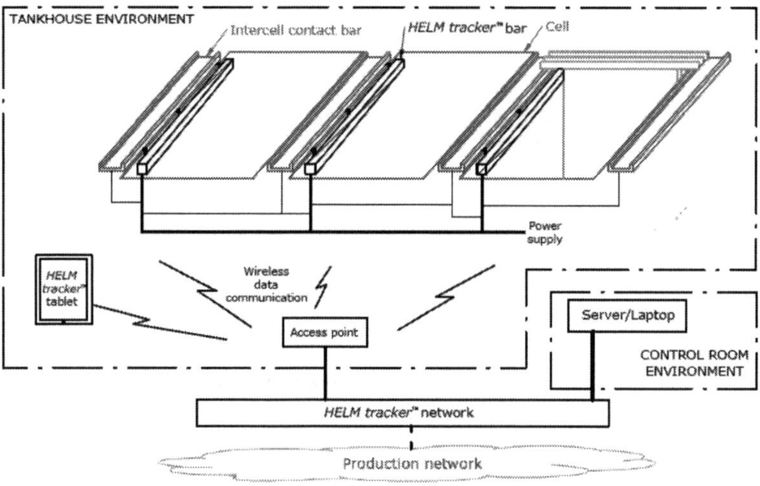

Figure 3. Arrangement of *HELM tracker* bars in cells, showing power wiring and wireless communications with the server.

The *HELM tracker* server collects and records cathode currents in a historian database. From here, the data is available for immediate monitoring, as well as longer term trending and analysis.

Operators can access the *HELM tracker* information through a web browser on a computer in the control room. However, *HELM tracker* also comes with a handheld tablet which can display all the same information. This allows an operator to walk around the tankhouse with the tablet and a short-clearing tool, and to respond as soon as a short or poor contact is identified. Simply displaying the current in each cathode allows the operator to detect these problems much more quickly than traditional methods, and time can be more usefully focused on their rectification.

In addition, the *HELM tracker* system is able to utilise spatial plant layout and to prioritise the current distribution issues to recommend an optimised sequence for fault rectification by the operator.

The system will allow much quicker rectification than in existing plants, where operators may go through the plant first with short-detection equipment, marking electrodes where problems are detected; and then return with short-clearing tools. As has been discussed, this change should allow increased current efficiency and either reduced power cost or increased production.

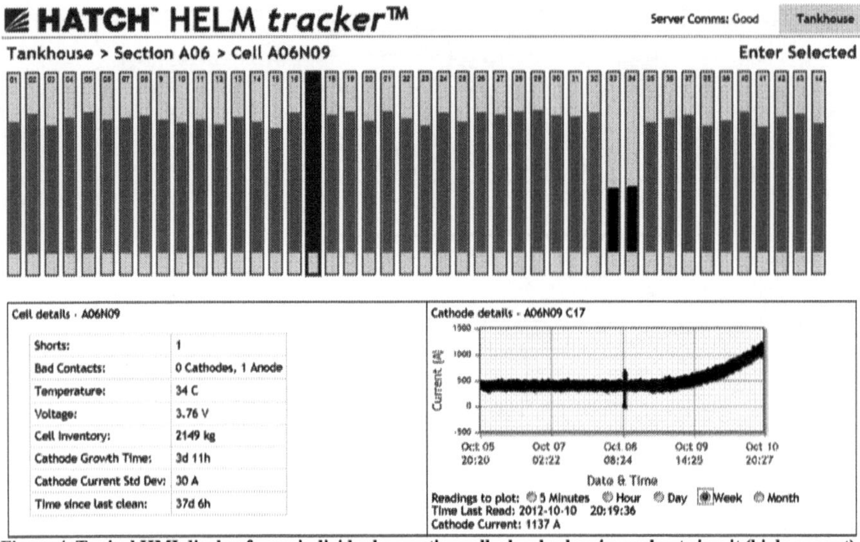

Figure 4. Typical HMI display for an individual operating cell, clearly showing a short circuit (high current) and a poor anode contact (adjacent low cathode currents).

Additional benefits to operating plants also include:

- Operators can spend less time working in the tankhouse searching for faults, and can spend more time rectifying them at the earliest opportunity. This may reduce labour requirements.
- In electrowinning plants with expensive titanium anodes, the *HELM tracker* system's earlier detection of shorts will mean that less damage is caused to the anodes, thereby prolonging their life.
- Fewer shorts mean that the deposition of metal on the cathodes is more uniform. As well as greater quantity of production, this system will allow greater quality.
- Fewer dendrites (uneven growths due to shorts) mean less damage to electrode bags and other cell dividers.
- Existing plants may be able to reduce electrode spacing if they have instant information about short circuits and increase plant capacity with minimal capital.

The benefits stated above are due solely to the qualitative early identification of shorts and bad contacts. Beyond this, the *HELM tracker* will collect quantitative data (primarily current and time), which can be exploited for further benefits. The system can identify the total charge (Amp hours) that has passed through individual cathodes, cells, and banks, to provide an indication of the overall inventory of plated production at any time. By detecting stripping and cleaning cycles, the system can not only identify which cells are next ready for harvesting, but also nominate the optimum time for uniform plate thickness.

The *HELM tracker* may further assist in tuning of various other plant parameters, such as temperature and electrolyte composition ratios, as well as tracking the performance of different cells and even individual cathodes and anodes over time.

Although we expect that measuring the currents in individual cathodes will give the most useful information to operators, the *HELM tracker* system also measures the voltage across each cell. This information may assist in assessing operating factors other than shorts or poor contacts, such as changes in cell voltage with; anode life; depolarising or polarising impact of additives and impurities; and anode or cathode bag blinding.

Development Project Overview

Hatch provides a range of specialized know-how and patented technologies to enable our clients' facilities to be safer, more environmentally friendly, and more competitive in terms of minimum capital and operating costs per unit of throughput. It is under this framework that the *HELM tracker* technology has been developed. Following 14 years of development and plant trials a commercialisation project is currently under execution. The commercialisation project has included the following stages:

- Concept development and laboratory test work
- Prototype development and laboratory test work
- Commercial design
- Commercial plant trials
- Finalisation of commercial product ready for sale

Hall effect measuring devices are widely used and the technology is well understood, however, there were a number of issues that had to be resolved to allow for this technology to be applied to the electrowinning/ electrorefining industry. The early stages of the commercialisation project built upon prior development works with a focus on refining the electronics design and compensation methodologies required to obtain an accurate electrode current measurement. Once the concept was defined, electronic and communication prototypes were developed for verification and optimisation. The prototyping phase provided further opportunities to provide a higher quality current measurement and improve the robustness of the design.

A key component of the overall *HELM tracker* package is the HMI, software and database historian that interprets, provides interactive displays and stores data for reporting and other functions. These aspects of the overall package were developed during the prototyping and commercial design phases. A key factor in the development of the software was the input of plant operators and management to ensure that the functionality meets the practical requirements of day-to-day operators and the reporting system is capable of providing plant management with sufficient information required to further improve operations.

The mechanical, electrical and communication aspects of the commercial design were developed to require minimal modifications to be made for each plant installation. This allows for a reduction in the cost of the units with improved efficiencies associated with fewer engineering hours, greater pricing power with common parts and less contingencies with reduced risk to operational function. This includes the ability to seamlessly switch between the standard 2.4 or 5 GHz WiFi communications system and the proprietary 900MHz wireless communication systems dependent on the specific plant requirements.

Figure 5 - The compact design is located against the cell wall under the electrode header bars preventing any interference with plant operations.

At the time of writing this paper, three plant trials are currently underway with promising early signs. The sites selected for the trial are spread between Australia and the United States of America in the production of copper in electrowinning and electrorefining plants. Following the completion of these trials in early 2013, it is expected that commercial units will be available for sale by the end of Q1, 2013. All trials to date have been for copper with the next step to include commercial trials in nickel, cobalt and zinc operations.

Applicability to Nickel and Cobalt

The *HELM tracker* system appears to be a very useful addition to new and existing electrowinning and electrorefining plants. Although plant trials that are underwayare only been carried out on copper electrowinning and electrorefining plants, there is every reason to believe that this technology would provide similar benefits in nickel and cobalt plants. Due to the nature of nickel and cobalt electrowinning plants the technology also stands to provide additional benefits:

- Nickel plants and some cobalt plants utilise anode or cathode bags. These bags are damaged by dendrites and nodules generated at the cathode. Early detection of high cathode currents would be expected to reduce bag damage and increase bag life. There is also a requirement for tight catholyte pH control which is prejudiced by bag damage.
- Nickel and cobalt electrolysis is more sensitive to current density variation and to in situ corrosion when bad contacts occur.
- Extending the operating life in existing chloride based and mixed sulphate/chloride systems of titanium catalytic anodes[4] for nickel and the future application of such anodes to purely sulphate systems

Future Developments

Following commercial trials, further development work is underway and will be implemented as required to improve the functionality of the *HELM tracker* system. This includes:

- Full integration with existing plant systems such as data historian, process control system and tank house management systems. Integration with these systems will allow for real time information to be part of automated solutions and for analyses of all relevant trended data to be carried out to assist with continuous improvement of operational decision making. The compatibility with the most common communication protocols in the process control industry will allow a smooth integration and a minimal impact on the existing infrastructure.
- To improve production and allow for keeping pertinent cathode data, it is intended to identify and track electronically every cathode in a tankhouse using RFID tagging or similar technology which is a natural fit with the *HELM tracker* technology.
- A conditional based notification system module using SMS (Short Message Service) will allow alerts and status to be sent to mobile phones, digital radio and email.
- To help the field technicians to quickly locate the problematic cathodes and to minimize exposure to chemicals, a visual identification mechanism linked to the *HELM tracker* system will be developed.
- Additional built-in sensing devices such as electrolyte temperature, concentration and flow are in development with data to be transmitted and available in similar fashion.

References

1. Outotec 2007, Copper SX-EW Technology, Outotec, Finland.

2. MIPAC CellView, viewed 10 October 2012, http://www.mipac.com.au/CellView.htm.

3. P.E. Aqueveque, E.P. Wiechmann, and R.P. Burgos, "Short-Circuit Detection for Electrolytic Processes Employing Optibar Intercell Bars" (Paper presented at the Industry Applications Conference, 2007 42nd IAS Annual Meeting, 23-27 September 2007).

4. S. Sandoval, C. Clayton, S. Dominguez, C. Unger, and T. Robinson, "Development and Commercialization of an Alternative Anode for Copper Electrowinning" (Paper presented at Copper 2010, Hamburg, Germany, 6-10 June 2010).

5. P.G.S. Garritsen, D.J. MacVicar, D.P Young, and J.G. Sojda, "Process for Electrowinning Nickel or Cobalt" (US Patent Application No. 178,819, 18 August 1980).

Ni-Co 2013

Pyrometallurgy: Solid-State Processing

Session Chair

Shijie Wang

Ni-Co 2013
Edited by: Thomas Battle, Michael Moats, Violina Cocalia, Harald Oosterhof, Shafiq Alam, Antoine Allanore, Rodney Jones, Nathan Stubina, Corby Anderson, and Shijie Wang
TMS (The Minerals, Metals & Materials Society), 2013

COBALT RECOVERY THROUGH SULPHATING ROAST OF CU/CO CONCENTRATE OF KATANGA MINING

K. Adham - Ph.D., P.Eng.
Hatch Ltd. - 2800 Speakman Drive - Mississauga, Canada

Keywords: Cobalt, Copper, Roaster, Fluidized Bed

Abstract

In 2006, Hatch was awarded a contract to supply fluidized bed technology for the Katanga Mining Limited (KML) copper-cobalt processing project in D.R. Congo [1]. The new roaster and the associated solids and off-gas treatment plants are part of the revitalization of an existing facility, which was historically one of the largest cobalt and copper producers in the world. The sulphide ore fed to the roasting plant is treated under sulphating roast conditions, followed by hydrometallurgical extraction. Equipment for the first roasting line has been installed and the plant commissioning completed. Hatch's Fluidization Technology Group (FTG) in supplied its proprietary equipment and technology to KML under a license agreement. Our offices in Canada and South Africa conducted most of the required engineering. Here, the process conditions for the sulphating roast, with emphasis on cobalt and copper recoveries are discussed.

Introduction

Design of the two new sulphating roaster lines included testwork on feed samples, performed in Johannesburg, South Africa. Most of the tests were conducted on the concentrate samples, in order to characterize their physical and chemical properties. Furthermore, a limited number of pilot scale runs were conducted, to improve the feed nozzle design, confirm the product's physical and chemical properties, and verify the cobalt and copper recoveries.

Cobalt and copper recoveries depend on the roasting conditions (e.g. temperature, residence time, excess oxygen) and are pivotal to the economical performance of the overall metal plant [2, 3]. Moreover, scale up from the observed recoveries in the pilot plant to the commercial plant, has a profound impact on the commissioning ramp up rate and the project's overall success [4].

Beside the highest possible cobalt and copper recoveries, the auto-thermal operation of those fluid bed roasters, were of prime importance in order to avoid the extra operating cost of auxiliary fuel consumption. In order to limit any fuel (coal) addition, the feed slurry was maintained at 75 wt% solids to minimize the energy loss to water vaporization. This resulted in a slurry that presented a significant challenge in the roaster feed system design due to its high viscosity and solid content.

Not all the sulphur content of feed concentrate can be converted to leachable sulphates; and a significant portion turns into SO_2/SO_3 gases. The presence of sulphur oxides in the roaster off-gas both necessitates a robust thermal design/monitoring of roaster shell (to avoid acid attack); and scrubbing of the roaster off-gas (with minimum loss of metal value).

Key Chemical Reactions

The principle desired chemical reactions taking place in the roaster are the sulphating of the various cobalt and copper sulphides present in the feed slurry. Such reactions make the Co/Cu values available to simple atmospheric leach, downstream of the roaster.

The exact composition of cobalt and copper species vary depending on the concentrates available from the mining operations. However, carrollite (Co_2CuS_4) can be expected in the typical Congo concentrates and from it, synthetic cobalt sulphide (CoS) can form at the elevated temperatures of roaster.

Copper speciation in the concentrate can range from chalcosite (Cu_2S) to chalcopyrite ($Cu_2S{\cdot}Fe_2S_3$) as in typical Congo concentrates [3]. The key simplified cobalt and copper sulphating reactions can be approximated as follows:

$$CoS + 2O_2 \rightarrow CoSO_4 \qquad \Delta H_{rxn} = -190\ \text{kcal} \qquad (1)$$

$$2Cu_2S + 5O_2 \rightarrow 2CuSO_4 \cdot CuO \qquad \Delta H_{rxn} = -407\ \text{kcal} \qquad (2)$$

$$4CuS + 7O_2 \rightarrow 2CuSO_4 \cdot CuO + 2SO_2 \qquad \Delta H_{rxn} = -532\ \text{kcal} \qquad (3)$$

The cobalt speciation and reaction is approximate, but with the right roasting conditions, almost all cobalt can be expected to substantially turn into a sulphate, which would be readily leachable with minimum net acid consumption. However, under the right cobalt conversion conditions, the copper species are only partially sulphated, with some turning into the oxide form. Therefore, in the subsequent atmospheric leach step, a net input of sulphuric acid is required for achieving the highest copper yield.

Historically, the cobalt selective sulphating roast temperature has been in the 650 to 725°C range [2, 3]. Figure 1 shows the cobalt-oxygen-sulphur equilibrium at 700°C, for different oxygen levels (provided in fluidizing air) and sulphur dioxide (product of roasting) partial pressures, as predicted by the commercial software HSC®.

At the typical SO_2 levels of 5 to 10 percent in the sulphating roaster, a very low level of excess oxygen is adequate for converting the cobalt to its sulphate form. However, the practical level of oxygen required more depends on the copper and iron oxidation criteria.

The aim of the fluid bed roaster is generally to maximize the cobalt and copper recovery, while minimizing the leached iron into the atmospheric acid extraction following the calcine cooler.

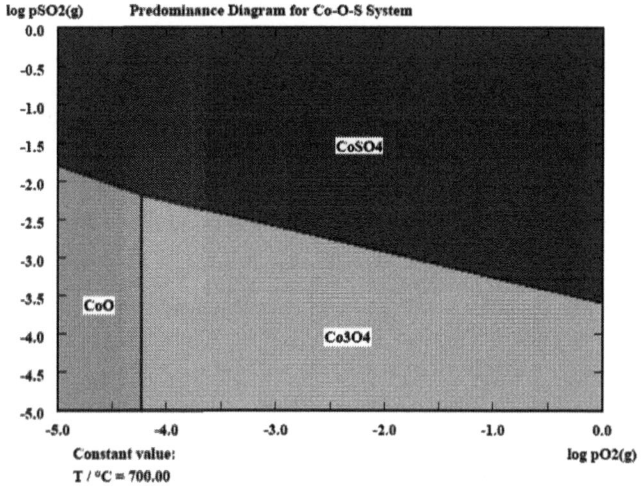

Figure 1: Predominance diagram for Co-O-S at 700°C

At the optimum conditions set by the cobalt selective sulphating roast, for example at 700°C, the copper recovery as sulphate is not complete, and inevitably results in some sulphur dioxide release. Figure 2 shows the copper-oxygen-sulphur equilibrium at 700°C, for different oxygen levels and sulphur dioxide partial pressures.

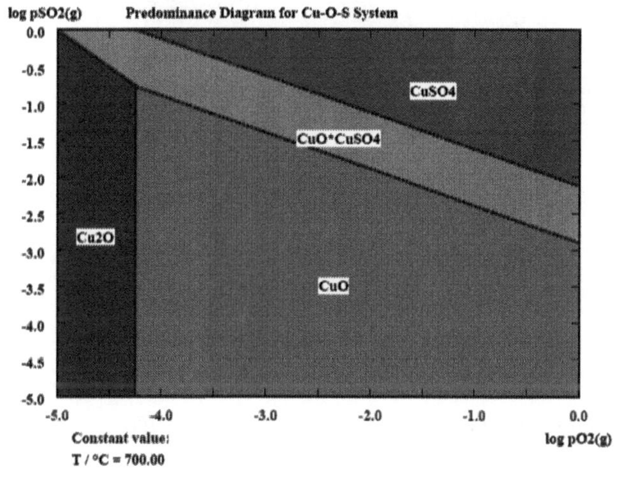

Figure 2: Predominance diagram for Cu-O-S at 700°C

Although not required for cobalt recovery, it is beneficial to have a higher level of oxygen overpressure (e.g. 20%-30% excess air) in order to create the least amount of copper oxides, and minimize the amount of sulphur loss from copper minerals to the off-gas. Otherwise, the copper oxide leaching downstream of the roaster will consume more sulphuric acid and have a slower leach kinetic. The overall sulphuric acid balance in the sulphating roast plant is an important aspect of operating cost/benefit of the metal complex.

In the fluid bed roaster, it is also very important to convert the iron sulphides to a form that is least soluble in the consequent atmospheric acid leach step; otherwise the cobalt and copper recovery circuits will struggle with iron control. The following approximate reaction shows the iron sulphide oxidation to a stable hematite form, while releasing sulphur dioxide to the roaster off-gas.

$$4FeS + 7O_2 \rightarrow 2Fe_2O_3 + 4SO_2 \qquad \Delta H_{rxn} = -583 \text{ kcal} \qquad (4)$$

Figure 3 shows the iron-oxygen-sulphur equilibrium at 700°C, for different oxygen and sulphur dioxide partial pressures, as predicted by the commercial software HSC®. It is clear that given the typical SO_2 concentration within the fluid bed sulphating roaster (5-10 volume %), the target amount of oxygen partial pressure (e.g. 20%-30% excess air as required for best copper recovery) should not be exceeded, in order to discourage the formation of soluble iron sulphate.

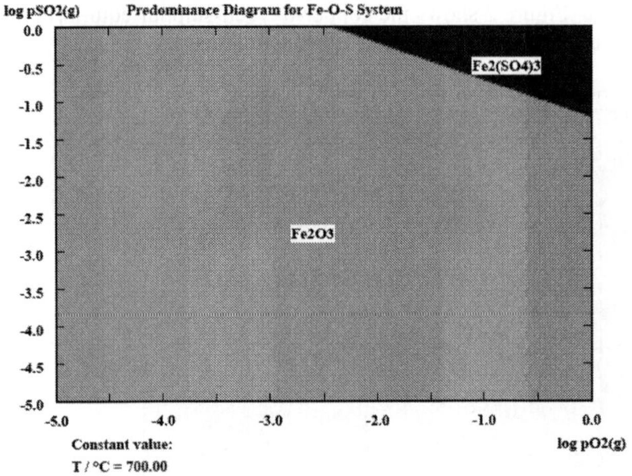

Figure 3: Predominance diagram for Fe-S-O at 700°C

All of the above mentioned reactions are exothermic, with transferable energy to the feed water evaporation and towards the product calcine and off-gas heating. However, depending on the naturally occurring dolomite content of feed, some endothermic magnesium carbonate dissociation is to be expected (see following reaction).

$$MgCO_3 \rightarrow MgO + CO_2 \qquad \Delta H_{rxn} = 28 \text{ kcal} \qquad (5)$$

Figure 4 shows the magnesium-oxygen-carbon equilibrium at 5% by volume oxygen concentration at one atmosphere pressure, for different carbon dioxide (product of calcination) levels and at a range of roaster temperatures.

Considering the minimum temperature of proven sulphating roast (600°C), the predominance diagram for Mg-O-C shows that substantial extent of calcination and MgO formation can be expected.

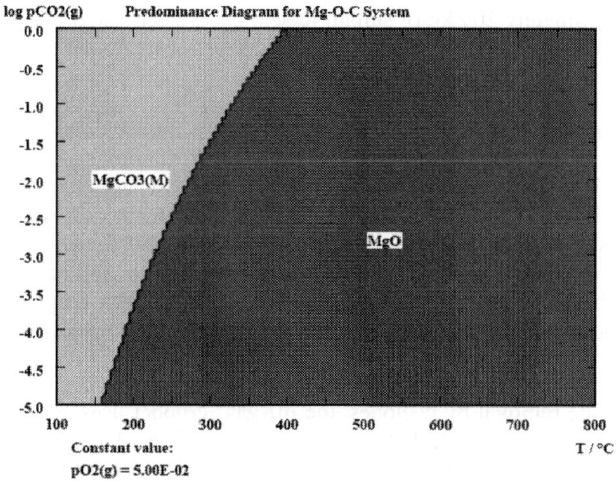

Figure 4: Predominance diagram for magnesium carbonate calcination

The decomposition of $MgCO_3$ to MgO and CO_2 gas, according to reaction 5, is important with regards to the thermal balance in the roaster, and the fate of released SO_2 from the Co/Cu sulphating (partial roast) and the iron (dead roast) reaction.

A high level of $MgCO_3$ can create a thermal demand on the roaster, resulting in the need for auxiliary fuel (e.g. coal). Furthermore, the dolomite content in the roaster product will be a net acid user in the downstream atmospheric leach step. On the other hand, the MgO inside the fluidized bed roaster can capture some of the released SO_2 values, and lower the amount of neutralizing agent (e.g. hydrated lime) required for the off-gas scrubbing.

Process Overview of the Roaster

Upstream of the roaster, the run-of-mine sulphide ore is milled and then concentrated in froth flotation cells. The 75 wt% solids roaster feed slurry contains approximately 40% copper and 4% cobalt (dry basis).

Roasting of the sulphide slurry feed produces a dried calcine product, in the form of oxy-sulphates of copper and sulphates of cobalt, which is highly amenable to low-temperature acid leaching. Cooled calcine product is then sent to leach tanks, where it is combined with the recycled spent electrolyte for leach. Product of leaching is then filtered prior to electrolysis.

Downstream processing of sour roast product leads to a significant net generation of sulphuric acid; approximately 200 kg of sulphuric acid can be generated for every ton of calcine product produced [3]. This is a large benefit driver for treatment of the open pit (oxide) ore, which requires the sulphuric acid.

The roasting area consists of a bubbling fluid bed roaster, a fluidized calcine cooler for product cooling to 150°C, and ancillary feed and product distribution systems. Shown in Figure 5 is a simplified equipment diagram. Solid product is discharged from the roaster via a pneumatic device (J-valve) and fed to the calcine cooler freeboard, where ambient air and water sprays (optional) maintain a bed temperature of 150°C, while fluidization maintains sufficient mixing, to avoid localized bed wetting. Roaster and cooler off-gases are passed through high-efficiency cyclones for solids removal. Dust removed from the off-gas is sent to the quench tank, where it is mixed with product discharged from the calcine cooler.

After coarse particulate removal by cyclones, the off-gas scrubber uses spent electrolyte for scrubbing to maximize copper/cobalt recovery for the fines. The use of water for scrubbing of off-gas dust would dilute the concentration of copper and cobalt in the spent electrolyte and have a negative impact on the acid balance in the plant.

Complete heat and mass balances were performed using the software METSIM®. The design case for the roaster was based on a solids feed rate of approximately 20 t/hr (dry basis). Based on research performed on samples of Katanga concentrate, a target bed operating temperature of 680°C was selected.

Based on thermodynamic data, the partial pressure of sulphur dioxide in the roaster bed and the bed gas temperature range have been selected to maximize production of soluble copper and cobalt and minimize production of soluble iron.

Heat input to the roaster is necessary during start-up and is achieved by three removable preheat burners. Once a bed temperature of about 700°C is achieved, oxidation of the sulphide feed slurry provides adequate heat to maintain temperature.

Residence time of solids in the roaster is in the range of 3 to 4 hours. Research and original pilot plant data performed on representative feed samples indicated a minimum retention time of 3 hours.

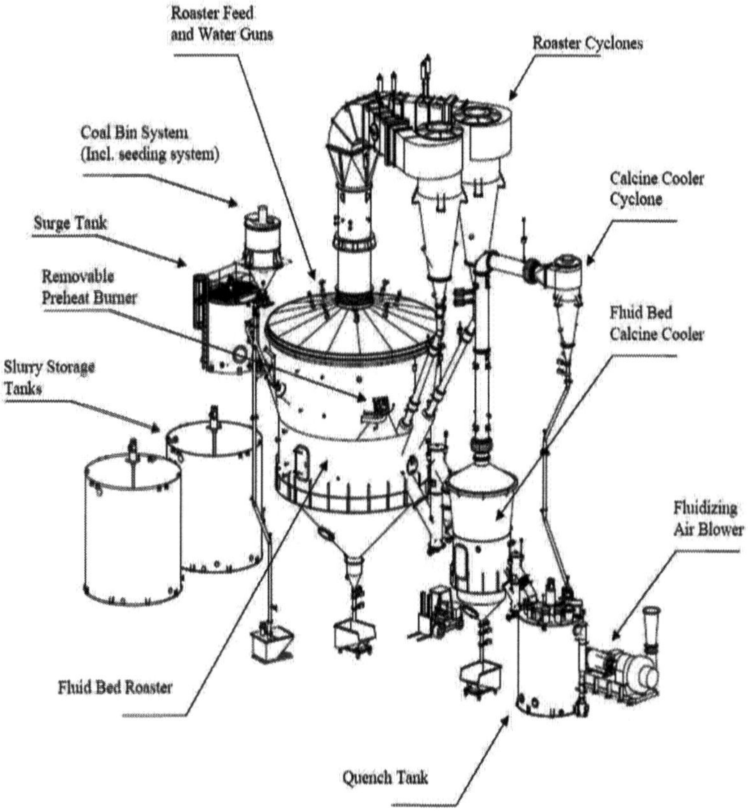

Figure 5: Sketch of fluid bed roaster and auxiliary equipment

Results of the METSIM® heat and mass balance provided the basis for sizing of the equipment. Based on the particle size distribution of the feed as well as particle and gas properties, a fluidization velocity of 0.7 m/s was selected to ensure complete fluidization in the bubbling regime, while minimizing solids elutriation to the cyclone.

Based on gas flowrate requirements and fluidization velocities, a roaster diameter of 8 m I.D. was selected. Based on retention time requirements, a bed height of 1.8 m was chosen. Roaster and cooler freeboard heights were selected based on the transport disengagement height for elutriated bed material.

The roaster feed system consists of four slurry feed ports, three operating and one standby, mounted to the roof of the roaster. Each feed port consists of a single nozzle used for atomization of slurry feed with plant air. Design includes control over atomization air flowrate and pressure and an installed purge air delivery system for the prevention of dust ingress to the gun.

Spraying of the feed slurry onto the bed from roof-mounted injection guns allows for significant evaporation of water in the freeboard. This is an important design parameter for the roaster, as the particle size distribution (PSD) of the bed should be significantly higher than that of the slurry feed. The bed PSD is a direct consequence of the design of the atomized feed system, and is controlled by the spray properties of the injection nozzles. As such, design of the nozzles was critical to ensure the solid particles had the desired size distribution prior to contact with the bed.

Conclusions

Hatch has designed two fluid bed roasters for the Katanga Mining Ltd. The KML sulphide ore concentrate has been successfully roasted under sulphating conditions (for Co/Cu recovery) and dead-roast (for Fe fixation); followed by hydrometallurgical extraction through atmospheric sulphuric acid leaching. The success of roaster operation depends on the proper feed atomization and the steady roaster temperature, excess air and residence time control. The subsequent electrowinning of paymetals also results in substantial amount of excess sulphuric acid, which can be used for other oxide ore leach operations.

References

1. K. Adham et al., Design of Copper-Cobalt Sulphating Roasters for Katanga Mining Limited in D.R. Congo, Copper 2010, Hamburg, Germany, 2010.

2. Theys, L.F. and Lee, L.V. (1958): Sulphate Roasting Copper-Cobalt Sulphide Concentrates: Journal of Metals: 134-136.

3. Thoumsin, F.J. and Coussement, R. (1964): Fluid-Bed Roasting Reactions of Copper and Cobalt Sulphide Concentrates: Journal of Metals: 831-834.

4. K. Adham, A. Tohn and L. May, A Success Path from Process R&D to Commercial Plants, Conference of Metallurgists of CIM, Montreal, Canada, 2011.

Ni-Co 2013
Edited by: Thomas Battle, Michael Moats, Violina Cocalia, Harald Oosterhof, Shafiq Alam, Antoine Allanore, Rodney Jones, Nathan Stubina, Corby Anderson, and Shijie Wang
TMS (The Minerals, Metals & Materials Society), 2013

Experimental Study on Reduction-Magnetic Separation Process of Low-grade Nickel Laterite Ore

CHEN Fa-tao, ZHANG Bo, LI Wen-cai, WANG Qiang, HONG Xin

(Shanghai Key Laboratory of Modern Metallurgy & Materials processing, Shanghai University, shanghai 200072, China)

Abstract: The reduction and magnetic separation of low-grade nickel laterite was studied at varing temperatures between 1325-1400℃and a weak magnetic field of 100 mT in order to acquire high grade concentrations. The influence of reduction temperature, percentage of CaO, and atomic ratio of C/O (carbon/oxygen) were investigated. According to the gas analysis, the reaction process was divided into two stages: the direct reduction had a greater advantage than the indirect reduction in the earlier stage when the temperature was low, and the direct reduction and the indirect reduction reacted together in the later stage when the temperature was high. The result indicated that reduction temperature of 1375℃, percentage of CaO of 12% and atomic ratio of C/O of 1.2, the grade of fine ore was improved up to 5.71%. At the same time, the recovery rate of nickel reached up to 96.35%.

Key words: low-grade, laterite, reduction-magnetic separation, nickel

1 Introduction

Nickel, known as "industrial vitamins", is a very important non-ferrous material. Under conditions of room temperature, nickel can make austenitic structure in steel be more stable, which leads to a steel with higher toughness and ductility. Nickel-based alloys not only have good performance in anti- corrosion and high temperature oxidation corrosion, but also have good mechanical strength and ductility [1], thus they are widely used in stainless steel, alloy steel, corrosion resistant alloys, batteries, electroplating, chemical industry, etc. Approximately 65% of the total metal nickel is consumed by the stainless steel industries. With the rapid development of the world's stainless steel industry, this will greatly promote and stimulate the consumption of nickel. Table 1.1 lists the global stainless steels yield in recent years [2].

Table 1.1 Global stainless steels yield /Kt

Region	2009	2010	Comparison,%	2010	2011	Comparison,%
Western Europe/ Africa	6449	7878	22.2	7878	7875	0.0
Central and Eastern Europe	237	340	43.2	340	387	14.1
America	1942	2609	34.4	2609	2486	-4.7
Asia (excluding China)	7472	9011	20.6	9011	8770	-2.7
China	8805	11256	27.8	11256	12592	11.9
Total	24904	31094	24.9	31094	32110	3.3

At present, identified land-based resources that average 1% nickel or greater containing at least 130 million tons of nickel, consist of sulfide ore (40%) and nickel laterite ore (60%). Table 1.2 lists the world mine production and reserves [3]. Historically, although laterite deposits are widely distributed, resource-rich, close to the surface, and large-scale, the smelting technology is complicated and costly, therefore approximately 60% of the world's production of nickel comes from sulfide ore. With the increasing depletion of sulfide ore, there is an increasing focus on the utilization of low-grade nickel laterite ore. With the development and improvement of the smelting technology, laterite has shown a great deal of economic potential, the proportion of nickel production from laterite is increasing, and it has become an important source of the metal nickel. According to statistics, the proportion of the total metal nickel from laterite ore will exceed 50%.

At present, the typical metallurgical processes for laterite ores are pyrometallurgical processes, hydrometallurgical processes, and a combination of processes of metallurgical and mineral separation. Pyrometallurgical processes can be divided into the process which produces ferronickel by reduction smelting and the process of sulfide smelting to produce nickel matte by the output of products. Hydrometallurgical processes can also be divided into ammonia-ammonium carbonate leaching and high-pressure acid leaching. The combination of process of metallurgical and mineral separation can acquire the valuable products when the ores are separated after reduction roasting. Pyrometallurgical processes are short, but require high energy consumption. Hydrometallurgical processes don't require melting and can selective recovery of valuable metals, but have the disadvantages of reagent requirements and protection of the environment [4]. The combination of process of metallurgical and mineral separation is suited to deal with any type of nickel laterite ore and has the advantages of a short process, low energy conservation, etc., thus it is worth further research.

In this study, the reduction and magnetic separation of low-grade nickel laterite was studied in order to acquire the high grade concentrations. The influence of reduction temperature, reduction time, percentage of CaO, and atomic ratio of C/O (carbon/oxygen) were examined.

Table 1.2 The world mine production and reserves

	Mine production		Reserves
	2010	2011	
Australia	170,000	180,000	24,000,000
Botswana	28,000	32,000	490,000
Brazil	59,100	83,000	8,700,000
Canada	158,000	200,000	3,300,000
China	79,000	80,000	3,000,000
Colombia	72,000	72,000	720,000
Cuba	70,000	74,000	5,500,000
Dominican Republic	---	14,000	1,000,000
Indonesia	232,000	230,000	3,900,000
Madagascar	15,000	25,000	1,600,000
New Caledonia	130,000	140,000	12,000,000
Philippines	173,000	230,000	1,100,000
Russia	269,000	280,000	6,000,000
South Africa	40,000	42,000	3,700,000
Other countries	99,000	100,000	4,600,000
World total(rounded)	1,590,000	1,800,000	80,000,000

2. Experimental

2.1. Characterization of the Materials

The chemical composition of the nickel laterite ore is listed in Table 2.1. The ore contained approximately 1.68 mass% Ni, 21.42 mass% Fe, 32.37 mass% SiO_2, and 22.95 mass% MgO, which is typical of low-grade nickel laterite ore. XRD analysis (Fig.2.1) shows that the laterite ore is mainly composed of goethite, garnierite, kaolinite and clinochrysotile. In this study, graphite is used as the reductant, which contained approximately 99.5 mass% C. Calcium oxide was used as a flux.

Table 2.1 Chemical compositions of nickel laterite ore (unit: mass%).

O	Fe	Si	Mg	Al	Ni	Cr	Ca	Mn	Co	S	Ti	Zn
44.37	21.78	14.60	13.86	2.02	1.63	0.75	0.41	0.38	0.08	0.06	0.05	0.03

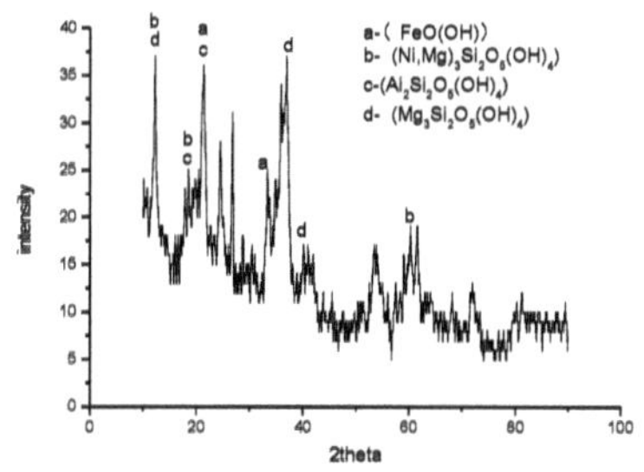

Fig.2.1 XRD pattern of the nickel laterite ore sample.

2.2. Experimental Methods and Processes

First, a large amount of free water in the laterite ore was taken off in the oven. After drying, the laterite ore was crushed and ground to 90 mass% passing 0.100mm. Then the ground laterite sample was mixed with graphite and calcium oxide to produce the pellets. The pellets were then heated at heating rate of 5℃/min in a high-temperature reactor shown in Fig.2.2. After reaching the given temperature, the pellets were kept at a given temperature for 300 minutes in order to make the material react fully. During the reaction, the reaction gas was detected by the gas analyzer. After roasting, the reduced pellets were cooled to room temperature under an Ar gas atmosphere. Lastly, the cooled pellets was crushed and ground to 95 mass% passing 0.100 mm and separated in a weak magnetic field of 100 mT to acquire the high grade concentrations. Fig.2.3 shows the process of the experiment.

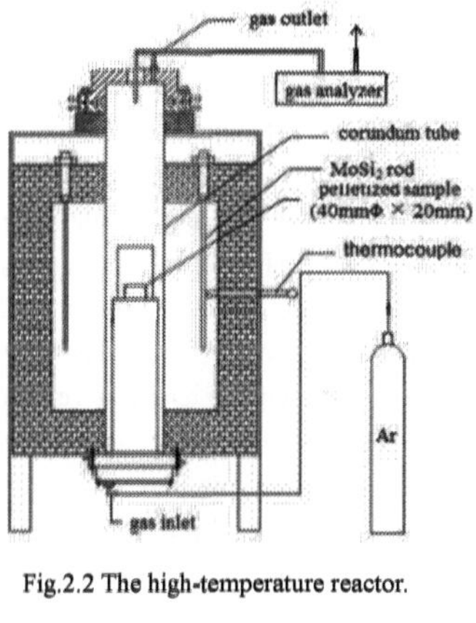

Fig.2.2 The high-temperature reactor.

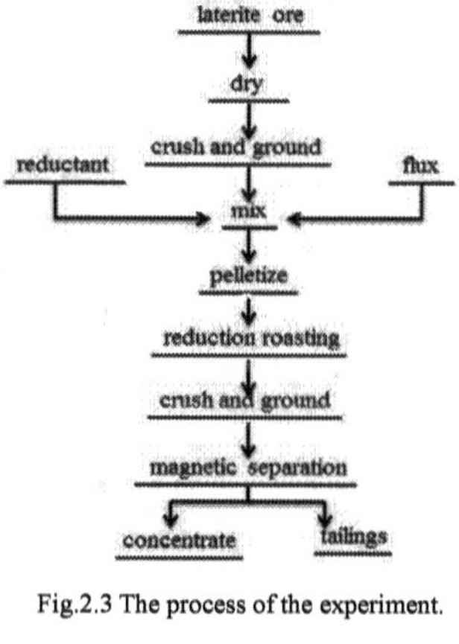

Fig.2.3 The process of the experiment.

3. Results and Discussion

In this study, the influence of reduction temperature, percentage of CaO, and atomic ratio of C/O (carbon/oxygen) are investigated in order to acquire the high grade concentrations. The experimental results are described in detail.

3.1 Beneficiation of Laterite Ore by Reduction and Magnetic Separation

3.1.1 Effect of Reduction Temperature on Reduction and Magnetic Separation

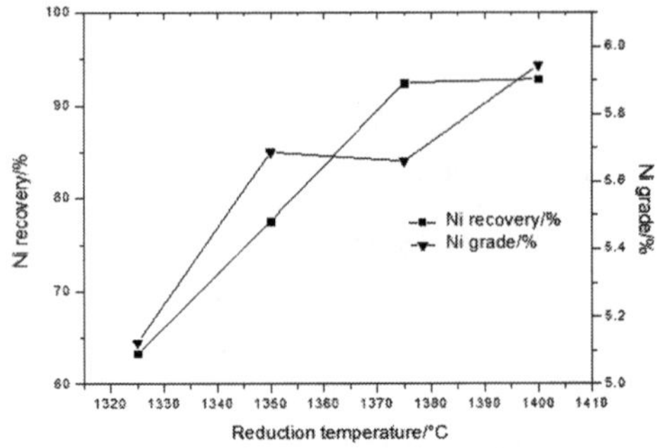

Fig.3.1 Effect of reduction temperature on reduction and magnetic separation.

The influence of reduction temperature on reduction and magnetic separation was examined by carrying out the experiment at 1325℃, 1350℃, 1375℃, and 1400℃ with the atomic ratio of C/O at 1.4 and the percentage of CaO at 12%.

The results are presented in Fig.3.1. It can be seen that the recovery and grade of nickel improves with increasing reduction temperature. When the reduction temperature was 1325 °C, the grade and recovery of the nickel were 5.12% and 63.36% respectively, while the reduction temperature was up 1400℃, the grade and recovery of the nickel were up to 5.94% and 91.65% respectively. During the experiment, the molten state of the material was observed when the reduction temperature was at 1400 ℃ and there was some ferronickel particles after reduction roasting at the same time. We conclude that the higher reduction temperature is beneficial to reduction and magnetic separation of laterite ore. This phenomenon is attributed to the fact that reduction temperature has an important impact on aggregation and grew up in the grains of ferronickel in the reaction. The larger the grains of ferronickel, the greater the degree of separation of valuable metal and gangue will be, which improves the separation of slag and iron and improves the recovery of valuable metals.

When the reduction temperature was very low, the growth rate of ferronickel crystal nucleus was also very low, which caused the diameter of ferronickel particles to be very small. The reason for this was that the temperature was low in the pellets, which led to the transfer and congregation process of iron and nickel metals to be very slow. The separation of ferronickel alloy and slag was very difficult. The recovery of nickel was much lower. When the temperature reached to about 1400℃, the molten slag and ferronickel particles began to form, the kinetic process of transfer was accelerated, and the bigger particles of ferronickel alloy came into existence and made the recovery rate of nickel maintain at a higher level. The bigger the molten slag was, the more difficult the separation of the metal and slag, and the higher the energy consumption. According to the targets of both efficiency and emission, the optimum reduction temperature was 1375℃. At this condition, the recovery rate of nickel reached 92.46%.

3.1.2 Effect of the Atomic Ratio of C/O on Reduction and Magnetic Separation

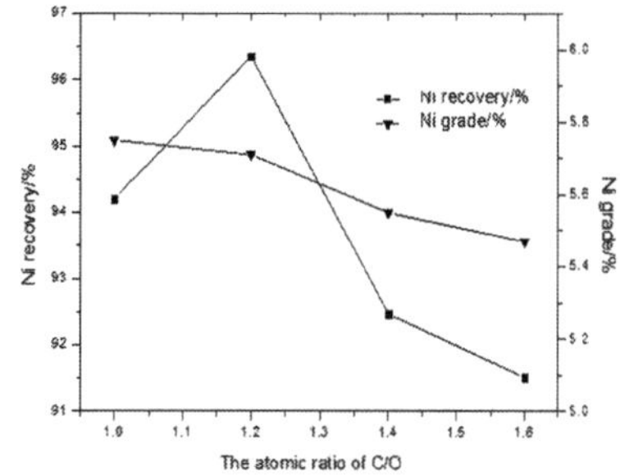

Fig.3.2 Effect of the atomic ratio of C/O on reduction and magnetic separation.

The influence of the atomic ratio of C/O on reduction and magnetic separation was carried out at the ratios of 1.0, 1.2, 1.4, and 1.6, with the reduction temperature of 1375℃ and the percentage of CaO of 12%.

The results are presented in Fig.3.2. With increasing atomic ratio of C/O, the grade and recovery of the nickel significantly increased and then decreased. When the atomic ratio of C/O was 1.2, the maximum grade and recovery of the nickel reached 5.71% and 96.35% respectively. Meanwhile, it was found that there were some ferronickel particles after reduction roasting when the atomic of C/O was low. This phenomenon was attributed to the fact that excessive residual carbon in the phase was detrimental to the agglomeration of the grains of ferronickel [5].

In this study, anthracite was a reducing agent, which mainly improved the reducing atmosphere inside the pellets which promoted reduction of iron oxide, nickel oxide and thus produced a large amount of ferronickel crystal nuclei inside the pellets which was good to the agglomeration of the grains of ferronickel. In addition, the carbon penetrated into ferronickel, which reduced the melting temperature of ferronickel. As a result, there was some metal molten phase in the pellets, which is advantageous for the promotion of agglomeration. However with increasing atomic ratio of C/O, the excessive residual carbon disperses inside the pellets, which increases the viscosity of the slag and hinders the metal phase aggregation. This reduces the recovery rate of nickel.

In order to have a clear understanding on the effect of excessive carbon on the recovery rate of nickel, we selected several groups with high atomic ratio of C/O to study. The results are presented in Fig.3.3. We found that the recovery rate of nickel was very low. According to the result, the optimum atomic ratio of C/O was 1.2.

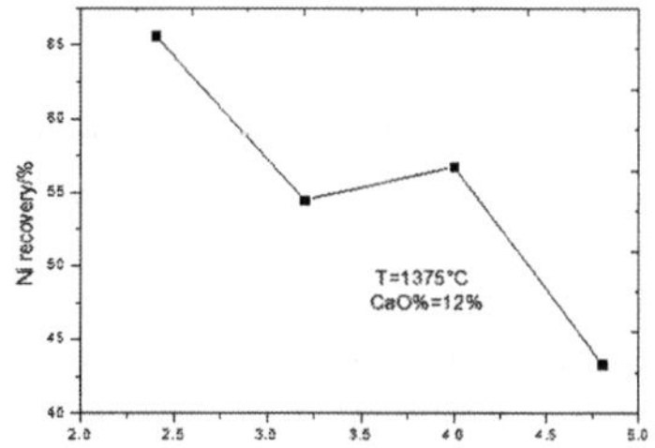

Fig.3.3 Effect of the atomic ratio of C/O in higher level on nickel recovery.

3.1.3 Effect of the Percentage of CaO on Reduction and Magnetic Separation

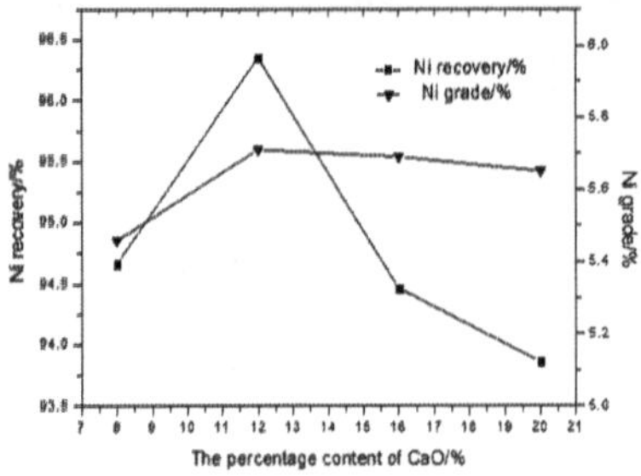

Fig.3.4 Effect of the percentage of CaO on nickel recovery.

The influence of the percentage of CaO on nickel recovery was carried out at 8%, 12%, 16%, and 20% with a reduction temperature of 1375℃ and atomic ratio of C/O of 1.2.

The results are presented in Fig.3.4. It can be seen that the recovery of nickel first increased and then decreased with increasing percentage of CaO. The maximum recovery of nickel was 94.45% when the percentage of CaO was 12%.

During the reduction process of laterite nickel ore, complex silicate containing nickel and iron oxide decomposed into simple nickel oxide at first. Then the simple oxide is reduced by the anthracite. The study shows that suitable additives can improve the reduction of laterite. During the reduction roasting, the right amount of calcium oxide can accelerate the decomposition of complex silicate containing nickel and iron: $CaO+NiO \cdot SiO_2=CaO \cdot SiO_2+NiO$ [6]. This improves the activity of NiO, and it is reduced to metallic easily. In addition, calcium oxide with several mineral can form some materials which have the lower melting temperature. As a result, there is some molten phase in the pellets, which is advantageous for the promotion of agglomeration. So the recovery rate of nickel increased at the beginning with increasing percentage of CaO. However the absorbing ability to calcium oxide for each mineral is different [7]. When there is excessive residual calcium oxide in the phase, the excessive residual calcium oxide disperses inside the pellets, which increases the viscosity of the slag and hinders the metal phase aggregation. This is detrimental to the agglomeration of the grains of ferronickel. This leads to the recovery rate of nickel decreasing with further increasing percentage of CaO. According to the result, the optimum percentage of CaO was 12%.

3.2 Characterization of the Fine Reduced Laterite Ore by Magnetic Separation

In order to find out the phase composition of fine laterite ore reduced with reduction temperature of 1375℃, the atomic ratio of C/O of 1.2, and the percentage of CaO of 12%, the samples were examined by X-Ray fluorescence(XRF), X-ray diffraction (XRD) and scanning electron microscopy (SEM) equipped with an energy dispersive spectrometer (EDS). The chemical compositions of the fine ore by reduction-magnetic separation was shown in Table 3.1, the XRD pattern of fine ore by reduction-magnetic separation was seen in Fig.3.5, and the SEM pattern of fine ore by reduction-magnetic separation was shown in Fig.3.6 and the chemical compositions of observed district in SEM pattern was shown in Table 3.2.

As shown in Table 3.1, the fine ore achieved effective separation between the valuable metals and gangue. The Ni and Fe grades were 5.71% and 75.02% respectively, and the corresponding Ni and Fe recoveries were 96.35% and 93.02% respectively. As seen from the XRD pattern of fine ore in Fig.3.5, the main material is ferronickel alloy and the rest is $(Mg,Fe)_2SiO_4$ and $CaSiO_3$, etc. in fine ore. As seen in SEM pattern of fine ore and the chemical compositions of observed district, we can find that a small amount of gangue embedded cloth in ferronickel alloy.

Table 3.1 Chemical compositions of fine ore (unit: mass%).

Fe	O	Ni	Mg	Si	Cr	Ca	Al	V	Mo	S
75.02	7.22	5.71	4.27	3.90	1.52	1.29	0.86	0.049	0.0418	0.039

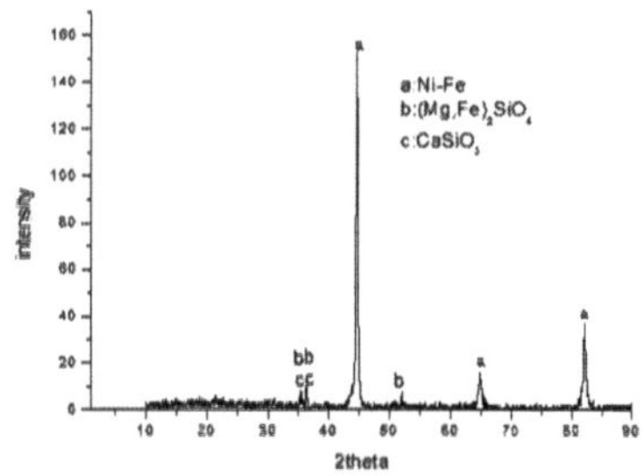

Fig.3.5 XRD pattern of the fine ore by magnetic separation of laterite reduced.

Fig.3.6 SEM pattern of the fine ore by magnetic separation of laterite reduced.

Table 3.2 Chemical compositions of observed district.

No.	Fe	Ni	Cr	Si	Mg	O	C
1	80.77	10.94	0	1.15	0.83	6.30	0
2	77.99	9.94	0	0	0	12.07	0
3	82.28	9.16	3.69	1.04	0	3.85	0
4	82.45	0	13.03	1.00	0	3.51	0
5	2.81	0	1.45	21.88	9.47	50.90	13.46

3.3 The Gas Analysis of Reduced Laterite Ore

During the experimental process, the gas of reduced laterite ore with reduction temperature of 1375℃, the atomic ratio of C/O of 1.2, and the percentage of CaO at 12% was detected by a gas analyzer, which helped clearly show the reaction process in the pellets. The change of volume fraction of CO, CO_2 with increase of the reduction temperature is shown in Fig.3.6.

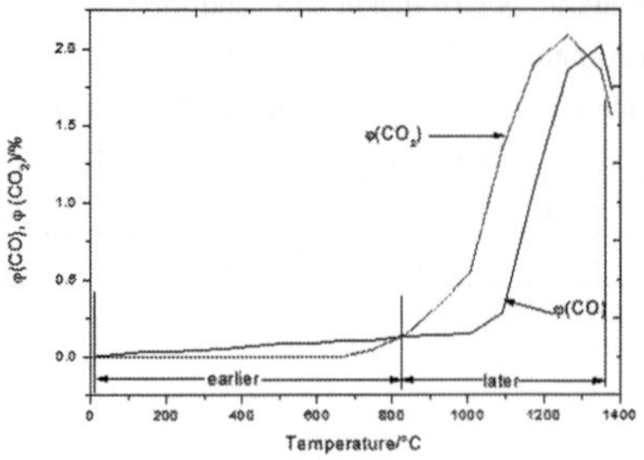

Fig.3.7 The change of volume fraction of CO, CO_2 with increase of the reduction temperature.

As shown in Fig.3.7, according to the gas analysis, there were two stages during the reaction process.

Early on, the volume fraction of CO was slightly more than the volume fraction of CO_2, but they were small in total. We considered that the direct reduction ($NiO+C=Ni+CO\uparrow$, $3Fe_2O_3+C=2Fe_3O_4+CO\uparrow$, $Fe_3O_4+C=3FeO+CO\uparrow$, $3FeO+C=2Fe+CO\uparrow$) had a greater advantage than the indirect reaction ($NiO+CO=Ni+CO_2$, $FeO+CO=Fe+CO_2$, $3Fe_2O_3+CO=2Fe_3O_4+CO_2$, $Fe_3O_4+CO=3FeO+CO_2$) during this stage when the temperature was lower. This phenomenon was attributed to the fact that diffusion of the gas became the limiting element of the reaction in the earlier portion. The reason for this was that there was almost no gap because of reducing agent and oxide close contact fully inside the pellets. And during this stage, the temperature was low in the pellets, which led to the diffusion of the gas was very slow.

Then, with increasing reduction temperature, the volume fraction of CO, CO_2 increased rapidly in the later portion. It was also found that the volume fraction of CO_2 was much larger than CO. This was due to the diffusion of the gas got faster and the reducing gas was in close contact with the oxides inside the pellets. This was the main reason that the volume fraction of CO_2 was much larger than the volume fraction of CO. There was excessive residual carbon, which promoted the boudouard reaction($C+CO_2=2CO$, $\Delta_rG^{\theta}=166550-171T$ J/mol). This was the primary reason that the volume fraction of CO_2 increased first before CO (Fig.3.7) in the later portion. It was concluded that there was direct reaction and indirect reaction at the same time and they promoted each other in the later portion.

4. Conclusion

For low-grade nickel laterite, the reduction and magnetic separation was studied at temperatures of 1325-1400℃ with a weak magnetic field of 100 mT in order to acquire the high grade concentrations . The influence of reduction temperature, the percentage of CaO, and the atomic ratio of C/O (carbon/oxygen) were investigated for reduction and magnetic separation. During the experimental process, the gas of reduced laterite ore was detected by the gas analyzer. The fine ore was characterized by X-Ray fluorescence (XRF), X-ray diffraction (XRD) and scanning electron microscopy (SEM) equipped with an energy dispersive spectrometer (EDS). The following words summarize this study.

(1) According to the gas analysis, the reaction process was divided into two stages: the direct reduction had a greater advantage in the earlier stage when the temperature was low, and there was the direct reaction and the indirect reaction at the same time in the later stage when the

temperature was high.

(2) The result indicated that optimum reduction temperature was 1375℃, the optimum percentage of CaO was 12%, and the optimum atomic ratio of C/O was 1.2, resulting in the grade of fine ore of 5.71%. At the same time, the recovery rate of nickel was 96.35%.

(3) The nickel and iron in the fine ore basically existed in alloy state, and a small amount of gangue embedded cloth in ferronickel alloy.

References

[1]. Kim J, Dodbiba G, Tanno H. Calcination of low-grade laterite for concentration of Ni by magnetic separation (Minerals Engineering, 23(4)(2002)), 282-288

[2]. ISSF-News and meetings.

[3]. U.S. Geological Survey (Mineral Commodity Summaries [EB/OL], 2012), 108-109.

[4]. Guo, X.Y., Li, D., Kyung, H.P., Tian, Q.H., Wu, Z., Leaching behavior of metals from a limonitic nickel laterite using a sulfation-roasting-leaching process (Hydrometallurgy, 99(2009)), 144-150.

[5]. Tetsuya Watanabe, Sadao Ono, Haruo Arai, et al. Direct reduction of garnierite ore for production of ferro-nickel with rotary kiln at Nippon Yakin Kogyo Co., Ltd., Oheyama Works (International Journal of Mineral Processing, 19(04)(1987)) , 173-187.

[6]. Bo li, Hua Wang, Yonggang Wei. The reduction of nickel from low-grade nickel laterite ore using a solid-state deoxidization method(Minerals Engineering, 24(2011)) , 1556-1562.

[7]. Yusuke Kobayashi, Hidekazu Todorok, Hitoshi Tsuji. Melting behavior of siliceous nickel ore in a rotary kiln to produce ferronickel alloys (ISIJ International, 51(01)(2011)), 35-40.

Ni-Co 2013
Edited by: Thomas Battle, Michael Moats, Violina Cocalia, Harald Oosterhof, Shafiq Alam, Antoine Allanore, Rodney Jones, Nathan Stubina, Corby Anderson, and Shijie Wang
TMS (The Minerals, Metals & Materials Society), 2013

STATE OF THE ART REFRACTORY CORROSION TEST WORK FOR THE NONFERROUS METALS INDUSTRY

D. Gregurek[1], A. Ressler[1], V. Reiter[1], A. Franzkowiak[2], A. Spanring[2], B. Drew[2], D.R. Flynn[3]

[1]RHI AG, TC Leoben, Magnesitstrasse 2, Leoben, A-8700, Austria
[2]RHI AG, Wienerbergerstrasse 9, Vienna, A-1100, Austria
[3]Stillwater Mining Company, Columbus, MT 59019, U.S.A.

Keywords: Stillwater Mining Company, Ca-ferritic slag, Magnesia-chromite bricks, Rotary kiln, Corrosion resistance test, Refractories, RHI

Abstract

This study is a result of a collaboration between Stillwater Mining Company and RHI AG aimed at the determination of refractory products to improve the lining life in the customer's vessel by using optimized corrosion testing methods at RHI's Technology Center.
As a first part of the work a complete chemical and mineralogical characterization as well as FactSage calculation was carried out based on provided slag samples. According to these results the rotary kiln tests were realized at different process temperatures with several selected magnesia-chromite- and alumina-chromia bricks. Based on the mineralogical investigation a high chemothermal load caused by the Ca-ferritic slag resulted in a deep-reaching infiltration of the brick microstructure and corrosion of main brick components. On the basis of this outcome the selection of suitable brick types will be described in detail in the following article.

Introduction

In order to improve the lining life for specific non-ferrous metal applications, test procedures must be developed to simulate the various demanding service conditions. Frequent changes in the processing conditions and routes in combination with the diversity of metal processing furnaces [1], requires a precise understanding of each individual system for appropriate refractory product selection [2]. The RHI Technology Center in Leoben, Austria combines practical corrosion testing methods and equipment such as an induction furnace, rotary kiln, cup test etc. together with the microscopic analyses to enable a comprehensive understanding of brick wear following pilot scale trials.
Within a direct cooperation project between Stillwater Mining Company and RHI AG a practical testing with different refractory brands and Ca-ferritic slag has been conducted at the Technology Center Leoben.
For that purpose 120 kg of slag was received from Stillwater Mining Company, for a detailed investigation and for additional rotary kiln tests to be carried out on different magnesia-chromite- and alumina-chromia brick brands. Prior to testing a complete chemical and mineralogical characterization of the slag was carried out at RHI lab facilities.

Slag characterization

In order to adjust the optimal testing conditions for the rotary kiln test, along with the complete chemical and mineralogical characterization, FactSage calculations have been carried out on slag samples received from the Stillwater Mining Company[3].

The slag represents a Ni,-Cu,- and sulphur- rich Ca-Fe-oxide slag (Ca-ferritic slag type). The main slag components include two different Ca-Fe-oxides (one of type C_2F) and Ni-Mg-Fe-(Cr)-(Co)-oxide of type magnesia-ferrite (see Figure 1).

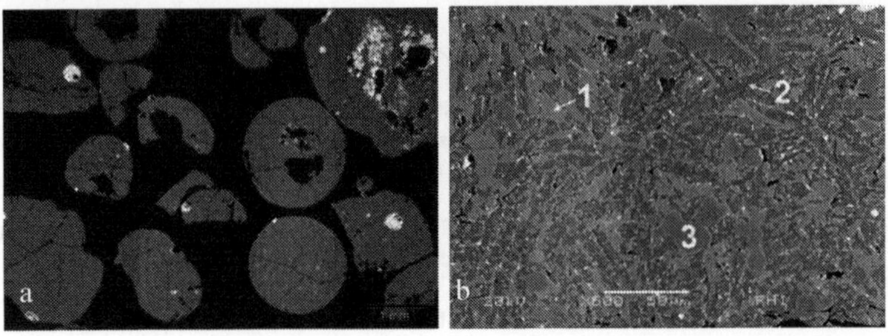

Figure 1. Exemplary microstructural overview & detail. (a) Reflected light microscopy. Granulated Ca-ferritic slag with Cu-(Fe)-Ni-(Pd)-sulphide inclusions (light phases). (b) Scanning electron microscope. Detail of slag granule. Ca-Fe-oxides (1 and 2). Ni-Mg-Fe-(Ca)-(Cr)-(Co)-oxide of type magnesia-ferrite (3).

In the slag different Cu-(Fe)-Ni-(Pd)-sulphide inclusions were found - mainly Ni-sulphide (Ni_3S_2; heazelwoodite), Cu-sulphide (Cu_2S; chalcocite) and Pd, Pt enriched Cu-Fe-Ni-sulphide. The melting temperature determined with a hot stage microscope is between 1350°C and 1410°C. Additional calculations of liquidus and solidus temperatures were carried out by help of the software package FactSage 6.0. Under the oxidising conditions the calculated solidus and liquidus temperatures are 1057°C and 1111°C respectively 1481°C and 1424°C for the slag samples slightly different in chemical composition.

Testing Facilities and Experimental Work

In the rotary kiln, with 92mm in diameter, up to six different brick brands can be installed and tested simultaneously (see Figure 2). The furnace is heated using a propane-oxygen gas mixture. The first rotary kiln test with Ca-ferritic slag was carried out between 1500°-1550°C[4]. The second rotary kiln test was performed at 1650°C[5].

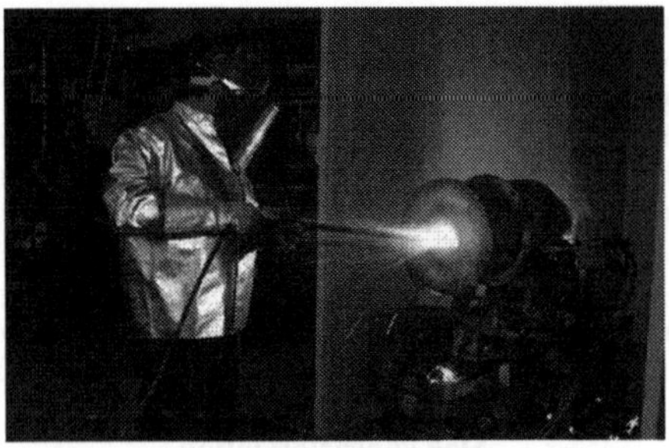

Figure 2. Rotary kiln test.

Both tests were carried out for 20 cycles. During testing 1.5 kg of slag were used per cycle.

In agreement with Stillwater Mining Company, five magnesia-chromite brands (i.e., RADEX FM5, RADEX OX6 COMPACT, RADEX FG B, RADEX FGLI and newly developed RADEX VFG), as well as one alumina-chromia brand (i.e., RESISTAL RK30SR) were selected for both furnace tests. The RADEX FM5 (type MCr 50, ISO 10081-2) represents a standard direct-bonded magnesia-chromite brick brand consisting of fused magnesia and chrome ore. RADEX OX6 COMPACT is a high-quality dense magnesia-chromite brick (type MCr 50, ISO 10081-2) based on magnesia-chromite co-clinker (i.e., OXICROM sinter) and chrome ore. RADEX FG B and RADEX FGLI are also high quality magnesia-chromite bricks (type MCr60) based on fused magnesia-chromite and chrome ore. The newly developed RADEX VFG brick brand contains fused magnesia-chromite and has a higher Cr_2O_3 content (Type MCr 40). RESISTAL RK30SR is an alumina-chromia brick (type ACr60/30 ISO 10081-4) based on Cr-corundum and chromium oxide.

The macroscopic overviews of selected bricks after testing, in cross section, are shown in Figure 3. The results from both trials, following evaluation of the wear area/depth in terms of the final cross-sectional profile, are presented in Tables I and II. The mineralogical investigations were performed on polished sections using a reflected light microscope and a scanning electron microscope combined with an energy-dispersive X-ray analyser (see Figures 4-6).

Results of corrosion testing

Rotary kiln test at 1500°-1550°C

All tested brick samples show crack formation due to thermal shock during kiln operation (Fig. 3). In case of the magnesia-chromite brick brands there is either minimal or no wear visible. The high wear observed for the alumina-chromia brick RESISTAL RK30SR can be explained by crack formation and loss of the brick parts due to high thermal shock from the beginning of testing.

Table I. Results of the first rotary kiln test

	Rotary kiln test (1500°-1550°C)		
Brick brand	Wear area*	Wear depth*	Infiltration depth**
	(cm^2)	(mm)	(mm)
RADEX FM5	0	0	13-17
RADEX OX6 COMPACT	0	0	7-10
RADEX FG B	7	5	10-13
RADEX FGLI	0	0	10-14
RADEX VFG	0	0	13-15
RESISTAL RK30SR	65^x	5^x	>20

*based on macroscopic wear evaluation from the final cross-sectional profile
**based on mineralogical investigation
^{x}not representative due to high thermal shock

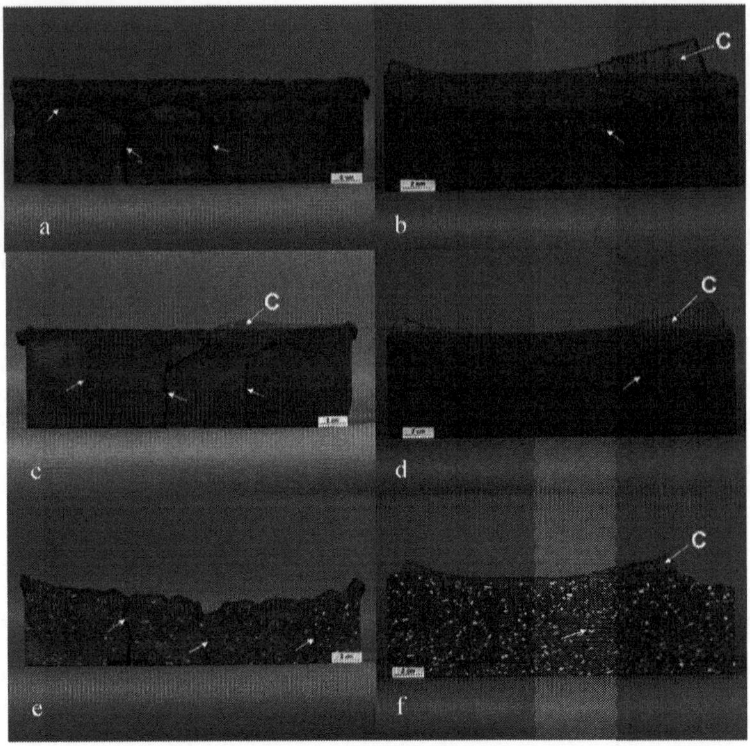

Figure 3. Selected exemplary magnesia-chromite RADEX VFG (a and b), RADEX OX6 COMPACT (c and d) and alumina-chromia RESISTAL RK30SR (e and f) bricks after the rotary kiln test at 1500°-1550°C (left side: a, c, e) and 1650°C (right side: b, d, f). Slag coating (C). Crack formation (arrows).

The main microstructural changes to the investigated magnesia-chromite bricks, after the rotary kiln test, are summarized below (exemplary see RADEX VFG, Fig. 4).

The immediate brick hot face was covered with a 1–2 mm thin reaction zone. Within this area the magnesia component is completely dissolved, only some chromite relics are still visible. Below the reaction zone a deep reaching infiltration and corrosion of the brick microstructure can be observed. The lowest infiltration of the brick microstructure was observed for RADEX OX6 COMPACT and the highest for RADEX FM5 (see also Table I).

In the infiltrated and partly completely degenerated brick microstructure (0-5 mm from the hot face) the single brick components cannot be distinguished any more. The high supply of Fe-oxide resulted in formation of a low melting Ni-rich magnesia-wuestite ((Ni)-Fe-Mg-oxide). The second brick component chromite, as well as, chromite precipitations are corroded. Due to corrosion of chromite Ca-Mg-Fe-Al-Cr-oxide and Ca-Fe-Cr-Al-oxide formed.

In addition, vertical cracks partly filled with Cu-Fe-Ni-sulfides formed in almost all brick brands. In case of RADEX FM5, RADEX FGLI and RADEX OX6 COMPACT cracks also formed parallel to the hot face at the interface between the infiltrated and non-infiltrated brick microstructure.

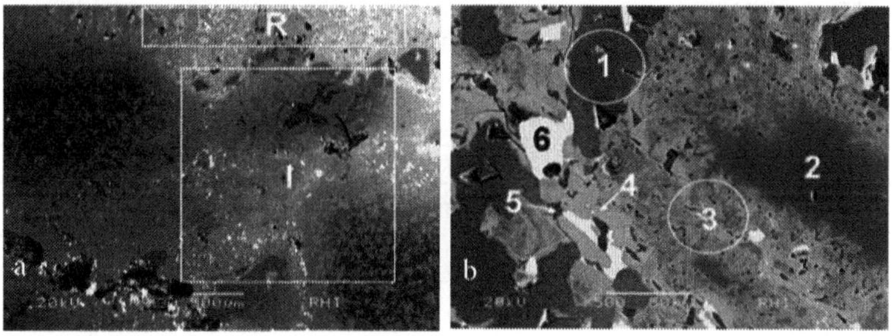

Figure 4. Exemplary microstructural overview & detail. Immediate brick hot face. RADEX VFG. (a). Reaction layer (R). Infiltrated and corroded brick microstructure (I). (b) Approx. 2 mm from hot face. Corroded magnesia (1) and chromite (2). Ca-Mg-Fe-Al-Cr-oxide (3). Ca-Fe-Cr-Al-oxide (4). Dicalcium silicate (5). Cu-Fe-sulphide (6).

In case of the alumina-chromia brick RESISTAL RK30SR (Fig. 5) infiltration of the microstructure can be traced over the whole polished section (0-20 mm from the hot face). Nevertheless the highest microstructural changes were observed in the area 0-2 mm from the hot face. In that brick area pore filling infiltration, recrystallization of Cr-corundum bearing matrix, corrosion of Zr-mullite and formation of Mg-Fe-Al-Cr-oxide took place. The immediate hot face is covered with a 1-2 mm thin reaction zone. In the infiltrated brick microstructure cracks formed parallel but also vertical to the hot face can be observed. Cracks are partly filled with slag.

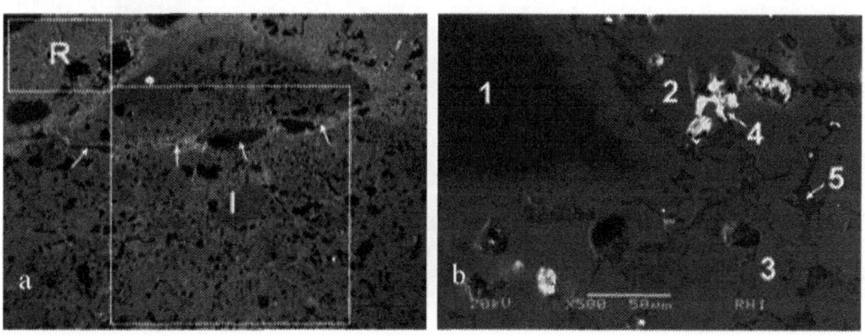

Figure 5. Exemplary microstructural overview & detail. Immediate brick hot face. RESISTAL RK30SR (a). Reaction zone (R). Infiltrated and corroded brick microstructure (I). Cracks formed parallel to the hot face partly filled with slag (arrows). (b) 1 mm from hot face. Degenerated, recrystallized and infiltrated brick microstructure. Fused alumina (1). Cr-corundum (2). Mg-Fe-Al-Cr-oxide (3). Zr-oxide left after the corrosion of Zr-mullite (4). Phosphorous-rich Na-Ca-Al-silicate (5).

Rotary kiln test at 1650°C

In the case of the magnesia-chromite brick brands in similar testing at 1500°-1550°C there is either minimal or no wear visible. Nevertheless in comparison to the first rotary kiln test the infiltration depth is significantly higher (see Table II). The highest infiltration depth has been observed for RADEX FM5 (Fig. 6). The alumina-chromia brick RESISTAL RK30SR show in comparison to the magnesia-chromite bricks higher wear. The infiltration depth cannot be exactly verified on the cross section.

Table II. Results of the second rotary kiln test

	Rotary kiln test (1650°C)				
Brick brand	Wear area*	Wear depth*	Infiltration area**	Infiltration depth*	Infiltration depth**
	(cm^2)	(mm)	(cm^2)	(mm)	(mm)
RADEX FM5	2	4	54	28	18-20
RADEX OX6 COMPACT	2	3	41	20	18-20
RADEX FG B	0	0	43	21	16-18
RADEX FGLI	0	0	38	20	14-15
RADEX VFG	0	0	46	24	15-18
RESISTAL RK30SR	10	7	x	x	>20

*based on macroscopic wear evaluation from the final cross-sectional profile
**based on the mineralogical investigation
x cannot be determined macroscopically on the cross section

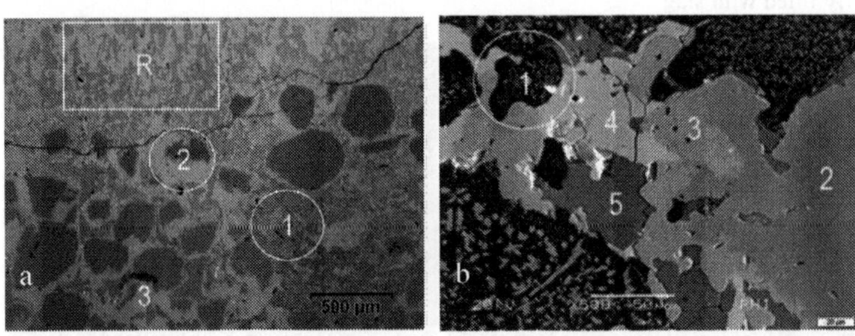

Figure 6. Exemplary microstructural overview & detail. RADEX FM5. Immediate brick hot face. Severe corrosion and degeneration of the brick microstructure. (a) Reaction zone (R). Corroded magnesia (1) and chromite (2). Ca-ferritic infiltrate (3). (b) Approx. 2 mm from hot face. Corroded magnesia (1) and chromite (2). (Ca)-Mg-Al-Cr-Fe-oxide (3). Ca-(Cr)-Al-Fe-oxide (4). Sulphur rich Ca-silicate (5).

Based on the mineralogical investigation the brick microstructure is highly degenerated especially within the area 0-20 mm from the hot face. Infiltration, corrosion of both brick components e.g. magnesia and chromite, formation of a low melting Cu-Ni-rich magnesia-wuestite can also be observed after the second rotary kiln test. Nevertheless the microstructural changes are much more pronounced in comparison to the first rotary kiln test at 1500°-1550°C. Additionally due to decomposition and oxidation of CuFeNi-sulphide sulphur attack and formation of sulphur bearing phases can be observed at the end of infiltration.

In case of the alumina-chromia brick RESISTAL RK30SR below the several mm thick reaction zone the highest microstructural changes were observed in the area 0-2 mm from the hot face. In that brick area pore filling infiltration, recrystallization of Cr-corundum bearing matrix, corrosion of Zr-mullite and formation of (Na)-Ca-Al-silicate of type gehlenite ($Ca_2Al_2SiO_7$) took place. Infiltration of the microstructure can be traced over the whole polished section (0-20 mm from the hot face).

Thermochemical calculations

The interactions of different combinations of refractory materials with the Ca-ferritic slag have been described thermochemically with regard to potential reaction products formed at the interface and the solubility of the refractory components in the slag.

The calculations have been conducted for a temperature of 1550°C and 1650°C as well as oxidising conditions with the software package FactSage 6.0. The phase constitution of a system is calculated under the assumption that the defined elements or compounds of the system react or partially react to reach a state of chemical equilibrium. Reaction kinetics are not considered in the thermochemical calculations.

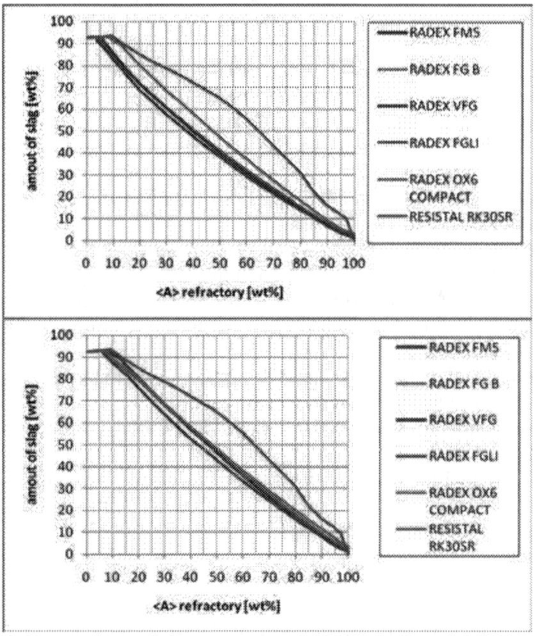

Figure 7. Comparison of slag amount at 1550°C (left) and 1650°C (right).

The equilibrium phases and phase compositions in dependence of the refractory/slag ratio have been determined. For this purpose a variable <A> has been introduced which is defined by the mass ratio of the refractory material to the total mass of refractory plus slag. <A> = 0 represents the composition of the slag with no contact to the refractory material and <A> = 100 represents the composition of the refractory material with no slag contact.

The slag amount in dependence of <A> has been compared for the different refractory bricks (see Figure 7) at 1550°C and 1650°C. The alumina-chromia brick shows a considerably higher solubility compared to the magnesia-chromite brands that can be ranked with regard increasing solubility:

- RADEX FGLI
- RADEX VFG
- RADEX FG B
- RADEX FM5
- RADEX OX6 COMPACT

From the different investigated magnesia-chromite bricks the RADEX FGLI shows the best result and the lowest solubility rate. But the difference of the various brands with exception of RADEX OX6 COMPACT is low.

Conclusion

During this cooperative work with Stillwater Mining Company a variety of tests were conducted at RHI's research facilities, including a complete characterization of the Ca-ferritic processing slag and two rotary kiln tests at 1500°-1550° and 1650°C.

All tested magnesia-chromite bricks show macroscopically either no or minimal wear after both rotary kiln tests. The possible explanation for the latter is the trial duration of ~20 cycles which is standard testing method at the TC Leoben due to tightened testing methods. Nevertheless the following observations can be summarized comparing both tests:

- significantly higher infiltration of the brick microstructure at 1650°C at the same number of cycles;
- sulphur attack and formation of sulphur bearing phases at the end of infiltration due to decomposition and oxidation of CuFeNi-sulphide;
- severe degeneration of the brick microstructure due to chemothermal load by Ca-ferritic slag. The microstructural changes are much more pronounced in comparison to the first rotary kiln test;
- in comparison to the magnesia-chromite bricks the alumina-chromia brick RESISTAL RK30SR shows significantly higher wear;
- magnesia-chromite bricks are better withstanding Ca-ferritic slags.

According to the obtained results diverse options can follow, including refractory proposals for field trials, executed on customer's site. This leads to an improved lining life in the vessel entailing technical and economical benefits.

References

1. W.G. Davenport et al., *Extractive Metallurgy of Copper* (Kidlington, Oxford OX5 1GB, UK: Elsevier Science Ltd., 2002), 432.

2. G. Routschka, *Refractory Materials* (Essen: Vukan-Verlag GmbH, 2004), 512.

3. V. Reiter, D. Gregurek, "Investigation of slag samples received from Stillwater Mining Company, Columbus, Montana, U.S.A." (Internal RHI report, Technology Center Leoben, Austria, 2012).

4. D. Gregurek, "Investigation of basic and non-basic bricks after the rotary kiln test carried out for Stillwater Mining Company, Columbus, Montana, U.S.A." (Internal RHI report, Technology Center Leoben, Austria, 2012).

5. D. Gregurek, "Investigation of basic and non-basic bricks after the second rotary kiln test carried out for Stillwater Mining Company, Columbus, Montana, U.S.A. " (Internal RHI report, Technology Center Leoben, Austria, 2012).

Ni-Co 2013

Ores and Processing

Session Chairs

Corby Anderson
Phillip Mackey

Ni-Co 2013
Edited by: Thomas Battle, Michael Moats, Violina Cocalia, Harald Oosterhof, Shafiq Alam, Antoine Allanore, Rodney Jones, Nathan Stubina, Corby Anderson, and Shijie Wang
TMS (The Minerals, Metals & Materials Society), 2013

MINERALOGICAL CHARACTERIZATION OF COBALTIC OXIDES FROM THE DEMOCRATIC REPUBLIC OF CONGO

Y. Vanbrabant[1], C. Burlet[1], P. Louis[2]

[1]Royal Belgian Institute for Natural Sciences, Geological Survey of Belgium, Jenner street, 13 – BE-1000 Brussels, Belgium, Email: yves.vanbrabant@naturalsciences.be
PEL Consult, Belgium[2]

Keywords: asbolane, cobalt oxides, Democratic Republic of Congo, heterogenite, lithiophorite

Abstract

The largest cobalt ore reserves are located in DRC, the Democratic Republic of Congo. Most of the cobalt is observed as black cobaltic oxide minerals: heterogenite [$HCoO_2$] and asbolane [$(Ni,Co)_2$-$xMn(O,OH)_4.nH_2O$] which are hardly differentiable since they exhibit similar macroscopic habit and textures. These minerals are frequently observed in similar environment (oxidized horizon of ore deposits) and they are commonly poorly-crystallized limiting their study with XRD. Their chemical composition is also not very well-constrained since they exhibit significant chemical substitutions with cations as Cu, Co, Ni and Mn. The difference in mineralogy and chemical composition of the different cobalt minerals will have an impact on their hydrometallurgical treatment. We compared the natural chemical variability of heterogenite, with that of asbolane samples from DRC. These minerals are compared by EDS and Raman microspectrometry techniques. We show that Raman microspectrometry is a quick and reliable tool to discriminate asbolane from heterogenite.

Introduction

Cobalt bearing oxides are an economically important ore extracted in the Democratic Republic of Congo (DRC) as a world-class source of cobalt [1]. This commodity has become a critical resource in developed countries and its availability remains strongly dependent on African production (Figure 1). The cobalt extracted in DRC is mainly used in rechargeable batteries (mobile devices, hybrid cars), alloys, magnets, tyre adhesives and catalysts for the petroleum and chemical industries. Cobalt fields of application in new technologies are so wide that is now a strategic material in the modern industry [2].

The main cobalt oxide ore extracted in DRC is called 'heterogenite'. This name generally describe a black powdery or botryoidal oxide encountered in the upper part of the oxidized horizon of Cu-Co bearing formations of the Neoproterozoic Lufilian fold-and-thrust belt in the Katanga Province (DRC) [3]. Nowadays, heterogenite is still intensively operated into formal mining sites where it is usually extracted in association with Cu-minerals. It is also mined in numerous informal small-scale sites where the miners (diggers) work in appalling conditions with basic equipment (heel bars), no safety measures and no environmental follow-up [4].

Historically, GECAMINES was the largest producer with 50-60% of world cobalt mine production until the end of the 1980's. Cobalt was mostly produced in the hydrometallurgical plants of Shituru and Luilu in Katanga, whose main production focused on copper cathodes. Cobalt was present in the ore at a ratio Cu/Co of between 10 and 20 to one. Most of the extracted ore was sold overseas, mostly to China, for further treatment or was treated in Katanga by new hydrometallurgical plants (SOMIKA, BOSS MINING and CHEMAF). Clear distinction criterions are needed, as well as quick and reliable determination tools, since the mineralogical form of the cobaltic oxides (heterogenite or asbolane, or even both) has a substantial impact on the reducing agent consumption during hydrometallurgical treatment. This task is however complex since heterogenite and asbolane share similar petrographic textures, their chemical composition is variable and unclear (especially for asbolane) and both forms are commonly XRD amorphous.

In this paper, we first describe the discrete mineralogy of 'heterogenite' ore. We then summarize the past and current nomenclature available for heterogenite and asbolane minerals. The current unclear definition of asbolane is compared with the case of lithiophorite. The distinction criterions between these three phases are then presented with their respective Raman signatures and with their EDS chemical composition. A special attention is given to the Mn-Co-Ni and Mn-Co-Cu content for samples of different mining sites in DRC, along with asbolane samples from New Caledonia. The metallurgical implications of the existence of these two phases are finally discussed in terms of best practices.

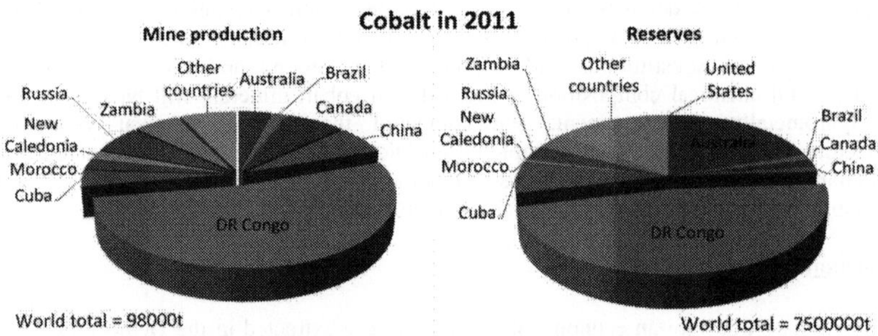

Figure 1. Estimated World mine production and reserves for cobalt in 2011 (source: USGS commodity summary 2012) [1]

Heterogenite Ore and its Discrete Mineralogy

The two cobalt bearing oxides found in 'heterogenite' ore are: heterogenite sensus-stricto [$HCoO_2$: ideal composition] [5] and asbolane [$(Ni,Co)_{2-x}Mn(O,OH)_4.nH_2O$: unclear composition]. Those are hardly differentiable minerals since they exhibit similar macroscopic habit (black mineral) and textures (botryoidal morphology, stalagmite or powdery). These minerals are frequently observed in similar environments (soil and oxidized horizon of ore deposits) and they are commonly poorly-crystallized which hinder their investigation with traditional X-Ray diffraction techniques.

The chemical composition of the three minerals is also not very well-constrained since the natural species exhibit significant chemical substitutions with cations as Cu, Co, Ni and Mn that expand the chemical composition spectra. Since the asbolane crystallography is not yet clearly defined, this phase is usually not referenced in XRD databases. However, its crystallographic structure is likely close to that of lithiophorite [$(Al,Li)MnO_2(OH)_2$] and Llorca and Monchoux [6] in their study of Co oxide ore from New Caledonia found occurrence of lithiophorite associated with asbolane and concluded that lithiophorite and asbolane occupy a continuous field.

History and Nomenclature of Heterogenite

Heterogenite ("ετερογενής", literally "of a different nature"), was named for the first time by Frenzel in 1872 in order to distinguish the cobalt-, nickel- and manganese ore found at the Maasen mine in Schneeberg, (Germany) from the other black oxides [7]. Since their discovery, these cobalt-bearing oxides have posed a mineralogical problem that is particularly difficult to resolve. For more than a century, material obtained from various deposits has been studied. Differences in the state of crystallinity of the material, due to variability in the copper, nickel and hygroscopic water content, determine a series of "varieties" with different aspects and properties. As such, more than 20 different chemical formulae have been proposed, and several names were given to them (including boodtite, heubachite, lubumbashite, mindigite, schulzenite, stainierite, transvaalite, trieuite and winklerite) [8].

In 1962, Max Hey [9] came to the conclusion that most of these minerals share the same formula, CoO.OH, analogous to that of manganite (MnO.OH) and goethite (FeO.OH). This was later supported by the work of Deliens [10], who studied the southern Katanga heterogenites and clarified this nomenclature by suggesting the use of only the name heterogenite as the general name for the oxyhydroxide of cobalt. Later, Deliens and Goethals [11] identified two polymorphs for this mineral, namely heterogenite-2H and heterogenite-3R.

In 2012, Burlet et al. showed that strong spectroscopic evidence proved that the mineral structure could be better described by the chemical formula $HCoO_2$ instead of CoO.OH [5]. The mineral should be thus considered as a cobaltic acid [12] instead of a cobalt oxy-hydroxide [3]. Heterogenite was in consequence categorized as a layered oxide with a delafossite-type structure [5].

Methodology

An extensive mineralogical-petrological study of heterogenite, in its mineralogical sense, has been conducted at the Royal Belgian Institute for Natural Sciences (RBINS) in its Geological Survey Department through two projects, namely TRACE financed by the Belgian Sciences Policy (2008) and GECO-projects sponsoed by the Belgian Foreign Affairs Office (2007-2012). 120 samples of heterogenite were studied through EDS, microprobe, EBSD, XRD and Raman equipment combined with more traditional petrographic observations along with backscatter SEM images. These samples are partly from mineralogical collections of the Royal Museum for Central Africa (RMCA) and RBINS. The other part of samples was recently collected during field campaigns in Katanga into industrial mining sites as well as informal artisanal extraction shafts.

For the sake of conciseness, only a subset of 13 heterogenite samples was reexamined during the present study. This selection covers the chemical variability in heterogenite samples as observed during the above-mentioned projects. 9 asbolane and 4 lithiophorite samples from the mineralogical collection at RBINS, as well asbolane samples collected in the field in Katanga, completes the studied set. The origin of the asbolane samples is divided between the Lufilian arc in Katanga and New Caledonia, while the lithiophorite samples come from different countries (Germany, Poland, USA). It is also worthwhile mentioning that few samples exhibit both heterogenite and asbolane phases. This point will be further detailed in the petrographical section.

Two main analytical tools were applied during this study, namely Raman microspectrometry and standardless EDS chemical analyses. All samples were prepared as polished sections in order to conduct clear observations under a scanning electron microscope as well under the Raman microscope. This approach allows accurate repositioning when passing between both techniques and chemical information can be therefore compared with the Raman response and vice versa. However the use of polished sections is not mandatory when using Raman technique, since spectra can be acquired even of raw samples, making this technique interesting for rapid evaluation of ore samples. As a reminder, both techniques provide information at the scale of a 'point', that is in practice at µm-scale. For instance the EDS chemical compositions are usually representative for a single mineralogical phase. Such measurement contrasts with bulk chemical analyses, which represents an average composition of all existing phases. The main analytical parameters for both techniques are summarized in Table I.

Table I. Applied Analytical Parameters

Raman microspectrometry	
Equipment/Manufacter	Senterra/Brucker
Laser wavelength	532 nm (green)
Lase power	usually 2 mW or 0.2 mW for asbolane
Optics	50 X-long distance
Aperture	50 X 1000 µm
Acquisition time	3 X 10 sec, 5 X 100 sec for reference spectra
Spectral window	114->4400 cm^{-1}
Standardless EDS	
Equipment/Manufacter	APOLLO 10 Silicon Drift /EDAX
Spot size	7
High voltage	30 kV
Magnification	between 200 and 500X
Acquisition time	50 sec

Experimental Results

Raman Signature of Phases

The Raman responses of mineralogical phases reflect discrete vibrational states of constituting molecules resulting from the inelastic scattering of an exciting source (e.g. laser). The recorded spectra usually consists of two main components, namely a broad and usually bell-shaped signal corresponding to the fluorescence of sample and the second component including several narrow peaks (bands) representing the Raman effect. The peak positions represent a key feature during the comparison between mineralogical phases. For instance, unknown phases are usually compared with Raman signature of a reference mineral on the peak positions. The latter are expressed as a Raman shift with respect to the monochromatic existing source and the unit is cm^{-1}. By contrast, the absolute intensity of a single peak is disregarded since it can be influenced by numerous factors, such as the fluorescence and sample surface state. The relative intensity of two peaks can however be analyzed since it assumes that the absolute intensity of peaks is controlled by the same processes. In this paper, only a qualitative comparison between reference spectra of asbolane, heterogenite and lithiophorite is realized on crystallized phases. The validation of the spectral response of heterogenite was conducted on well-crystallized samples and comparisons were confirmed by microcrystallographic information from EBSD mapping.

Experience has shown that valuable Raman responses on heterogenite samples can also be acquired even on XRD amorphous specimens. We consider that the comparison of the chemical composition of an amorphous phase with that of a reference phase combined with similar Raman responses allows a positive determination of an unknown or amorphous phase.

Figure 2 shows reference spectra of heterogenite, asbolane and lithiophorite. All Raman responses are different, but they share nevertheless a similar pattern of peaks composed of two groups. The first one gathers 4 peaks located between 376 and 649 cm^{-1}. The second group shows three small-amplitude peaks with Raman shifts between 938 and 1217 cm^{-1} and with similar intensities. In addition to these two groups, an additional very weak peak is observed at 3484 cm^{-1} for asbolane and 3458 cm^{-1} for lithiophorite.

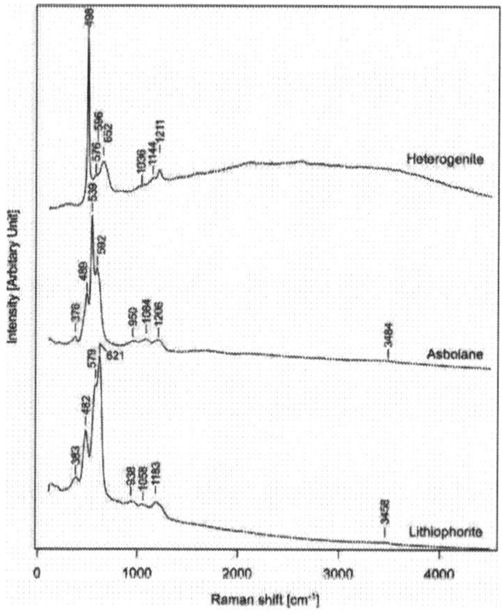

Figure 2. Reference Raman spectra of heterogenite, asbolane and lithiophorite.

The distinction between the three phases can therefore be readily realized on the respective intensities of individual bands from the first group. The peak at 498 cm^{-1} (e.g. with the small Raman shift) in heterogenite sample is the most intense, while the middle peak of asbolane (@ 539 cm^{-1}) and the peak with the higher shift (621 cm^{-1}) of lithiophorite have the highest intensity.

Similar patterns were recorded on amorphous phases and a good correlation is observed between the respective spectra and the EDS chemical fields of the three studied phases as described below.

EDS Geochemical Compositions

The chemical variability within the studied samples was examined through the determination of geochemical composition of the three phases in 798 points with EDS standardless measurements. The list of the quantified elements is: C, O, Na, Mg, Al, Si, P, S, Cl, K, Ca, Ti, V, Mn, Fe, Co, Ni, Cu, Zn. The main elements observed in heterogenite are Co, O with significant amount of Cu. Asbolane samples show a composition dominated by Mn, O, Co and different amounts of Cu. Finally, lithiophorite includes Mn, O and Al. EDS technique is however unable to detect light elements, such as H which is a major component into the three studied phases. The same limitation exists for Li in lithiophorite.

Since standardless EDS spectra only provides (semi-)quantitative chemical compositions the measured concentrations must be regarded with some cautions. Nevertheless the main advantages of this technique are its speed and the recording of all elements in a single test, which allows a rapid assessment of the chemical variability. The relative concentration in the different phases can be represented as triangle plot of Ni, Mn and Co (Figure 3.a) and Cu, Mn and Co (Figure 3.b). These representations show that heterogenite, asbolane and lithiophorite minerals occupy different chemical fields.

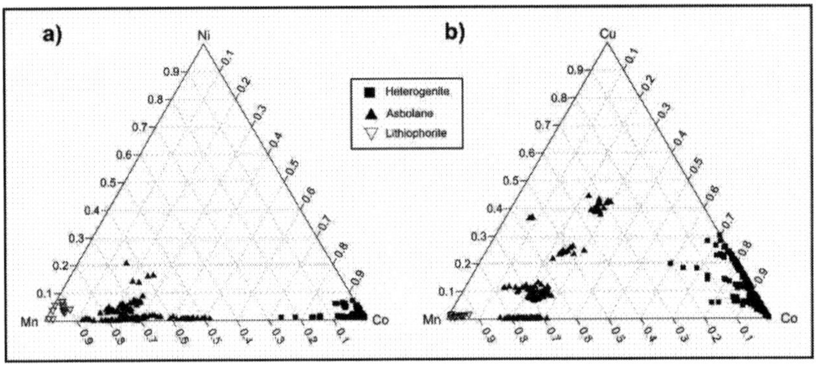

Figure 3. Triangle plots of relative concentration (wt%) for the 3 studied phases. a) Ni-Mn-Co ; b) Cu-Mn-Co

Heterogenite, asbolane and lithiophorite samples show globally small amounts of Ni (see Figure 3.a), which is usually below 1 wt%, that is close to the detection limit of the EDS detector. Studied asbolane specimens from the Democratic Republic of Congo (Katanga) show even lower Ni-concentration, while samples from New Caledonia are Ni-rich. This difference reflects the different geological setting of this mining province, since asbolane minerals from the Lufilian Arc (Katanga) generated from the alteration of folded and thrust sedimentary rocks, while this phase in New Caledonia formed during the alteration of ultrabasic rocks which contain larger amount of Ni. Heterogenite from DRC commonly does not show significant concentration of Ni. Only two mining sites (Likasi and Shinkolobwe) have provided samples with Ni amounts larger than the detection limit.

Heterogenite samples are the richer Co-phase, whereas asbolane specimens occupy an intermediate position with significant concentrations of Mn and Co. The relative concentration of Mn/(Mn+Co) in asbolane reaches a maximum value of 0.84. This ratio can decrease down to ~0.3 due to a significant increase in Cu concentration (see Figure 3.b). In this case, the phase could be regarded as a cupro-asbolane. Raman spectra acquired on this amorphous phase show the signature of asbolane with some specificities. By contrast, the increase of Cu concentration in heterogenite is not necessarily associated with an increase of the Mn/(Mn+Co) which remains for most samples close to zero. Only a few measurements show Mn contamination in heterogenite samples. Based on our set of samples it seems that no intermediate phase can be observed between asbolane and heterogenite.

Petrological Observations

Even if heterogenite and asbolane are distinct phases in a mineralogical sense, it is worth mentioning that both phases can coexist at a microscopic scale in the same sample. Figure 4 shows an example of an asbolane matrix associated with a band of heterogenite. This observation reinforces the statement that both phases constitute distinct minerals without an intermediate form. In addition, the occurrence of heterogenite and asbolane within the same sample implies that a precise characterization at a microscopic-scale of Co-ores is required in order to define the best ore processing method.

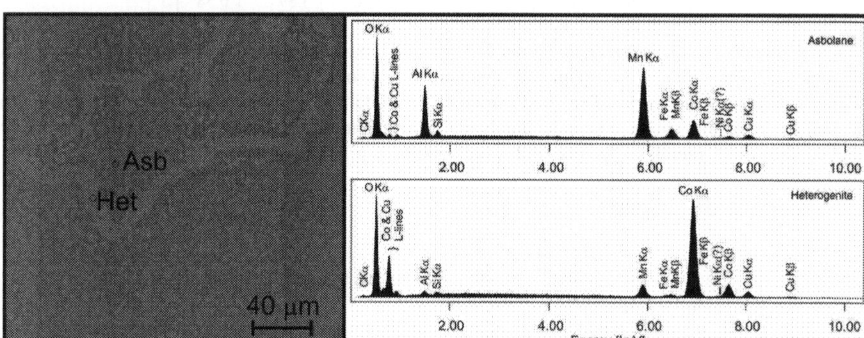

Figure 4. Example of coexisting heterogenite (Het) and asbolane (Asb), with their respective EDS spectrum.

Metallurgical Implications

Pyrometallurgical Treatment

Until mid of the 1990s, most of the cobalt produced in the DRC was produced in GECAMINES as cobalt cathodes in their hydrometallurgical circuits at the Shituru and Luilu plants. There was nevertheless also some pyrometallurgical production of impure cobalt in their plant of Panda near Likasi. The electric furnaces were fed by some cobalt-rich ore and by different cobalt/copper materials which were difficult to treat hydrometallurgically. Limestone and coal were added to improve slag characteristics. After melting and reduction, a metallic phase and slag were produced. The metallic phase was allowed to cool slowly and two phases were produced: a red alloy containing circa 85 % Cu and a white alloy with circa Co 40-45 %, Cu 15-20 %, Fe 20 – 35 %; remaining was Si and other minor impurities [13]. Mn went nearly quantitatively found into the slag. No one cares whether the cobalt in the ore was as heterogenite or as asbolane.

After the sharp decrease in cobalt production from GECAMINES, some new plants were erected which treated locally the collected 'heterogenite' ore in quite efficient small electrical furnaces. There was no separation between the white and the red alloys. In fact, the improperly called 'white alloy' presently produced had a relatively dark purple color. Hereunder are data for typical pyrometallurgical operation of an electric furnace in the Democratic Republic of Congo.

Table II. Typical feed and "white alloy" composition in electric arc furnace operation

	Co	Cu	Fe	Ni	Mn
Co in feed (%)	5.93	4.12	5.97	0.63	0.56
Co in metal (%)	31.14	23.5	29.0	3.59	0.58

The pyrometallurgical operation presents different advantages compared to the direct exportation of cobalt ore. It allows a reduction, by a factor 5 – 6, of the weight to be transported. It also allows substantial reduction of reagent during hydrometallurgical operations, in particular due to the reduction of the Mn / Co ratio in the "white alloy" and due to the reduction of the content of acid consuming minerals in the ore (dolomite, calcite, e.g.).

Hydrometallurgical Treatment

Unlike the pyrometallurgical processing of cobalt ores, the mineralogical form of the cobaltic oxides has a substantial impact in hydrometallurgy:

- Both asbolane and heterogenite require a reducing agent for their dissolution. The specific consumption is nevertheless substantially higher for asbolane than for heterogenite due to required consumption for Mn^{+3} or Mn^{+4} reduction.
- Dissolved manganese has to be removed from solution

Different agents can be used for the reduction leaching of asbolane and heterogenite: SO_2, sodium sulphite, bisulphite or metabisulphite ($Na_2S_2O_5$), $FeSO_4$, metallic iron scraps or powdery copper.

If we consider reduction by $FeSO_4$, we have the following reactions:

For asbolane:

$$(xMnO_2 . yCuO . zCo_2O_3) + (2x + 2z)FeSO_4 + (2x + y + 3z)H_2SO_4 \rightarrow xMnSO_4 + yCuSO_4 + 2zCoSO_4 + (x + z)Fe_2(SO_4)_3 + (2x + y + 3z)H_2O \quad (1)$$

For heterogenite:

$$Co_2O_3.H_2O + 2FeSO_4 + 3H_2SO_4 \rightarrow 2CoSO_4 + Fe_2(SO_4)_3 + 4H_2O \quad (2)$$

The ferric sulphate produced during heterogenite or asbolane leaching can be regenerated as ferrous sulphate by using different types of reducing agent. If using SO_2 as reducing agent, we have:

$$Fe_2(SO_4)_3 + SO_2 + 2H_2O \rightarrow 2FeSO_4 + 2H_2SO_4 \quad (3)$$

Manganese sulphate can be removed from solution by different ways as discussed by Ferron and Louis [14].

Selective precipitation par SO_2 / O_2 or SO_2 / air.

Surprisingly, SO_2, which is an efficient reducing agent that can directly or indirectly be used for the reduction leaching of asbolane or heterogenite becomes a very strong oxidizing agent when mixed with oxygen, pure oxygen or air [15]. Very promising results were obtained at laboratory and on pilot plant scale for selective manganese removal from industrial solutions.
Global manganese precipitation reactions are:

$$2MnSO_4 + SO_2 + O_2 + 3H_2O + 3CaCO_3 \rightarrow Mn_2O_3 + 3CO_2 + 3\ CaSO_4.2H_2O \quad (4)$$

or

$$MnSO_4 + SO_2 + O_2 + 2H_2O + 2CaCO_3 \rightarrow MnO_2 + 2CO_2 + 2CaSO_4.2H_2O \quad (5)$$

Removal of manganese from solution by solvent extraction

Manganese can also be removed from $CoSO_4$ or $CoCl_2$ solution by using DEHPA (Di-(2-ethylhexyl) phosphoric acid) in the solvent extraction (SX) technique [16]. Along with Mn, other impurities will be co-extracted too: Cu, Ca, Zn. To remove the Ca loaded on the organic phase, the use of HCl is required for scrubbing and stripping, and HCl-resistant materials of construction. This problem can nevertheless be solved by using a reverse SX configuration [16].
It must be noticed that the SX technique requires 3 times the reagent consumption for manganese removal:

- NaOH for manganese loaded on DEHPA
- Acid for stripping
- Lime, NaOH or Na_2CO_3 for effluent neutralisation and manganese precipitation.

Table III. Extraction pHs (pH50) for different metals using DEHPA

	Zn	Ca	Mn	Cu	Co
pH 50	1.70	1.72	2.71	2.80	3.70

Selective cobalt precipitation leaving manganese in the effluent

If the Mn/(Mn+Co) ratio is not too high in the cobalt bleed from the main Cu hydrometallurgical circuit, it is possible to have a fairly good selective cobalt precipitation as hydroxide or basic sulphate after acid neutralisation, iron and copper removal [16]. This is due to the difference of solubility products of cobalt and manganese hydroxides: with pS values for Co, and Mn hydroxides of respectively 15.4, and 12.7 (pS = -log [solubility product]). The cobalt precipitation will be done at a pH around 8.2, with close to 98% cobalt precipitation and 5-20% manganese co-precipitation. An even better selectivity can be reached by doing the precipitation in two steps counter-currently. This is probably, in most cases, the easiest and cheapest way to remove Mn from Co solutions.

Conclusions

The 'heterogenite' ore deposits in the Katanga Province (Democratic Republic of Congo: DRC) is composed of two main mineralogical Co-oxides phases, namely heterogenite [$HCoO_2$] and asbolane [$(Ni,Co)_2$-xMn$(O,OH)_4.nH_2O$: unclear composition]. These minerals cannot readily be distinguished from each other since they exhibit similar macroscopic habits, the same microscopic features and both are frequently observed as amorphous phases. In addition, their chemical composition is highly variable. The issue on the nature of asbolane also requires some comparisons with the better-defined lithiophorite that can also be found in some Co-oxide ore deposits, such as in New Caledonia.

Our observations on a set of heterogenite and asbolane samples from DRC combined with samples from other localities shows that each phase, even under in amorphous form, can be readily distinguished by Raman microspectrometry. This technique is therefore attractive during ore deposit characterization campaigns or during the follow-up extraction operations where it is important to distinguish the main Co-containing phase(s). The main advantage of this technique is its speed since no sample preparation is required during the collection of Raman spectra, that usually last few tens of seconds. The method provides information at a µm-scale and several points are thus required to fully characterize ore batches composed of different mineralogical phases. Our petrographical observations show also that asbolane and heterogenite mineralogical phases can coexist at a µm-scale as two distinct phases into 'heterogenite' ore.

The distinction between heterogenite and asbolane from our sample set can also be conducted on a chemical base showing that heterogenite represents the richer Co-phase with variable Cu concentrations. By contrast, only Mn traces are usually observed in heterogenite minerals from DRC except in few samples, but always in lower concentration than in asbolane. The latter shows variable Mn/(Mn+Co) ratios between 0.85 and 0.3 and the decrease of this value is related to enrichment into Cu.

Most of asbolane and heterogenite samples from DRC show only traces or small concentrations of Ni. This observation contrasts with Ni-rich asbolane samples from New Caledonia studied in this paper or those described by Llorca and Monchoux [6]. The composition of the primary rocks from which Ni and Co elements were mobilized and then concentrated into alteration horizons can explain this difference in Ni-content. In New Caledonia, primary ultramafic rocks represent the main Ni and Co source, whereas folded and thrusted sedimentary formations form the bedrock of the mining district of the Katanga Province.

The mineralogical form of the cobaltic oxide, asbolane or heterogenite, has nearly no impact on the pyrometallurgical treatment of this type of ore. On the contrary, the Mn content in asbolane has a serious impact on its hydrometallurgical treatment. It will require more reagents for leaching and a Mn removal step. This could be done by selective Mn precipitation by SO_2/O_2, by SX or by selective Co precipitation as hydroxide.

Acknowledgments

The first two authors (YV & CB) thanks the Belgian Foreign Affairs Ministry for financing the GECO-project and the Belgian Science Policy Ministry for sponsoring the TRACE project. We are also gratefull for the support of our colleagues at the University of Lubumbashi (DRC), especially Prof. F. Lubala and Mr. L. Kipata. This study would not have been possible without the help of Mme. Ch. Lange (Katanga Mining Limited) and Mr. J. Minsiomi (Boss Mining SPRL).

References

1. USGS, *Cobalt Statistics and Information, yearly commodity summary* (2012) http://minerals.usgs.gov/minerals/pubs/commodity/cobalt/mcs-2012-cobal.pdf

2. Hawkins M., *Why we need cobalt, Applied Earth Science (*Transactions of the Institution of Mining & Metallurgy Section B 110 (2), 2001), 66–71

3. Burlet C., Vanbrabant Y., Goethals H., Thys T., Dupin L., *Raman spectroscopy as a tool to characterise heterogenite (CoO.OH), (Katanga Province, Democratic Republic of Congo)* (Spec. Acta Part. A, Vol. 80, 1, 2011), 138-147

4. Malzato M., *Environmental Liability in the Mining Sector : Prospects for Sustainable Development in the Democratic Republic of the Congo, In: J.P. Richards* ((Ed) Mining, Society and Sustainable World, Springer XXV, 2010) 289-321

5. Burlet C., Vanbrabant, Y., Goethals, H, *Heterogenite ($HCoO_2$) delafossite crystal structure and polytypic differentiation* (Spec. Acta Part. A., 2012), submitted

6. Llorca S. and Monchoux P., *Supergene cobalt minerals from New Caledonia* (Canadian Mineralogist, Vol.29, 1991), 149-161

7. Frenzel A. (Journal für praktische Chemie, 1872) 404

8. Buttgenbach H., *Les minéraux de Belgique et du Congo belge* (Dunod, Paris, Liège H. Vaillant-Carmanne SA, 1947)

9. Hey Max, *Cobaltic hydroxide in nature*, (Mineralogy Magazine 33, 1962), 253-259

10. Deliens Michel, *Les Oxydes hydraté de cobalt du Shaba Méridional* (Doctorate thesis - University of Louvain, Belgium, 1972)

11. Deliens M., Goethals H., *Polytypism of heterogenite*, (Mineralogical Magazine 39,1973)

12. Delaplane R., Ibers J., Ferraro J., Rush J., *Diffraction and Spectroscopic Studies of the Cobaltic Acid System* $HCoO_2$-$DCoO_2$ (J. Chem. Phys. 50, 1969), 1920

13. Cobalt Monograph, Centre d'Information du Cobalt, Brussels, 1960, p 30 seq.

14. Ferron J., Louis P., *Alternative options for the cobalt-manganese separation in hydrometallurgical acidic circuits*, Hydrometallurgy of Nickel and Cobalt 2009, 39th Annual Hydrometallurgical Meeting, Sudbury, Canada, 2009, 335-350

15. Ferron J. and D. Turner, "Purification of Cobalt-Containing Ore-Leaching Solution by Removal of Iron and Manganese with Oxidation by the O2-SO2 Gas Mixture", WO Patent 2000056943, 2000

16. Louis P., *"Cobalt Solution Purification: By Solvent Extraction or by Selective Precipitation?"*, Hydrometallurgy of Nickel and Cobalt 2009, 39th Annual Hydrometallurgical Meeting, Sudbury, Canada, 2009, 255-270

Ni-Co 2013
Edited by: Thomas Battle, Michael Moats, Violina Cocalia, Harald Oosterhof, Shafiq Alam, Antoine Allanore, Rodney Jones, Nathan Stubina, Corby Anderson, and Shijie Wang
TMS (The Minerals, Metals & Materials Society), 2013

POLYMET MINING CORPORATION'S NORTHMET PROCESS DEVELOPMENT

David Dreisinger[1], Mike Ounpuu[2], Dan Imeson[3], Niels Verbaan[3]

[1]PolyMet Mining Corp.
Suite 390 - 3600 Lysander Lane
Richmond, Canada V7B 1C3

[2]Consultant
22 Crestwood Ave
Peterborough, Canada K9J1M8

[3]SGS Minerals Services – Lakefield Site
PO Box 4300
185 Concession Street
Lakefield, Canada K0L 2H0

Keywords: NorthMet, Copper, Nickel, Platinum, Flotation, PLATSOL™

Abstract

Recent process development work for the recovery of copper, nickel, cobalt and precious metals (Au, Pt, Pd) from PolyMet's NorthMet deposit is described. The process uses a number of flotation process steps combined with hydrometallurgical treatment to produce a range of possible products including copper and nickel flotation concentrates for custom smelting, mixed nickel and cobalt hydroxide for refining and a precious metal precipitate for treatment.

Introduction

PolyMet Mining Corp. (PolyMet) is advancing the development of the NorthMet Project. The NorthMet deposit is located in northern Minnesota, adjacent to the historic Iron Range. The deposit was discovered in the 1960's and consists of a large, magmatic, disseminated sulphide, polymetallic deposit with values in Cu, Ni, Co, Zn, Au, Ag and Platinum Group Metals (PGM). The shallow, tabular orebody may be mined by open-pit methods with minimal pre-strip and a low waste:ore strip ratio.

The heart of the processing strategy revolves around the acquisition of the Erie Plant from Cleveland Cliffs. The Erie plant acquisition allows for PolyMet to proceed with a "brownfields" processing strategy with an existing 100,000 tpd crushing and grinding plant and associated infrastructure (roads, power, railway, water supply) including waste disposal facilities. The Erie plant was built in the 1950's for approximately $355 Million USD and operated until 2001 processing taconite ore. The acquisition of this plant and related infrastructure has provided most of the heavy equipment needed for production, reduced initial capital costs and simplified permitting. Initial plans are to use approximately 1/3 of the nominal capacity by processing 32,000 ton/day of ore.

Figure 1 below shows the NorthMet site in northern Minnesota, adjacent to the Mesabi Iron Range. Figure 2 provides an overview of the Erie plant site with the tailings area in the background. Figure 3 shows the indoor view of the Erie plant milling circuit with rod and ball mill lines.

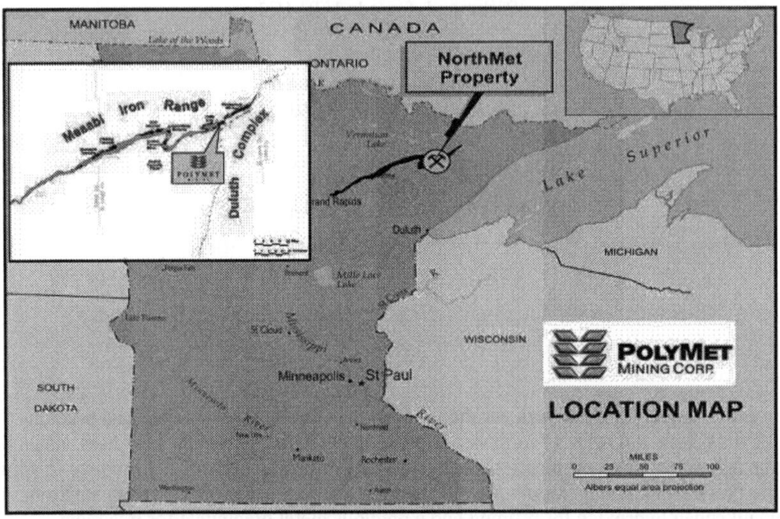

Figure 1. Location of the NorthMet Property in Northern Minnesota

Figure 2. The Erie Plant Site

Figure 3. The Erie Plant Milling Circuit with Parallel Rod and Ball Mill Lines

The original processing strategy at NorthMet [1] involved flotation of a bulk concentrate containing base and precious metals followed by PLATSOL™ hydrometallurgical processing. The PLATSOL™ process was developed to treat NorthMet bulk concentrate. The process directly extracts base and precious metals into solution followed by production of a PGM precipitate for refining by others, copper cathode via SX-EW, a gypsum product and a mixed hydroxide precipitate (MHP) containing nickel and cobalt.

Further work has been performed at SGS Minerals (Lakefield site) designed to further simplify the processing of NorthMet ore. This has involved two programs of testing. The first program tested the use of a "split cleaner" flotation flowsheet designed to produce a higher grade copper-nickel concentrate and a low grade pyrrhotite-containing concentrate. The higher grade copper-nickel concentrate is then processed to produce separate copper and nickel concentrates. The second program focused on processing the separate nickel concentrate or the separate lower grade pyrrhotite concentrate via PLATSOL™ in order to extract valuable metals. Importantly, the recovery of a PGM precipitate, an enriched copper concentrate and a mixed hydroxide precipitate of nickel and cobalt were produced following the PLATSOL™ treatment. This testing has given maximum flexibility to the production of various products from the NorthMet orebody.

Development of the Split Cleaner Flotation Flowsheet

Approximately 6 tonnes of split drill core were received at SGS Minerals (Lakefield). The entire sample was a single composite identified as C9. The composition of the sample was selected to be representative of the NorthMet ore. A series of seven open circuit batch tests and two locked cycle tests (LCT) were carried out to establish the flotation kinetics of the C9 composite and to optimize process variable such as regrind targets, reagents dosages, and reagent addition points in preparation of the pilot plant campaign.

The pilot plant only included the frontend of the circuit without the Cu/Ni separation stage due to the very low mass recovery into the combined CuNi 3rd cleaner concentrate (see Figure 4). The plant was operated for approximately 50 hours comprised of one 7-hour long commissioning run and a 43-hour extended run. The pilot plant flowsheet including reagent addition points and dosages is shown in Figure 4. Sodium isopropyl xanthate was used as a collector. Copper sulphate was used as an activator in the conditioning step prior to the flotation scavenger circuit. Carboxymethylcellulose (CMC) was used in combination with regrinding to clean the final concentrates.

The metallurgical performance of the pilot plant was excellent with high overall recoveries of the value metals, consistent with previous flotation testing programs since 1998. Flotation plant surveys were taken at regular intervals during the pilot plant to generate an overall calculated mass balance. This is summarized in Table I. The CuNi 3rd cleaner concentrate is enriched in copper and nickel and precious metals compared with the historical bulk concentrate. The pyrrhotite 3rd cleaner concentrate contains relatively low concentrations of base and precious metals but accounts for approximately 5% incremental recovery of copper and nickel, 22% sulfur recovery and 10-14% precious metal recovery. If the two 3rd cleaner concentrates were to be combined (which was not done during testing), the combined concentrate composition would again approximate the bulk concentrate composition referred to in earlier testing [1] with an overall recovery of 93.9% Cu, 63.2% Ni, 87.9% S and 75-80% of the precious metals. Note that 30-40% of the Ni in the NorthMet ore is as a silicate mineral phase that does not float.

Table I. Split Cleaner Flotation Pilot Plant Calculated Mass Balance.

Product	Wt.%	Assays (% or g/t)						Distribution (%)					
		Cu	Ni	S	Pt	Pd	Au	Cu	Ni	S	Pt	Pd	Au
CuNi 3rd Clnr Conc	1.48	18.2	3.41	27.7	2.41	10.5	1.33	89.1	58.0	66.1	65.1	69.4	61.3
Po 3rd Cleaner Conc	0.53	2.81	0.85	25.5	1.43	4.59	0.89	4.8	5.2	21.8	13.8	10.9	14.3
Combined Conc	2.01	14.1	2.74	27.0	2.15	8.97	1.21	93.9	63.2	87.9	78.9	80.3	75.6
Scavenger Tails	98.0	0.02	0.032	0.08	0.012	0.045	0.008	6.1	36.8	12.1	21.1	19.7	24.4
Feed	100	0.30	0.086	0.61	0.05	0.22	0.03	100	100	100	100	100	100

During the course of the pilot plant operation, the various concentrates were accumulated and stored for subsequent batch testwork. The CuNi 3rd cleaner concentrate was batch tested for Cu/Ni separation using the locked cycle methodology. The flowsheet involved regrinding, pH adjustment to + 11.5 with lime with aeration and then up to five cleaning stages (Figure 5). The three products from the separation flowsheet are a copper rougher tail, a copper first cleaner scavenger tail and a copper concentrate. Combining the flowsheets shown in Figure 4 and Figure 5 together gives the flexibility to produce up to four separate concentrates. Table II summarizes the composition of the concentrates collected during the SGS pilot campaign and subsequent locked cycle testwork. The copper concentrate grade is excellent at 28.2% Cu with relatively low Ni (0.57%). The copper first cleaner scavenger tails and the copper rougher tails have relatively high values of nickel and copper, along with precious metals. The pyrrhotite concentrate still represents a low value material for processing.

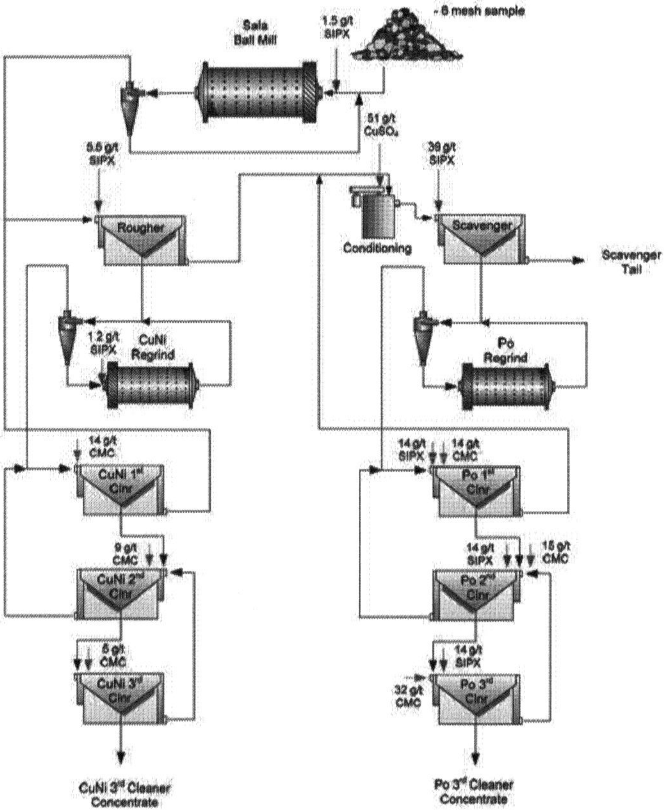

Figure 4. SGS Split Cleaner Flowsheet for Pilot Testing.

In summary, the split cleaner flowsheet with subsequent copper-nickel separation provides maximum flexibility for further sale of concentrate coupled with hydrometallurgical processing. Under one scenario, all the concentrates could be combined to provide a bulk concentrate for hydrometallurgical extraction via PLATSOL™. This approach would conform to the historical process flowsheets presented for NorthMet. Alternate scenarios would involve performing limited PLATSOL™ processing of selected concentrates (from the group Cu 1st Cleaner Scavenger Tails, Cu Rougher Tails and Po 3rd Cleaner Concentrate). The testing of the PLATSOL™ process on these materials, in combination with copper concentrate enrichment for copper recovery is described in the following section.

Table II. SGS Product Analysis for Concentrates Obtained from C9 Composite

Product	Assays (%)					Assays (g/t)		
	Cu	Ni	Co	S	Fe	Pt	Pd	Au
Cu 5th Cleaner Concentrate	28.2	0.57	0.038	30.2	32.7	1.59	10.7	1.71
Cu 1st Cleaner Scavenger Tails	8.92	6.65	0.32	19.6	27.0	3.66	12.0	5.81
Cu Rougher Tails	3.63	7.00	0.33	17.4	25.7	2.91	7.05	0.58
Po 3rd Cleaner Concentrate	3.40	0.89	0.063	25.2	41.1	1.36	4.21	0.80

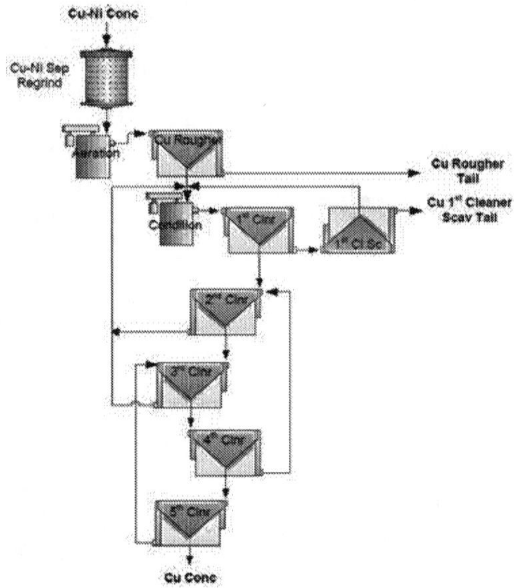

Figure 5. SGS Copper-Nickel Separation Flowsheet

Hydrometallurgical Processing of the NorthMet Flotation Concentrates

The hydrometallurgical testing program focused on two aspects. First, the PLATSOL™ treatment of pyrrhotite or Cu-Ni concentrates was tested and second, the use of copper metathesis (precipitation on a less noble sulphide) to recover copper dissolved in the PLATSOL™ process solution was studied. Figure 6 shows the flowsheet for the use of PLATSOL™ followed by copper concentrate enrichment. The chemistry and application of PLATSOL™ has been discussed previously [1]. The novel aspect of this flowsheet relates to the insertion of the copper concentrate enrichment step. The purpose of this step in the flowsheet is to (1) recover copper from the PLATSOL™ solution so as to avoid having to use the copper SX-EW technology at NorthMet and (2) to enrich the copper concentrate by metathetic copper precipitation and removal of iron, nickel and cobalt. The nickel and cobalt that is metathetically displaced from the copper concentrate will report to the mixed hydroxide precipitate product for sale. The metathetic copper concentrate enrichment step will therefore maximize the recovery of copper to a saleable copper concentrate with minimum nickel/cobalt content and maximize the saleable portion of the nickel and cobalt produced as high quality mixed hydroxide precipitate.

The metathesis reactions involved in the copper concentrate enrichment process are shown below.

$$CuFeS_2 + CuSO_4 = 2CuS + FeSO_4 \quad (1)$$
$$CuFe_2S_3 + 2CuSO_4 = 3CuS + 2FeSO_4 \quad (2)$$
$$Fe_7S_8 + 7CuSO_4 = 7CuS + S + 7FeSO_4 \quad (3)$$
$$NiS + CuSO_4 = CuS + NiSO_4 \quad (4)$$
$$CoS + CuSO_4 = CuS + CoSO_4 \quad (5)$$

A continuous pilot plant test of the flowsheet (Figure 6) was established at SGS Minerals. The first four steps in the process, PLATSOL™, PGM cementation, copper concentrate enrichment and copper precipitation were included in the continuous run. The copper depleted solution from copper precipitation was collected and a series of batch tests were used to complete the remaining flowsheet steps. The step by step results of the testing are summarized below.

PLATSOL™ Processing [1]

Two PLATSOL™ campaigns were carried out without interruption. Table III lists the main parameters of the two campaigns. Two different feed materials were used: campaign C1 used a nickel concentrate and campaign C2 used a pyrrhotite flotation. Due to limited feed availability the autoclave retention time was increased to ~ 120 min in the second campaign. Retention time calculations are based on an autoclave working volume of 33L (at 225 °C). Autoclave discharge (ACD) pulp was filtered on filter pans without thickening or flocculation and residue recycling was initiated as soon as sufficient leach residue cake was available. Filter cakes were repulped in ACD PLS and adjusted to target pulp density to reach target 100% solids recycling.

The switch over from campaign C1 to C2 was performed "on the fly" without stoppage. The solids recycling routine was not stopped in between, meaning that campaign C1 leach residue was recycled in the early stages of campaign C2. PLS from campaign C1 was collected 2 hours into campaign C2 (so 1 turnover) before a second collection vessel for C2 PLS was commissioned. Head assays of the two flotation products are included in Table IV and the composition of the synthetic autoclave dilution liquors is included in Table V. The pH of both liquors was adjusted to 2.

Table III. PLATSOL™ Autoclave Conditions

Campaign	Duration (h)	Conc	Dilution Liquor	Average T (°C)	O_2 Pressure (psig)	Feed % Solids	Solids Recycle (%)	Cl (g/L)	Ret. Time (min)
C1	12	NiCon	ADL1	224.3	100-110	9.5	96	8.5	64
C2	15	PoCon	ADL2	225.4	100-110	9.2	111	9.6	119

Table IV. Flotation Concentrate Assays

Campaign	Sample	Assay (%)												Assay (g/t)		
	Type	Ni	Cu	Fe	Co	Al	Mg	Cr	Ca	Zn	Si	S	S^{2-}	Au	Pt	Pd
C1	Ni Con	3.44	5.66	34.7	0.18	1.82	1.91	0.07	1.16	0.06	5.68	24.4	23.3	0.9	3.35	10.3
C2	Po Con	0.8	2.17	32.4	0.04	1.39	2.07	0.04	0.84	0.07	5.21	25.3	23.2	0.62	0.97	3.32

Table V. Autoclave Dilution Liquor Composition (Synthetic Solution)

Campaign	Sample Type	Analysis (mg/L)							
		Ni	Cu	Fe	Co	Mg	Ca	Zn	Cl
C1	ADL1	20,600	0.6	1.9	880	4000	590	390	10400
C2	ADL2	5,290	0.3	3.1	210	4000	770	390	10400

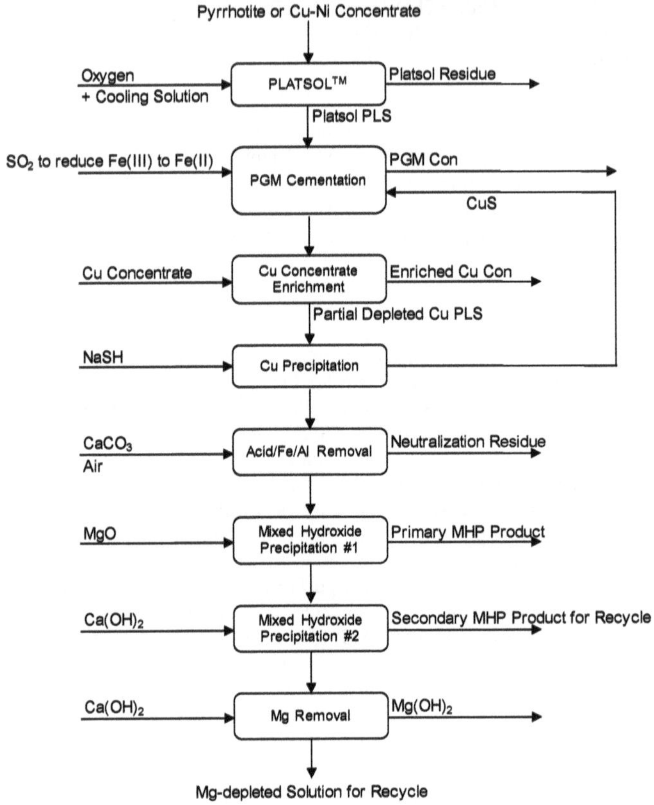

Figure 6. Hydrometallurgical Flowsheet for PLATSOL™ Leaching of Pyrrhotite or Cu-Ni Concentrates followed by Copper Concentrate Enrichment

Metal extractions were calculated by an assay-based Si-tie method. By "tying" the residue and feed for the element to the ratio of Si assay in residue and feed, a correction is made for weight gain or loss and since Si only dissolves to minor extent (3.1% in C1 and 2.1% in C2), the effect of dissolved Si is minor. Nevertheless a correction for dissolved Si is included in the calculation. Results of this calculation are included in Table VI for each campaign. For the residue assay an average of the final 3 grab samples have been used. The percentage of sulphide oxidation is also calculated through this method and resulted in sulphide oxidation levels of 95.5% and 97.4% in campaigns C1 and C2 respectively. Base metal extractions were 97.0% Ni and 99.1% Cu and 98.1% Co in campaign C1 and 95.5% Ni, 99.0% Cu and 96.7% Co in campaign C2. Precious metal recoveries were somewhat lower: 91.0% Au, 87.6% Pt and 92.0% Pd in campaign C1 and 84.0% Au, 94.2% Pt and 95.9% Pd in campaign C2. Recoveries for all metals would be expected to be higher with increased recycle of residue in the "Recycle PLATSOL" configuration [1].

Table VI. PLATSOL™ Metal Extraction

Campaign	Extraction (%)													
	Ni	Cu	Fe	Co	Al	Mg	Cr	Ca	Zn	Si	S^{2-}	Au	Pt	Pd
C1	97.0	99.1	-0.4	98.1	25.5	33.8	10.1	-66.4	97.4	3.1	95.5	91.0	87.6	92.0
C2	95.5	99.0	3.7	96.7	45.0	61.4	-13.2	-12.3	99.1	2.1	97.4	84.0	94.2	95.9

The SGS autoclave has six compartments. Samples were taken across the autoclave toward the end of each of the campaigns. The profiles for base metals, silica and sulfur were virtually constant meaning that most of the reactions of these species occur in the first compartment of the autoclave. The PGM extractions though show the expected increase across the autoclave. This is shown in summary form in Table VII.

Table VII Profile Extractions for Precious Metals for Campaigns C1 and C2

Compartment	Campaign 1 Extraction (%)			Campaign 2 Extraction (%)		
	Au	Pt	Pd	Au	Pt	Pd
1	58.0	73.5	74.3	69.1	73.2	71.1
2	59.9	71.4	76.1	72.4	83.1	84.1
3	78.6	75.5	83.6	74.6	89.4	91.5
4	74.8	83.5	89.6	74.6	91.5	93.4
5	83.2	89.5	94.0	76.8	92.9	94.8
6	85.1	91.2	95.3	79.0	92.9	94.2

Precious Metal Precipitation

The filtrate from the PLATSOL™ autoclave discharge was treated first for precious metal recovery. The precious metal precipitation circuit consisted of a preheat tank, SO_2 reduction tank (PGM1) and two precipitation tanks (PGM2 and 3). The preheat tank was used to preheat PLS from ambient to 95°C. In PGM1 ferric iron was reduced by SO_2 gas to form ferrous iron. The addition of SO_2 was controlled by online ORP measurements and hooked up to a solenoid valve that allowed SO_2 gas to flow into the reactor when the online ORP reading was above the programmed set point (400 mV versus Ag/AgCl). In tank PGM2 synthetically produced CuS solids were pumped in to maintain a target 10 g/L CuS concentration. Dissolved PGMs precipitated onto the CuS in tanks PGM2 and PGM3 and were filtered straight onto Buchner filters. Filtered PGM3 solids were re-pulped in PGM3 filtrate and recycled to PGM2 to reduce the flow of fresh CuS into the circuit. The target level of 10 g/L CuS was maintained throughout both campaigns of which ~90% was recycled CuS.

The results of the precipitation were excellent with <0.01 mg/L of Au, Pt, Pd detected in the final product solution. The composition of the precious metal PLS solution, the synthetic CuS and the final precipitates recovered are shown in the tables below.

Table VIII. Precious Metal Feed Liquor Compositions

Campaign	Analysis (mg/L)														
	Ni	Cu	Fe(T)	Fe^{2+}	Co	Al	Mg	Cr	Ca	Zn	Si	Cl	Au	Pt	Pd
C1	23000	7500	1970	50	1100	820	4800	21	540	480	430	9620	0.05	0.18	0.72
C2	11000	4800	5500	78	540	1900	6600	32	670	520	350	10700	0.04	0.20	0.63

Table IX. Synthetic CuS Assay

Analysis (%) or g/t														
Ni	Cu	Fe	Co	Al	Mg	Cr	Ca	Zn	Si	S	S^{2-}	Au	Pt	Pd
0.004	63.2	0.01	<0.002	<0.008	<0.003	<0.004	<0.04	<0.001	0.06	31.1	29.9	0.07	<0.02	0.4

Table X. Precious Metal Precipitate Assays

	Analysis (%) or g/t													
	Ni	Cu	Fe	Co	Al	Mg	Cr	Ca	Zn	Si	S	S^{2-}	Au	Pt	Pd
Cake 1	0.0	61.2	0.1	<0.002	0.03	0.01	<0.006	<0.05	0.0	0.1	30.9	25.7	4	25	67
Cake 2													5	28	85
Cake 3	0.1	42.7	0.2	0.004	0.58	0.02	<0.006	<0.05	0.0	0.2	25.9	20.3	11	66	170
Cake 4	0.0	59.6	0.6	<0.002	0.02	0.01	<0.004	<0.04	0.0	0.2	30.9	28.1	16	90	244

Copper Concentrate Enrichment

The solutions from precious metal precipitation (now free of precious metals) were used in a program of copper concentrate enrichment. Copper concentrates from SGS flotation programs on NorthMet ore, corresponding to the PLATSOL feed source concentrate (either Cu-Ni concentrate or pyrrhotite concentrate) were obtained and used for the test (Table XI).

Table XI. Copper Concentrate Composition for Copper Enrichment Tests

Campaign	Analysis (%) or g/t														
	Ni	Cu	Fe	Co	Al	Mg	Cr	Ca	Zn	Si	S	S^{2-}	Au	Pt	Pd
C1	0.38	30.5	33.5	0.018	0.09	0.47	<0.004	0.07	0.038	1.23	32.7	30.5	1.32	1.13	5.76
C2	0.64	30.5	31.5	0.025	0.15	0.36	<0.006	0.05	0.056	1.21	31.1	29.8	1.6	1.44	9.24

In campaign C1 the target ratio of 0.8 ton of copper concentrate over nickel concentrate was set and in campaign C2 the target was set at 1.0 ton of copper concentrate over pyrrhotite concentrate. The flow of solution was then matched to an addition of copper concentrate. Copper enrichment was performed in a series of three tanks (CuE1, CuE2, CuE3) at ~95 °C in the first tank, cooling to 50-60 °C in the last tank. Fresh concentrate was added continuously to CuE1 tank along with a recycle of solids recovered by filtration of CuE3 discharge. The nominal recycle rate was 1 t of recycle solids per t of fresh concentrate. After some initial adjustment of retention time, a total retention time of 7.1 hours was used.

The copper concentrate enrichment process was monitored by following the ORP (mV vs Ag/AgCl) of the slurry and correlating with the soluble copper content in solution. Figure 7 shows the relationship is very strong, especially in tank CuE3. By targeting a low ORP in the CuE3 tank it was possible to maintain a low residual copper in solution (about 250 mg/L Cu).

Solid composition of the copper concentrate was analyzed over time in each of the two campaigns. The results (Table XII) indicate that the copper concentrate was slightly enriched in copper content and depleted in iron, nickel and cobalt. The calculated values were 12.5 and 8.6% copper enrichment, 18.6 and 11.1% iron removal, 5.6 and 29.1% nickel removal and 1.8 and 20.0% cobalt removal for campaigns 1 and 2. More than 95% of the copper from the PLATSOL™ autoclave solutions was removed during the enrichment process, validating the process concept.

Table XII. CuE3 Solid Assays

Campaign	Analysis (%) or g/t														
	Ni	Cu	Fe	Co	Al	Mg	Cr	Ca	Zn	Si	S	S^{2-}	Au	Pt	Pd
C1	0.33	26.5	30.4	0.02	0.21	0.66	<0.004	0.1	0.062	1.95	31.4	29.8	Nss	Nss	Nss
	0.31	31.2	24.3	0.02	0.11	0.39	<0.004	0.06	0.045	1.1	31.0	30.9	1.3	1.1	5.2
	0.39	30.7	30.3	0.02	0.09	0.33	<0.004	<0.04	0.043	0.9	31.6	31.6	1.7	1.5	6.4
C2	0.39	30.7	30.3	0.02	0.09	0.33	<0.004	<0.04	0.043	0.9	31.6	31.6	1.7	1.5	6.4
	0.52	30.5	28.5	0.02	0.12	0.38	<0.004	<0.04	0.049	1.11	32	32	1.6	1.3	7.7
	0.55	29.7	29.4	0.03	0.14	0.41	<0.004	<0.04	0.054	1.23	32.7	31.3	1.6	1.3	8.5

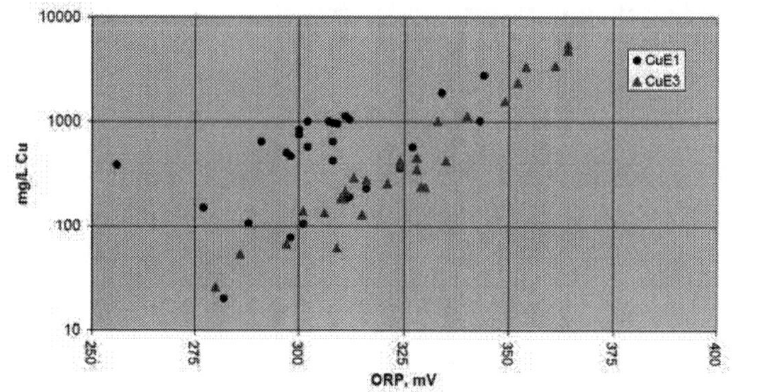

Figure 7. Soluble Copper versus ORP (mV versus Ag/AgCl) for CuE1 and CuE3 Tanks.

<u>Copper Removal with NaSH</u>

Copper removal from the copper enrichment solutions was performed by adding a 37.5 g/L NaSH solution to precipitate CuS. The residence time was set at approximately 2 hours using a two tank arrangement. The original control strategy was to ratio the flow of NaSH solution to the flow of feed solution. However, this strategy was replaced by ORP control, as visualized in Figure 8 below. As long as the ORP of the second tank (NaHS2) was kept below 250 mV, the copper in the final solution was maintained at less than about 10 mg/L in solution. The CuS recovered from this process would normally be recycled to the precious metal precipitation process. However, in this case the duration of the pilot plant was too short to allow for this integration.

Batch Test Results on Iron Removal, Mixed Hydroxide Precipitation and Magnesium Removal

The NaHS filtrate collected during the pilot plant was treated in a series of successive bulk batch tests consisting of iron/aluminium precipitation, 2-stage MHP precipitation and magnesium removal. All of these processes were operated in batch form. Continuous testing of these unit operations has been carried out previously in the context of the PLATSOL™ processing of NorthMet bulk concentrate. The results are summarized below.

Iron/acid removal, at a pH of 3.8 and with continuous air sparging to oxidize ferrous iron, led to iron levels of <2 mg/L Fe and 5 mg/L Fe for campaigns C1 and C2, respectively. Aluminum levels were <2 mg/L Al in both campaigns. Due to the nature of batch tests, nickel and cobalt losses were probably slightly inflated at 5% Ni and 4% Co in campaign C1 and 8% Ni and 4% Co in campaign C2.

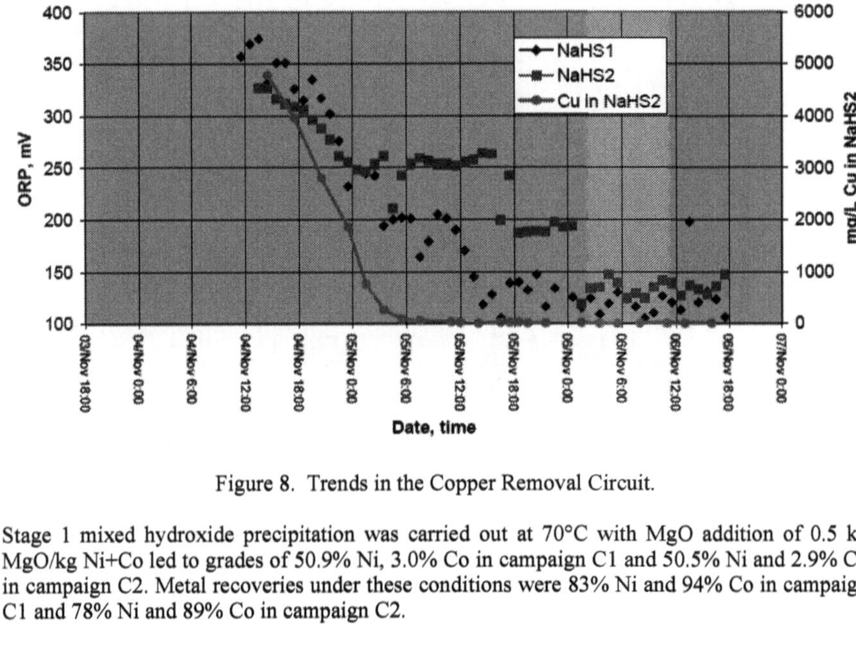

Figure 8. Trends in the Copper Removal Circuit.

Stage 1 mixed hydroxide precipitation was carried out at 70°C with MgO addition of 0.5 kg MgO/kg Ni+Co led to grades of 50.9% Ni, 3.0% Co in campaign C1 and 50.5% Ni and 2.9% Co in campaign C2. Metal recoveries under these conditions were 83% Ni and 94% Co in campaign C1 and 78% Ni and 89% Co in campaign C2.

Stage 2 mixed hydroxide precipitation was carried out 65°C with controlled addition of hydrated lime to reduce nickel levels to <10 mg/L. A pH of 7.5 (campaign C2) to 7.8 (campaign C1) was determined to reduce nickel to target levels. Equivalent cobalt values were <0.3 mg/L Co in both campaigns. The composition of the precipitate ranged from 20.8% Ni to 21.9% Ni and from 0.29% Co to 0.38% Co. Mg co-precipitation was very low (2 – 2.5% Mg in precipitate).

Magnesium removal of MHP Stage 2 filtrate targeted 50% removal at 50°C. In campaign C1, this was accomplished at pH 8.6 and in campaign C2 at pH 8.3.

Conclusions

The processing of the NorthMet deposit of PolyMet Mining using a combination of flotation and hydrometallurgical unit operations has been presented. The development of a split-cleaner flotation circuit followed by a copper-nickel separation process allows for generation of up to four concentrate products;

- Cu 5th Cleaner Concentrate
- Cu 1st Cleaner Scavenger Tails
- Cu Rougher Tails
- Po 3rd Cleaner Concentrate

The Cu 5th cleaner concentrate represents a high grade (~30% copper concentrate) while the Cu 1st cleaner scavenger tails and the Cu rougher tails are nickel rich and may individually or in combination be processed at a custom nickel smelter facility. Alternately these nickel rich materials could be processed via PLATSOL™. The pyrrhotite 3rd cleaner concentrate is low grade and unlikely to be toll smelted. This material is ideally suited to PLATSOL™ process treatment.

The split-cleaner flowsheet was demonstrated by treating 6 tonnes of NorthMet ore through an SGS pilot plant. An overall recovery of 93.9% of the copper and 63.2% of the nickel reported to the bulk and pyrrhotite concentrates along with 75-80% of the precious metals.

The PLATSOL™ treatment of a nickel rich concentrate and a pyrrhotite 3rd cleaner concentrate was demonstrated. The PLATSOL™ autoclave process was combined with precious metal precipitation, copper concentrate enrichment, copper removal, iron/aluminum removal, mixed hydroxide precipitation of nickel and cobalt and finally magnesium removal. The copper concentrate enrichment process proved to be a novel addition to the NorthMet flowsheet. The copper in solution from the autoclave processing was recovered by metathesis on the copper concentrate in the enrichment process. The solvent extraction and electrowinning of copper process applied to precious metal free PLATSOL™ solutions is no longer required by using this metathesis process route. This development provides maximum operational flexibility for treatment of the NorthMet ore.

Acknowledgements

The authors would like to thank PolyMet Mining Corp. for permission to publish this work.

References

1. Dreisinger, D.B., Molnar, R., Mezei, A., Baxter, K. and Wardell-Johnson, M., "The Recovery of Nickel and Cobalt from the NorthMet Deposit Using Bulk Flotation and Chloride Assisted Pressure Leaching", Proceedings of ALTA 2010 Ni-Co-Cu Forum, Perth, May, 2010.

Ni-Co 2013
Edited by: Thomas Battle, Michael Moats, Violina Cocalia, Harald Oosterhof, Shafiq Alam, Antoine Allanore, Rodney Jones, Nathan Stubina, Corby Anderson, and Shijie Wang
TMS (The Minerals, Metals & Materials Society), 2013

Talvivaara Nickel Mine – from a project to a mine and beyond

Marja Riekkola-Vanhanen[1], Lauri Palmu[1]
[1]Talvivaara Mining Company Plc, Ahventie 4 B 47, FI-02170 Espoo, Finland

Key words: sulphide ores, bioleaching, bacteria, hydrometallurgy

Abstract

The first commercial application of bioleaching in Europe is the Talvivaara Sotkamo Mine in Finland. The ore contains low grade sulphide minerals. Utilization of the deposit has been extensively studied for over 20 years. Bioheapleaching technology was chosen for the extraction of nickel from the ore based on its favorable capital, operational costs and good performance data obtained in a large on-site pilot trial.

Building of the industrial scale plant was started in 2007 and metals recovery started in 2008. In full production the annual nickel production will be approximately 50,000tpa. In addition, the mine will also produce zinc, copper and cobalt.

Talvivaara will start to recover uranium from the PLS solution. A novel solvent extraction process will be used to extract uranium for the yellowcake production.

An update of the existing operation, uranium extraction and the plans for the production expansion and upgrading beyond the 50,000tpa will be discussed.

Introduction

The Talvivaara Deposits are located approximately 350 km south of the Arctic Circle. The climate is characterized by extreme seasonal changes in temperature and light. Average daylight hours in summer (June to August) are 20 hours whereas in winter (November to March) this is reduced to seven hours. Winter temperatures range from 0 to minus 20 °C with exceptional short-lived cold spells of below 30 °C. The average annual snowfall is 0.7 m. Talvivaara deposits, Kuusilampi and Kolmisoppi comprise one of the largest sulphide nickel resources in the world with 1121 Mt in measured and indicated resource categories (Table 1).

Table I: Total mineral resources at 0.07% nickel cut-off in 2010.

Category	Mt	Nickel %	Zinc %	Cobalt %	Copper %
Measured	432	0.23	0.50	0.02	0.13
Indicated	689	0.23	0.50	0.02	0.13
Subtotal	1,121	0.23	0.50	0.02	0.13
Inferred	429	0.20	0.47	0.02	0.12
Total	1,550	0.22	0.49	0.02	0.13

Talvivaara's extraction technology is bioheapleaching, which is widely used for other metals, notably copper and gold. The company has demonstrated the viability of bioheapleaching technology for the extraction of nickel on site using Talvivaara ore (Riekkola-Vanhanen, 2007; Riekkola-Vanhanen, 2010). The leaching process generates heat and is therefore suitable in subarctic conditions. The planned annual nickel production of 50,000 tonnes is anticipated to be reached in 2014 – 2015. As by-products the mine will also produce approximately 90,000 tonnes of zinc, 15,000 tonnes of copper and 1,800 tonnes of cobalt.

Deposit Geology

Talvivaara is situated in the southern part of the Early Proterozoic Kainuu Schist belt. The Ni-Cu-Co-Zn mineralisations are hosted almost entirely by high-grade metamorphosed and intensively folded black schist. About 90 % of the ore is hosted by black schist and the rest by metacarbonate rocks. The main mineral assemblage in the black schist is quartz, micas, graphite and sulphides. The valuable metal containing sulphides are pyrrhotite, pyrite, sphalerite, pentlandite, violarite and chalcopyrite. The average content of the ore using the cut-off of 0.07% of nickel is 0.23% Ni, 0.50% Zn, 0.13% Cu, 0.02% Co, 10.3% Fe and 8.4% S. The distribution of nickel in different sulphides is pentlandite 66%, pyrrhotite 33% and pyrite 1%. The distribution of cobalt is pentlandite 11%, pyrrhotite 26% and pyrite 63%. All copper is in chalcopyrite and zinc in sphalerite.

The Talvivaara deposit consists of two ore bodies, Kuusilampi and Kolmisoppi, about 3 km apart. Known dimensions for the Kuusilampi deposit are length 2600 m, width 40 – 1000 m and depth 600 m, and for Kolmisoppi 1500 m, 30 – 350 m and 300 m respectively. The metal distribution in the deposits is homogeneous.

Project History

Geological work in Talvivaara area commenced in the early 1900's when geological mapping of bedrock was conducted by Geological Survey of Finland (GSF) and continued through 1951. Between 1951 and 1962, various Finnish companies carried out exploration activities in the area. The result of this phase led GSF to commence detailed exploration in the area in 1977. As a result of the work done by GSF, two polymetallic deposits, Kuusilampi and Kolmisoppi, were established. Outokumpu acquired the deposits from the Finnish State in 1985 and the mining license covering the Talvivaara Deposits was granted to Outokumpu in 1986. Between 1989 and 1992, Outokumpu focused on further geological and metallurgical work on the Talvivaara Deposits.

Despite extensive exploration in the Talvivaara area that established the presence of large deposits, it was concluded at the time that exploitation of the relatively low-grade deposits would not be economically viable using conventional techniques. Therefore, the Talvivaara deposits remained unexploited until the Talvivaara Project acquired the rights to the deposits in February 2004 and continued the work by focusing on sampling for metallurgical purposes and investigating areas for the pilot trial.

With the deposits Talvivaara was assigned the right to use all exploration data and research documentation relating to various process options studied. Bioheapleaching seemed to be the most economical option. Construction of a 17,000 tonnes on-site pilot heap was started in May 2005 and initial bioheapleaching in August that year. A metals recovery pilot using the PLS (pregnant leaching solution) from the process was run in 2006 at the OMG plant in Kokkola. The environmental permit was granted in March 2007 followed by the commencement of the construction phase of the project. In April 2008 ore mining started at Kuusilampi open pit and in July 2008 bioheapleaching was initiated. The first metal sulphides were produced at the plant in October 2008.

Talvivaara Black Schist Ore Bioleaching Tests

Talvivaara's application of the bioheapleaching technology has its origins at Outokumpu Research, where it has been developed in conjunction with Talvivaara ore since 1987. All the rights relating to the accumulated research and development data on bioheapleaching were transferred to Talvivaara in connection with the acquisition of the mining licenses in February 2004.

In order to validate the bioheapleaching process on a larger scale, an on-site pilot study was commenced at Talvivaara in May 2005 (Riekkola-Vanhanen, 2007; Puhakka et al., 2007). Heap irrigation was started in August 2005 and metals recovery in November 2005. The pilot heap was constructed of 17,000 tons of Talvivaara ore. The particle size used was 80% < 8 mm. In laboratory tests the temperature remained at the room temperature level. The size of the pilot was large enough to generate heat and the rise of temperature could be observed. The rise was due to the oxidation of the large quantity of pyrrhotite and pyrite in the ore. The temperatures measured inside the heap varied from 30 °C to nearly 90 °C.

The primary leaching of the 8 meters high pilot heap (heap pad of 50 x 100 m) took one and a half years and yielded nickel and zinc recoveries of about 90%. The heap was reclaimed during the coldest time in winter 2007 and moved to a new leaching pad. The temperature of the ore was under 0 °C when irrigation of the secondary heap was started at the end of February. The temperature started to rise immediately and was at the level of 80 °C in April and started to decrease gradually after that to the level of 20 – 40 °C, where it remained to the end of November 2008.

The main reason to install a secondary leaching phase is to get better recoveries of copper and cobalt. Copper is in chalcopyrite and the main part of cobalt in pyrite. As sulphide minerals have semiconductor properties, galvanic interactions appear when there is an electrical contact among mineralogical phases. During dissolution of a mineral assembly of different sulphides, those minerals that have the highest rest potentials behave as cathodes which means that they are galvanically protected and their leaching is hindered until the minerals with lower rest potentials have been leached. The electrochemical potentials of chalcopyrite and pyrite are higher than the ones of pyrrhotite, pentlandite and sphalerite (Riekkola-Vanhanen and Heimala, 1993). The leaching recoveries from the pilot plant were substantially better than anticipated based on laboratory and other smaller pilot results and confirmed the viability of the bioheapleaching of the Talvivaara ore. It was assumed that most of nickel and zinc will be leached in the primary heap within one and one-half years. Nickel recovery reached 80% within 400 days. The corresponding zinc recovery was 80% in 480 days. After secondary leaching the recoveries of nickel and zinc were 99%, the ones of copper and cobalt 22 and 35%. The pilot plant showed that the assumed recoveries for nickel and zinc can be reached (Puhakka et al., 2007).

Production Process

Overview

The mining method selected for use is large scale open pit mining. The crushed (80% < 8mm) ore is agglomerated and conveyed to the heap pad. After 13 to 14 months of bioleaching on the primary pad, the leached ore is reclaimed, conveyed and re-stacked onto the secondary heap pad. After secondary leaching, the barren ore will remain permanently in the secondary heaps. In the metals recovery process, metals are precipitated as sulphides from the PLS using gaseous hydrogen sulphide (Figure 1). The resulting products are intermediates to be transported for further processing in refineries operated by the company's customers.

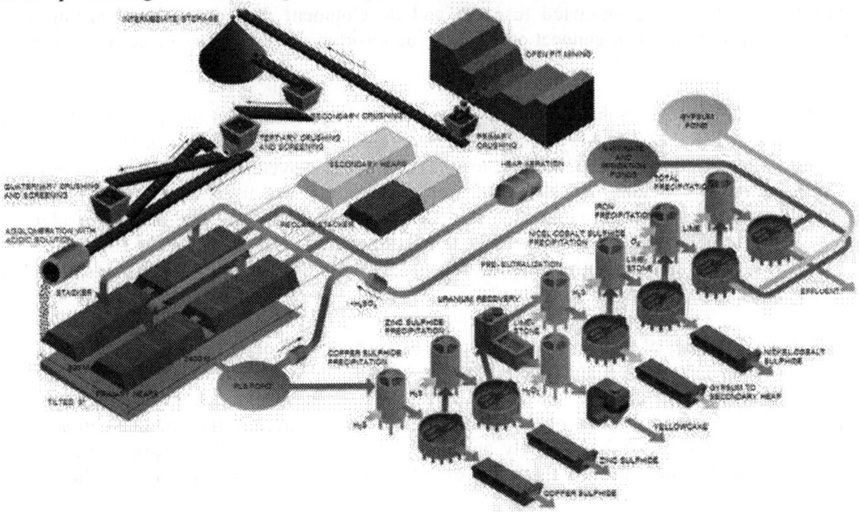

Figure 1. Process flow sheet of Talvivaara Sotkamo process.

Open Pit Mining

Planned annual ore production is approximately 24 million tonnes. Sufficient areas to extend the pit will be prepared in subsequent years, normally a year prior to when mining is scheduled to commence. Any overburden or moraine not required for road, perimeter wall or other construction, is stockpiled for later use in rehabilitation. Ore and waste are extracted using conventional large scale open pit drill and blast methods. A fleet of self-propelled diesel hydraulic track drill rigs is used. Fragmentation by blasting is preferred to crushing because of lower costs. It is also beneficial to create more fracturing at blasting stage which increases the surface area and therefore improves leaching solution entry and consequently leaching efficiency.

Materials Handling

During the process, the ore is crushed and screened in four stages into p80<8mm. After primary crushing, the ore is conveyed to the fine crushing station where it is crushed and screened in three phases. All material <10 mm continues to agglomeration. It takes place in a rotating drum, where PLS is added to the ore in order to consolidate the fine particles with coarser particles. This preconditioning step makes the ore permeable to air and water for bioheapleaching. After agglomeration, the ore is conveyed and stacked from eight to ten meters high on the primary heap pad. After 13 – 14 months of bioleaching on the primary pad, the leached ore is reclaimed, conveyed and re-stacked onto the secondary heap pad, where it is leached further in order to recover metals from those parts of the primary heaps, where leaching solution has had poor contact. Such areas include, for example, the slopes of the heaps and areas between channels formed by the circulating leaching solution. Also the leaching of copper from chalcopyrite and cobalt from pyrite continue in the secondary leaching. After secondary leaching, the barren ore is expected to remain permanently in the secondary heaps. The heaps will be covered air- and water-tight and revegetated.

Bioleaching

The bacteria used in the Talvivaara process grow naturally in the ore and the bioleaching process is technically viable by creating optimal conditions for growth of the bacteria. The designed parameters for bioheapleaching are based on laboratory test work, experience from other mine sites and pilot heap operation.

The primary heap comprises of two heap pads, a lower section and an upper section and a 40 m wide corridor between them reserved for a service area, pipelines and two fixed conveyor lines. The heap pads are 2400 m long and 400 m wide, the entire surface area is 210 hectares. Each section is further divided into two sectors. The heap pads are designed so that in the longitudinal direction they are horizontal and in the cross direction they slope toward the PLS ponds. The cross slopes vary from 2% to 3.5 %. The heap pad is equipped with piping, laid about 1.5 meters from the bottom of the pad, through which low-pressure fans supply air to the stacked ore. From the top, the heap is irrigated with leaching solution, which is collected from the bottom of the heap. Primary heap is irrigated at a rate of 5 l/m^2xh. The pH value of the irrigation solution is adjusted to 1.7 – 2.0 and the solution is circulated evenly on the heap surface. PLS is collected with a drainage system and discharged to the PLS collection ponds. About 10 – 20 % of the solution is pumped to the metals recovery plant and the rest of the solution is circulated back to the heap. The amount of solution is kept constant by adding process water and the pH value is adjusted. Several physico-chemical and microbiological process parameters, such as the pH of the irrigation solution, the rate of irrigation, the rate of removal of the PLS, and the rate of aeration are continuously monitored and modified in order to enhance and speed up the metals recovery to the solution.

The primary heap pad is constructed as a dynamic pad. This means that after the first stacking cycle, the ore is reclaimed and conveyed to the secondary leach pad for further leaching. While the reclaimer moves ahead reclaiming the ore, the stacker follows behind it, stacking new ore from agglomeration. This cycle goes on continuously.

The secondary leaching pads are constructed on top of waste rock dumps. The benefit of the arrangement is multiple lining systems and reduced earthwork quantities. It also reduces the final footprint of the operation and final rehabilitation costs. Both heap pads and waste dumps are lined with bentonite and plastic layers to prevent leakages.

Metals Recovery

In the metals recovery process, metals are precipitated from the PLS using gaseous hydrogen sulphide and pH adjustment (Figure 2). Solid waste is formed in pre-neutralisation, iron removal and total precipitation. In pre-neutralisation the acidity in the feed and the acid formed in copper and zinc sulphide precipitation are neutralized. At the same time a considerable part of aluminium in the solution is precipitated. The underflow from the thickeners is filtrated on belt filters. The inert precipitate is transported to the waste rock area as a filter dry product. Iron is removed from the solution as goethite and/or hydroxide which are pumped to the gypsum pond. The final precipitation is the outlet for manganese and magnesium in the process. Remaining iron and gypsum also precipitate in this step. After iron removal the solution goes to the final precipitation where the pH value is raised to a value of 10 with slaked lime. The remaining metals precipitate as hydroxides. The underflow from the thickeners is pumped to the gypsum pond. The slurry has to be carefully analyzed to ensure its suitability for the gypsum pond. The overflow from the thickener is recycled back to the heap leaching circuit.

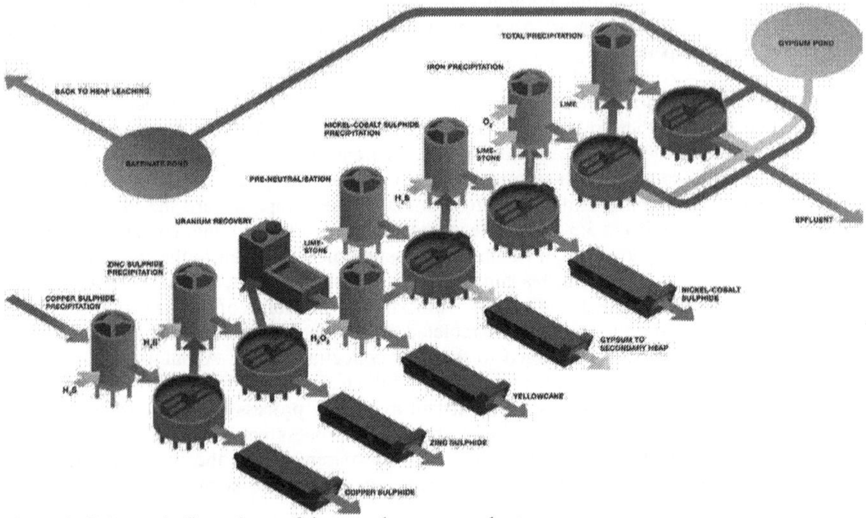

Figure 2. Schematic flow sheet of the metal recovery plant.

The resulting products of metals recovery are intermediates, copper and zinc sulphides and a mixed nickel cobalt sulphide. These intermediates are transported for further processing in refineries operated by Talvivaara's customers.

Water Management

The water management includes all relevant pipelines, ponds and pump stations related to the processes outside the metals recovery plant area. It also includes surface water management, effluent treatment and other surface water construction projects.

Water management plays an important role at the operation. The most important component is the recycling of the leaching solution from the irrigation pond to the heap and thereafter to the PLS pond. From the PLS pond, approximately 90 per cent of the solution is recycled back to irrigation to increase the metal grade, while about 10 – 20 percent is lead to metals recovery. After metals precipitation, the remaining solution goes into the raffinate pond for pH adjustment and is reused to irrigate the heaps.

Production Ramp-up

Year 2008 marked the transformation of Talvivaara from a project to an operating mine. During the year, the construction project went through its most intensive periods, providing work for some 2,000 contractor employees during the peak installation months in late summer and early autumn. As a result of this sizeable and multidimensional effort, the project reached all its operational milestones on time enabling start-up of production processes in stages beginning with the first blast of ore in April. Some final construction and installation work remained to be completed in the following years. In total the amount of work required to bring the mine into full operation was approximately 3,500 man-years.

Talvivaara's focus has remained firmly on production ramp-up throughout the years 2009 - 2012. Optimization of the already operating equipment and processes continued, and pre-requisites for full-scale production were fulfilled through the commissioning of the second production lines.

The scalability of the production processes and progress in ramp-up were confirmed by the production volumes achieved in 2011: 16,087 t of nickel (2010: 10,382 t, 2009: 735 t) and 31,815 t of zinc (2010: 26,462 t, 2009: 3,133 t). Although the production fell short of the originally budgeted figures, the ramp-up trend seen has been encouraging with a relatively steady increase in nickel production.

The metal plant operations have faced a series of technical challenges, ranging from a hydrogen plant failure to insufficient hydrogen sulphide capacity caused by installation faults in the hydrogen sulphide generator. Also, hydrogen sulphide emissions forced production levels to be restricted for several months because of the odour discharges. A break-through in controlling the odours was made through the use of hydrogen peroxide as the odour controlling chemical.

Mining has performed well. The crushing processes in materials handling have caused the biggest problems in the process. The primary crusher had to be fitted with a new mantle. The fine crushing circuit had to be redesigned and more crushers had to be added.

The installation and commissioning of the primary heap reclaiming and secondary heap stacking systems have represented a major challenge. Both systems were started up in autumn 2010 and the secondary stacker has since then been in production with good results. Commissioning of the primary heap reclaiming equipment has however been slower, resulting in reduced overall crushing and stacking output.

In bioheapleaching the building of the first heap sector proceeded much too slowly due to difficulties in crushing. Irrigation and aeration could only be started when about 25m of heap had been built. The agglomerates started to break apart during that time. The particle size was also smaller than planned. When irrigation and aeration were started, there were difficulties in getting an even airflow through the heap. With improved crushing systems the bioheapleaching started to progress according to expectations. The primary heap was fully stacked for the first time in November 2010, and secondary leaching started with good results. In process development, particular attention has been paid to improve aeration. As a result, nickel grades in leach solution have increased especially in the newer heap sections, reaching levels well above 3 g/l.

The quality of zinc and nickel/cobalt sulphides has been very good. Nickel/cobalt sulphide contains about 50% of nickel and 1.5% of cobalt, the moisture content is under 20%. Zinc sulphide contains about 64% of zinc and its moisture content is about 10%. The impurity levels in both sulphides are very low. Copper concentrations have just started to rise in the PLS and saleable amounts of copper sulphide production started in 2012.

Personnel

The total number of employees at Talvivaara is about 500. The personnel are mostly recruited locally from the Kainuu region, where Talvivaara is the largest provider of new job opportunities. In its recruitment process, Talvivaara has sought to maintain a representative staff age structure, in spite of the exceptionally vigorous rate of recruitment. Although the mining industry has conventionally been male dominated, Talvivaara seeks to hire employees representing both genders. This has however proven difficult due to the limited number of female applicants.

Sustainable Development

Talvivaara has continued to develop its operations in line with its sustainable development policy which emphasizes continuous improvement and operational excellence. With respect to safety issues Talvivaara's goal is a safe and healthy working environment. The company has continued to develop its safety culture based on zero accident philosophy. The company is also committed to continuous improvement in environmental efficiency, operational risk management and reduction of environmental impact.

Open and frequent communication is of outmost importance when building a long-term, trusting relationship with the nearby residents and communities, and Talvivaara has focused increasingly emphasis on its community programme during the last years. Open days for visitors and meetings with the local residents have been continued. A new, locally focused internet site was launched in the beginning of this year to provide up-to-date environmental information and a discussion and feedback channel for the local residents (http://paikanpaalla.fi/).

Business Development

Extraction of Uranium as a By-product

Talvivaara ore contains a low concentration of uranium which will be recovered from the existing production process as a uranium yellow cake by-product. The pregnant leach solution (PLS) from the existing metals recovery plant will enter the uranium plant with a uranium concentration of approximately 20 ppm. A novel solvent-extraction process will be used to selectively extract the uranium from the PLS into an organic solution. Sodium carbonate will be used to strip the uranium from the organic phase back into an aqueous phase. The yellowcake will be precipitated from the aqueous phase with hydrogen peroxide, settled in a thickener and then dried before being packaged in steel drums.

The low uranium concentration in the PLS, the high volume throughput and environmental considerations were the primary factors in designing the processes and selecting the major equipment.

The uranium recovery plant is under final construction stage and commissioning of the new plant will start in early 2013. The plant will produce 350 tonnes of uranium per year, when the full nickel production is reached.

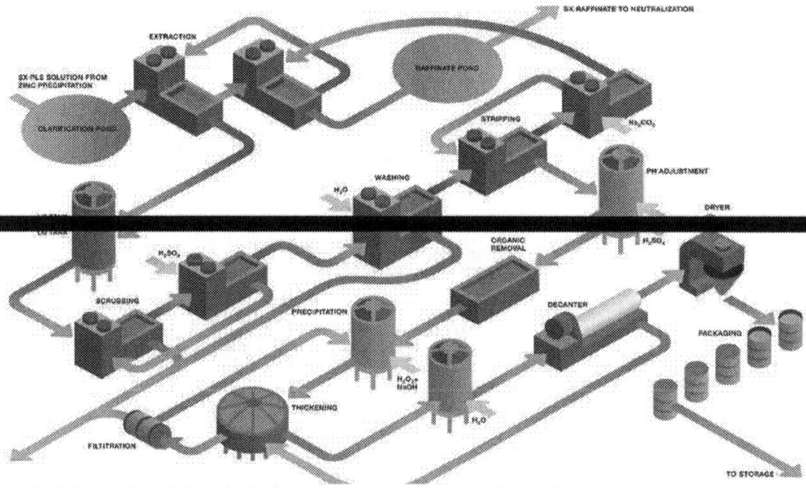

Figure 3. Schematic of the uranium recovery plant process flow sheet

Expansion Beyond 50,000 tpa Nickel

Following the announcement of upgrade of mineral resources in 2010 Talvivaara announced the plans for expansion of the mine and production up to approximately 100,000 tpa nickel. Further plans are to upgrade the nickel production from nickel sulphide to metallic nickel, where the nickel cathode production would account for a substantial part of the expanded production. The production expansion would also increase the zinc, cobalt, copper and uranium production, where the option is to produce also cobalt cathodes.

The mine and production expansion and new refinery process is currently under focus on the Environmental Impact Assessment program which is expected to be ready in the beginning of 2013. Conceptual, basic and detail designs for the new refinery process has been under work for the last 2 years.

The new process would include the pressure leaching of the nickel sulphide, cobalt solvent-extraction plant for separation of the nickel, solution purification and neutralizing plants, and nickel electrowinning tankhouse. Cobalt can be shipped out as cobalt salt solution from solvent extraction solution or optionally as cobalt cathodes from cobalt electrowinning plant. The expansion is designed to be build and commissioned in modular stages.

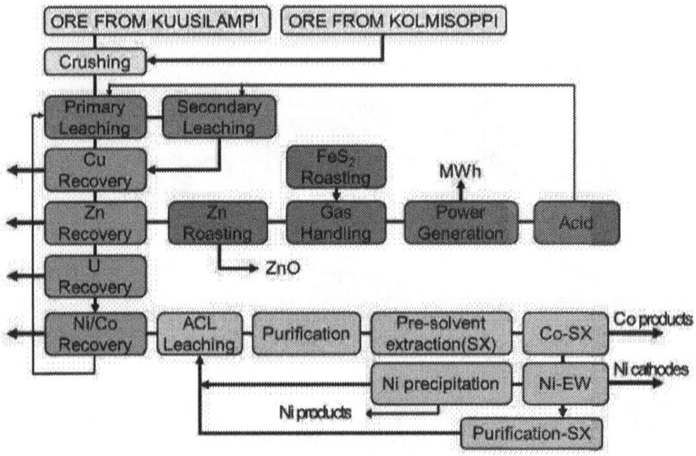

Figure 4. Block diagram of the Talvivaara Sotkamo process with the options for the production expansion.

Energy Strategy

Talvivaara has announced its new energy strategy that is to be self-sufficient for energy in the long term to cope the increased energy demands of the production expansions. The energy strategy focuses on environmental sound energy production including 60MW ownership of the new nuclear plant project and on-site energy production using local bioenergy, wind power and utilization of the energy generated from the new process plants.

Environment

Along with the present production and for the expansion projects extensive work and efforts has been put to environmental issues to minimize and improved effluent and emission levels.

REFERENCES

1. Puhakka, J.A., Kaksonen, A.H.,and Riekkola-Vanhanen, M. Heap leaching of Talvivaara black schist ore. In Biomining, ed. Rawlings,D.E. and Johnson,D.B. Springer-Verlag, 2007, 139-152.
2. Riekkola-Vanhanen, M. and Heimala, S. Electrochemical control in the biological leaching of sulphide ores. In Proceedings of the 10th International Biohydrometallurgy Symposium, ed. A.E. Torma, J.E. Wey and V.I. Lakshmanan, V.I. MS Warrendale, Pennsylvania, 1993, pp.561-570.
3. Riekkola-Vanhanen, M. Talvivaara black schist bioheapleaching demonstration plant. Advanced Materials Research Vols. 20-21, 2007. pp 30-33.
4. Riekkola-Vanhanen M. Talvivaara Sotkamo Mine – bioleaching of a polymetallic nickel ore in subarctic conditions. Nova Biotechnologica 10-1, 2010. pp 7-14.

Ni-Co 2013
Edited by: Thomas Battle, Michael Moats, Violina Cocalia, Harald Oosterhof, Shafiq Alam, Antoine Allanore, Rodney Jones, Nathan Stubina, Corby Anderson, and Shijie Wang
TMS (The Minerals, Metals & Materials Society), 2013

The sintering character of limonitic nickel laterite

Chao Wu, Hongxu Li*, Yu Chen, Chao Li, Zhiqian Zhang,

The Key Laboratory of Ecological and Circular Metallurgy, Ministry of Education, Metallurgical and Ecological Engineering School, University of Science and Technology Beijing, Beijing 100083,China
*Corresponding Author

Key words: laterite ore, limonite, dehydration, sintering, mineral phase transformation

Abstract

The world's proven nickel resources are about 113 million tons, less than 30% for sulfide, more than 70% for nickel laterite, while only approximately 40% of current production is derived from laterites in both limonite and saprolite. In this paper, the sintering character of limonitic nickel laterite with high iron and low silica was studied to offer technical support for producing Ni pig iron through sintering-blast furnace route. The results show that such minerals generally containing water up to 27~35%, pre dehydration is very necessary before sintering, The TGA (Thermogravimetric Analysis) and DTA (Differential thermal analysis) revealed removed rules of adsorbed water and crystal water in the minerals. XRD demonstrated that host minerals of the iron-enriched and magnesium depleted version of laterites are mainly ferric iron oxides, typically goethite, FeO(OH). Nickel embedded in the mineral is mainly as the pattern of NiO. The sinter strength of such minerals of sintering is poorer with drum index below 75%, additives can significantly improve the quality of sinter with the drum index increased to 77% or more. The paper also studied the material ratio, sintering temperature and other factors influencing the quality of sinter, while a series of crystal structure changes and mineralogical formation of sinter were also elucidated.

Introduction

Nickel has a wide range of uses in industry, it is not only the main alloying elements for the production of stainless steel and indispensable raw materials for alloy structural steel, but also basic raw materials and ingredients in a variety of nickel alloys. The world's proven nickel resources are about 113 million tons, less than 30% for sulfide, more than 70% for nickel laterite, while only approximately 40% of current production is derived from laterites in both limonite and saprolite. With the influence of the reduction of high grade sulfide nickel ore which is easy to exploit and the factor of nickel price, the nickel production amount from nickel laterite will increase. So extracting nickel from laterite nickel ore is gradually becoming common in the world [1]. In 2011, global stainless steel production has reached a new record once again for about 32.05 million tons, and China has ranked first for about 11.84 million tons. In recent years, strong demand in the global stainless steel has greatly stimulated the nickel market, resulting in the global nickel price increasing. As the world's largest stainless steel producer and consumer, China's demand for nickel is increasing.

The main treatment processes for laterite are hydrometallurgy and pyrometallurgy. Hydrometallurgy processes include: pressure ammonia leaching - hydrogen reduction, sulfuric acid selective leach electrowinning / hydrogen reduction, hydrochloric acid leaching and chloride leaching methods. Pyrometallurgy includes nickel matte and Ni pig iron process [3], which is commonly applied in China in recent years. The main technical routes are the following two: one is the rotary kiln - furnace process (RKEF) to produce high grade Ni pig iron; the other is blast furnace process to produce other Ni pig iron products. The main production steps are as follows: laterite nickel ore + quicklimedehydration, sintering, agglomeration → blast furnace smelting→refining and casting→ Ni pig iron [4]. Therefore, this paper makes the following research for the sintering and smelting mechanism of the nickel laterite.

Materials and Experiments

Nickel laterite materials analysis

Nickel laterite is nickel mineral changed by weathering, etching, enrichment, and then it turns loose clay-like, composed of iron, aluminum, silicon, and some hydrous oxides. The element and moisture content are rich in this kind of ore. The elemental composition and content (Table 1) can be obtained by XRF elemental analysis and chemical analysis.XRD demonstrated that host minerals of the iron-enriched and magnesium-depleted version of laterites are mainly ferric iron oxides, typically goethite, FeO(OH), and nickel embedded in the mineral is mainly as the pattern of NiO (Figure 1) .All of the samples came from Baosteel.

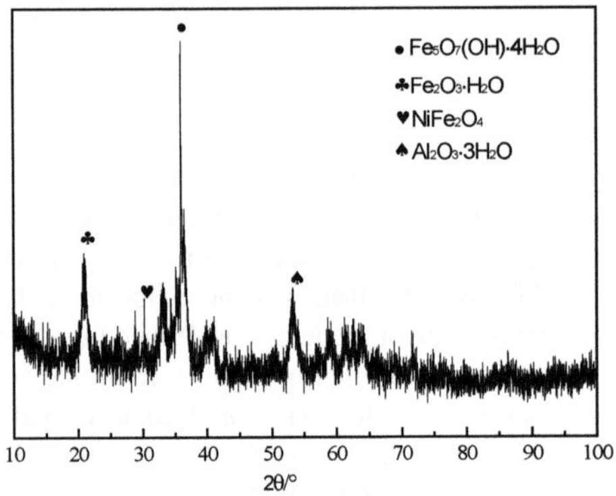

Figure 1 XRD analysis of raw materials

Table 1 Nickel laterite chemical composition (mass fraction %)

Fe_2O_3	Al_2O_3	SiO_2	NiO	MgO	Cr_2O_3	MnO	SO_3	CaO	P	Water content
70.98	11.19	5.95	2.12	1.91	1.51	1.05	1.12	0.16	0.01	29.54

Experiments

Pre dehydration

Nickel laterite's moisture content can be measured by thermogravimetry and differential thermal analysis. Due to the high water content, the ore is very sticky. This has a significant impact on the subsequent sintering process and the quality of sinter. So adding lime with stirring and mixing to dehydrate the ore is necessary as pretreatment for laterite sintering.

Roasting experiments

In the roasting experiment each material sample weighed 50g, roasting time was 90 minutes, the maximum temperature was 900 ° C. Each carbon amount selects 4%, 8%, 12% and 16% of the ore then install them in the crucible into the muffle furnace to determine the optimum carbon content.

Micro sintering experiments

In this micro sintering experiments laterite ore should be crushed to the needed granularity firstly: less than 8mm accounted for 90%. The reducing agent coke was screened to 200 mesh and the proportion is about 80%.. Conduct four groups totally: the additives fluorite content are 0% and 5%, respectively, the maximum temperature is seted of 1050 ° C (alkalinity to 1), and 1200 ° C (alkalinity to 2). When the furnace temperature up to the maximum temperature of the test requirement, the samples should be cooled to room temperature in the furnace after heat preservation for 10 min. A small amount of the sample was pulverized after the experiment for XRD analysis and then to determine the compressive property.

In a word, a series of crystal structure changes and mineralogical formation of sinter in the sintering process can be studied through chemical analysis, thermogravimetry (TG), differential thermal analysis (DTA) and X-ray diffraction analysis around the physical and chemical changes occurring in the whole process.

Results and Discussion

Pre dehydration

There are many carbonates like $CaCO_3$ and $MgCO_3$ in the sinter. In the sintering process, these carbonates must be decomposed sufficiently and combined with the other ingredients in the slag to compound into a new mineral in the sintering, otherwise, it will affect the mechanical strength and anti-pulverization ability of the sinter. If it is guaranteed that $CaCO_3$,the hardest decomposition of carbonate can be decomposed completely, several other carbonate decomposition can also be completed under the same conditions.The actual sintering process requires not only complete decomposition of $CaCO_3$,but also complete absorbtion of CaO by liquid and combination with other mineral, i.e. free CaO, which exists in the form of a white powder, will absorb water and digestion, affecting the sinter strength seriously . In order to ensure that the limestone decomposed completely and absorbed during the sintering , it can be replaced by a certain amount of quicklime to enhance the sintering process. Quicklime digests in the mixture and turns into hydrated lime $Ca(OH)_2$. This forms into gel particles which are widely dispersed in the mixtureand some strongly hydrophilic particles seize water on particles and surface of concentrate, so that capillary force has been generated between these particles and hydrated lime particles to increase the mixture consolidation tendency[5]. The quicklime is pulverized and mixed sufficiently into the nickel laterite sinter during the pre dehydration as a beneficial measure to enhance the sintering process, because it can improve the contact conditions of CaO and Fe_2O_3 and the formation of SFCA in the solid phase. But the amount should be appropriate, the active and non-overburning lime with a particle size less than 5mm is encouraged. Practice has proven that the production has raised 20% to 35% when adding 2% to 4% quicklime in the pre dehydration. In this experiment, we chose 5% quicklime with a particle size of 1mm. Infiltrate material for 9 days after pretreatment.

Dehydration law of the adsorbed and crystal water in the minerals

Usually the water of crystallization in the solid solution is easily decomposed from 120 ° C to 200 °C . However, the limonite which exists as the form of OH^- has a decomposition temperature up to 300°C due to the lattice transition during decomposition. Kaolin ($Al_2O_3{\cdot}2SiO_2{\cdot}2H_2O$) and Beidellite [$(Fe{\cdot}Al)_2O_3{\cdot}3SiO_2{\cdot}2H_2O$] in the gangue require a temperature of 500 ° C when their lattices enter into OH^-. The decomposition is an endothermic reaction.

We can get the removed rules of adsorbed and crystal water in the minerals combined with XRD and the mineral thermoanalysis as follows: the evaporation of free water causes the endothermic peak appearing at 102 °C in the DTA curve ,the corresponding mass loss in the TG curve is 2.72 mg, for 15.54% of the total weight. The next endothermic peak appears at 302.6 C due to the evaporation of crystal water, the mass loss in the TG curve corresponding to 4.41 mg, for 25.2% of the total weight, the mass loss in this stage has reached to 9.66% .From 400 °C to 650°C, hydroxyl in the mineral begins decomposing, endothermic peak has climbed to 507°C because of the decomposition of hydroxyl in kaolin and beidellite , the mass loss percentage is 4.34% while the total weight loss has peaked 29.54%. The curve change above 650°C is caused by the decomposition of the silicate minerals which will not change the mass . The hydroxyl has escaped substantially till 900°C. It can be obtained by thermal gravimetric analysis that water is substantially removed when the temperature has reached 900 °C, whereby the proportion of evaporated water is approximately 15.54%, and the total proportion of crystal water and hydroxyl is about 14%.

Figure 4 shows the nickel laterite thermal analysis when adding 5% quicklime. The first endothermic peak in DTA curve advanced to 86 °C, due to the preabsorbent and preheat, 8.4% of the total weigh has been lost. Other similar conclusions can be drawn from the figure.

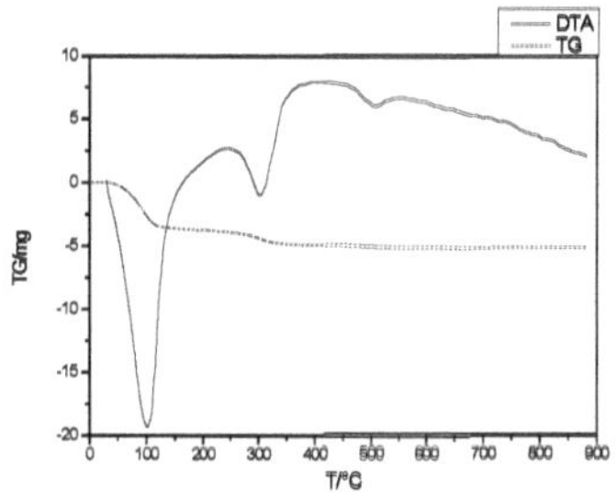

Figure 3 Thermal Analysis(Raw Ore)

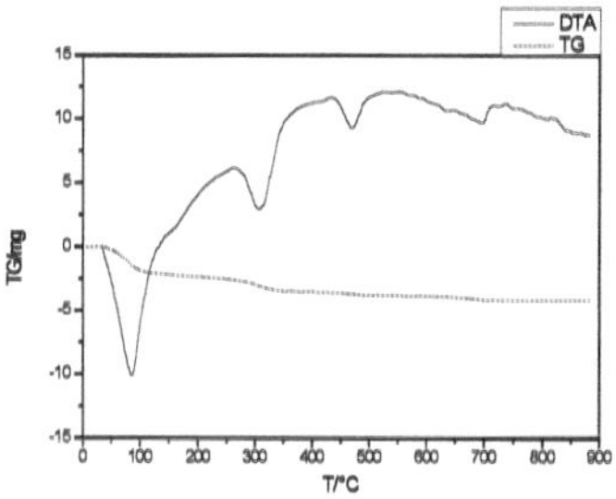

Figure 4 Thermal Analysis(Added 5% lime)

Sintered material ratio and temperature

The carbon in the mixture starts to burn when the temperature reaches to 700 °C, occurring the following four reactions:

(1) $C + O_2 = CO_2$ $\Delta H = -3350KJ/KgC$ (2) $2C + O_2 = 2CO$ $\Delta H = -9800KJ/KgC$

(3) $2CO + O_2 = 2CO_2$ $\Delta H = -23700KJ/KgC$ (4) $CO_2 + C = 2CO$ $\Delta H = 23700KJ/KgC$

The reaction (1) is prone to happen, the reaction (3) is easy to conduct in the low-temperature region, the reaction (2) and (4) is favored in the high temperature zone but their development has been limited in most time of the sintering process. Overall, the sintering exhaust gas is mainly CO_2 with a little CO and free oxygen [5]. The thermal decomposition, oxidation and reduction reactions of iron oxides occur before reflowing of the sintered material during the sintering process, so it must be controlled because of its great influence on the liquid ingredients and mineral composition of the sinter. Three types of iron oxide may exist in the sinter, Fe_2O_3 has a higher decomposition pressure, while the decompression pressure of Fe_3O_4 and FeO are low. The decomposition reaction of Fe_2O_3 occurs in the high-temperature region in the sintering process, the reaction is: $Fe_2O_3 = Fe_3O_4 + 1/2\ O_2$;

When there is SiO_2, Fe_3O_4 will conduct as follow: $2Fe_3O_4 + 3SiO_2 = 3(2FeO \cdot SiO_2) + O_2$;

In the area of the low temperature (300 °C) when CO existed, Fe_2O_3 is easy to reduce:

$3\ Fe_2O_3 + CO = 2\ Fe_3O_4 + CO_2$;

As a result, the free Fe_2O_3 in the sinter is little. The oxidizability of the sinter , the degree of oxidation of iron oxide in the sinter, has a great influence on the reductibility of the sinter. The high degree of oxidizability means there is much iron in the form of Fe_2O_3 morphology and the reproduction is good; Conversely, the reproduction is poor. Therefore, we should try to increase the degree of oxidation in the guarantee of sufficient strength to improve the reducibility of the sinter. Fuel ratio is the primary factor to affect the Fe_2O_3 content in sinter; increase the alkalinity of sinter is conducive to hinder the decomposition and reduction of Fe_2O_3; reduce the fuel particle size is not only helpful to the generation and bonding of the liquid phase, but also contributes to the slag oxidation. These steps all can reduce the fuel consumption to improve the content of Fe_2O_3.

The actual reduction in the sintering process depends on the sintered materials, technology and some other conditions, such as the amount of carbon, alkalinity and sintering temperature,etc. The roasting experiments and XRD analysis (Figure 5) have shown how the carbon amount impacts the composition changes of iron oxide (Figure 6). With the increase of the carbon amount, the degree of sinter reduction has been raised. Because the duration of high temperature is short during the sintering process, it is easy to form soluble substance between CaO and ore, which restricts the reduction process, resulting in the decline of the FexO amount in the sinter. While it is not welcome in blast furnace of the increasing of the FexO mass fraction, the development of the reduction reaction should be limited in the sintering process. Figure 6 can demonstrate that 4% carbon is more appropriate.

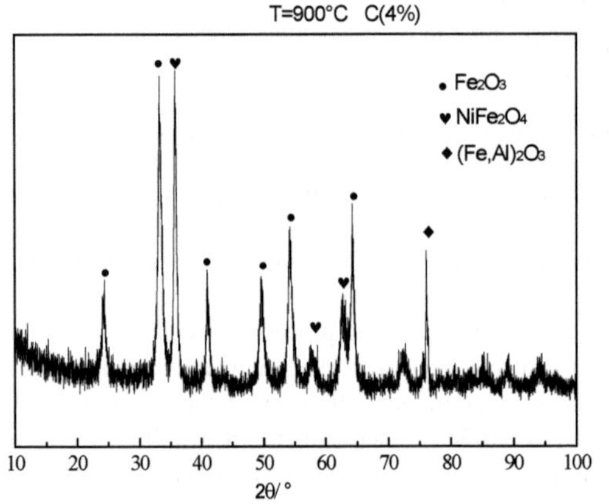

Figure 5 XRD of laterite after roasting

In order to meet the requirements of blast furnace smelting, the sintering process should develop new technologies. Low temperature sintering process is an advanced technology in the world, which can improve the sinter quality significantly. It is based on the" calcium ferrite (SFCA) theory", which requires that sinter basicity should be controlled to 1.8 ~ 2.0, with a sintering temperature of 1200 ~ 1280 °C,and the preparation of raw materials should be strengthened, particularly the particle size: 90% of nickel laterite should be less than 8mm; screening about 80% of the reducing agent coke to 200 mesh; and the proportion of lime and fluorite which is less than 1mm should be greatered than 80%.

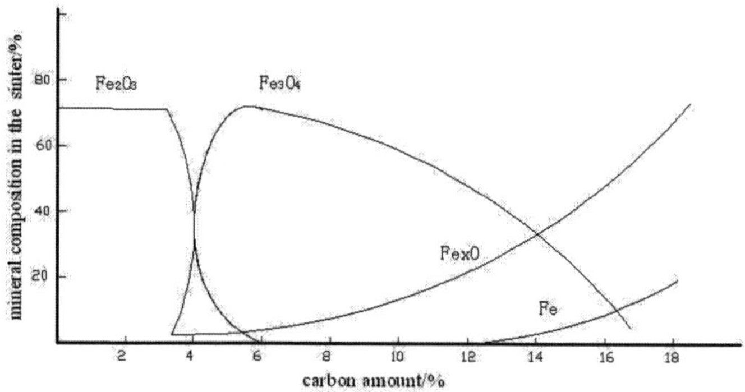

Figure 6 the changes of iron mineral composition in the nickel laterite during sintering

<u>Sintering Performance Analysis</u>

The sinter is the product of melt condensation and crystalline formation. Low melting point compounds have been generated by solid-phase reactions (Figure 7) occurring in the nickel laterite sintering process at first, and then they successively melt at the sintering temperature. The more fully the solid phase reaction is carried out, the more liquid phase will be generated, the higher the strength of the sinter will be. The calcium ferrite generation performance [6],which means the generation capability of composite calcium ferrite in the sintering process is an important indicator to detect the performance of the sinter. The calcium ferrite is the main and optimal bonding phase of high basicity sinter. The bonding phase of calcium ferrite will significantly improve the reducibility and strength of the sinter ,slag conditions and heat recovery rate, accelerate the slag-forming speed and the dissolution of the lime, create good dephosphorization conditions in the blast furnace, and shorten the refining cycle, which is very useful for improving the quality of steel. Therefore, it is wanted the calcium ferrite would generate as far as possible in the nickel laterite sintering process. Figure 8 shows the microstructure of nickel laterite sinter. There are many holes with different sizes and shapes, the main binding phase is acicular SFCA which is uniformly distributed. Sinter microstructure is an important factor in the reductive of sinter. When the limonite grains are small and dense, and have less binding phases among them, the sinter is easily reducible, while coarse grains exist or packed by silicate cement phase, it is difficult to restore. In addition, the sinter with high porosity, grain loose disseminated, as well as more number of cracks is easily restored. However, this sinter structure is amenable to be undermined, and has poor strength .It can be learned from the microstructure (Figure 8) and the compressive strength, that sample without fluorite is loose and brittle, with large porosity and low compressive strength . Adding fluorite increases the fluidity of the liquid phase in the sintering process, the microstructure of the sinter has been turned to dissolution structure, improving its uniformity and compressive strength [7].

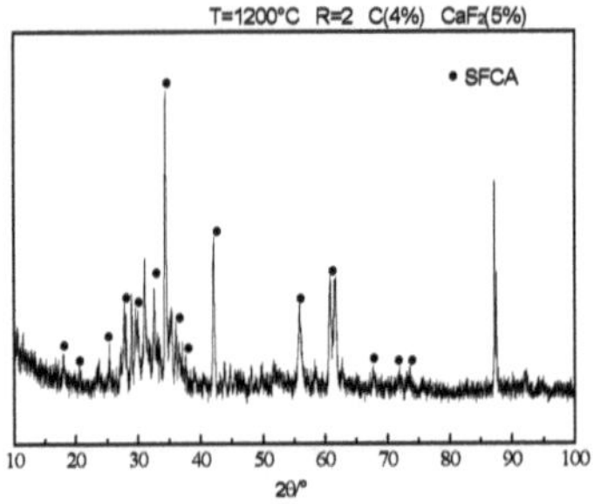

Figure 7 XRD results of sintering (1200 °C)

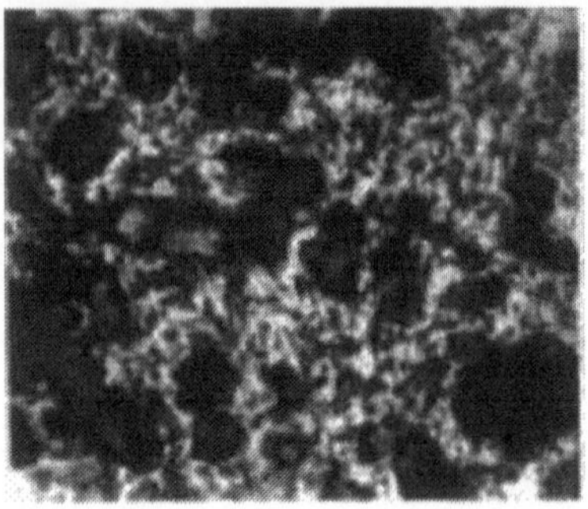

Figure 8 microstructure of nickel laterite after sintering

Conclusions

The host minerals of the iron-enriched and magnesiu- depleted version of laterites are mainly ferric iron oxides, typically goethite, FeO(OH). Nickel embedded in the mineral is mainly as the pattern of NiO. Due to the high water content , pretreatment that includes adding 5% lime with stirring and mixing to dehydrate before sintering is necessary. The thermal weight and differential thermal analysis can be obtained when the sample temperature is up to 900 °C, water is almost completely removed, evaporative water proportion is about 15.54%, water of crystallization, hydroxyl total proportion is about 14%. The appropriate parameters are as follows: the sintering temperature is 1200 °C, the alkalinity is 2, with 4% of the carbon amount and 5% of the additives fluorite. In nickel laterite sintering, producing high basicity sinter and avoiding excessive melting can promote the generation and conservation of the SFCA system, which has a significant importance to improve the strength and reducibility of the sinter.

REFERENCES

[1]Li Yanjun, Yu Haichen, Wang Dequan. "Nickel Llaterite Resources Status and Summary of Processing" [J]. Metal Mines, 2010, (11) :5-15.

[2] Cai Xieao. "Status of China special steel and Development Strategy" [J]. Metallurgy. 2009,5 (5) :10-12.

[3] Xiao Anxiong. "The investigation of world non-ferrous metal smelting plant from American metal journal(the third department - nickel laterite)" [J]. Chinese non-ferrous metallurgy, 2008, 8(4):1-12.

[4]Zhang Mei. "Laterite Nickel Pyrometallurgy Smelting Progress in China "[J]. Chinese Metal Bulletin. 2008, (35) :30-31.

[5] Wang Xiaoliu. "Ferrous metallurgy (iron-making part)" [M]. Beijing: Metallurgical Industry Press, 2008.

[6]Wu Shengli, Liu Yu, Du Jianxin, Mi Kun, Lin Hong. "Basic Characteristics New Concept of Iron Ore Sintering "[J]. Journal of University of Science and Technology Beijing. 2002,6 (3): 254-257.

[7] Si Xinguo, Lu Xionggang, Zhao Ermin, Zhang Yuzhu, Zhang Qingjun, Liu Limei. "the Compressive Strength Mechanism Study of Low Silicon Sinter" [J]. Steel. 2009, 8 (8) :11-14

Ni-Co 2013

Pyrometallurgy: Smelting

Session Chairs

Antoine Allanore
Rodney Jones

Ni-Co 2013
Edited by: Thomas Battle, Michael Moats, Violina Cocalia, Harald Oosterhof, Shafiq Alam, Antoine Allanore, Rodney Jones, Nathan Stubina, Corby Anderson, and Shijie Wang
TMS (The Minerals, Metals & Materials Society), 2013

NICKEL, COBALT AND COPPER RECOVERY FROM SEA NODULES BY DIRECT SMELTING PROCESS

K.K. Sahu, S. Agarwal, D. Mishra, A. Agrawal, N.S. Randhawa, K.M. Godiwalla and R.K. Jana
Council of Scientific and Industrial Research, National Metallurgical Laboratory
Jamshedpur - 831007, India

Key words : Sea nodules, reduction smelting, pilot scale smelting, cobalt, nickel, copper.

Abstract

A direct smelting process has been developed at CSIR-National Metallurgical Laboratory, to recover all four valuable metals such as Ni, Co, Cu and Mn. Initially sea nodules containing 1.1% Cu, 1.15% Ni, 0.76% Co, 22.3% Mn, 5.5% Fe and 13.1% SiO_2 was smelted in presence of coke to produce an alloy rich in Ni, Co, Cu and Fe. Various parameters such as amount of coke, quartz, smelting temperature and holding time etc. were optimized on different scale between 0.5-20 kg scale. Under optimized condition, the alloy of composition : 16.% Cu, 18.4% Ni, 1.2% Co 57.0% Fe and 3.6% Mn was produced on 20 kg scale using 7% coke with more than 90% metal recovery. The process was further up-scaled to 250 kg where similar recovery was achieved with 8% coke and 1 h holding time. The alloy and slag was further processed to recover different value added products.

Introduction

With the rapid depletion of land based mineral resources for nickel, copper and cobalt, several countries have explored the possibility of using deep sea manganese nodules. Although the accretion rate is very slow, only a few mm in a million years, however the present reserve of these nodules is substantial. Therefore, this can be an attractive resource due to the presence of significant amount of nickel, copper and cobalt.

India is completely dependent on imports to meet its requirements of cobalt and nickel and has launched an ambitious programme of sea bed exploration of minerals for some of these strategically important metals with a view to explore them for the national needs. The Indian program of work considers the manganese recovery option in addition to recovery of Cu, Ni & Co for better process viability & environmental consideration. Three process namely (i) combined pyro-hydrometallurgical process of reduction roasting and ammoniacal leaching, (ii) direct leaching process in ammoniacal medium with dissolved sulfur-dioxide as aqueous reducing agents and (iii) High temperature reductive sulphuric acid leaching, have been developed and extensively tested in India.

The direct reductive acid leaching process was developed by Hindustan Zinc Limited, Udaipur. Starch was used in the process as reductant during leaching. The separation process is based on selective sulphide precipitation followed by solvent extraction and electrolysis. The other two processes are based on ammonia leaching, developed at the Institute of Materials and Minerals Technology, Bhubaneshawar (IMMT-B) and National Metallurgical Laboratory,

Jamshedpur (NML). The IMMT-B process constitutes [1] direct ammoniacal leaching in presence of sulphur dioxide followed by demagnetization and solvent extraction to produce pure metal by electrolysis. The process is tested up to 500kg scale in order to prepare a pre-feasibility report. NML developed a process based on reduction roasting-ammonia leaching followed by metal separation using solvent extraction technique to produce pure metal by electrolysis. The process was tested on 100kg scale per day [2].

The above processes generate very dilute leach liquor as a result downstream operation requires very large equipments and higher energy. The ammoniacal leaching process also generate huge quantity of residue containing high moisture & sulphate / ammonia, which was further treated by NML to develop a process for the recovery of manganese as Fe-Si-Mn by electrothermal reduction-smelting route. However, due to low Mn/Fe ratio, standard grade Fe-Si-Mn alloy was produced by blending with high grade Mn-ore or by pre-smelting to remove iron. To obviate the above gap areas, direct reduction smelting of sea nodules was explored to recover Cu, Ni, and Co in the form of alloy and Mn as Fe-Si-Mn from Mn-rich slag by further smelting. The alloy containing high percentage of Cu, Ni, Co was subjected to matte formation followed by pressure leaching to generate concentrated leach liquor, which was suitably treated to recover metal by selective hydrothermal reduction / SX. The major advantages of the process are (a) enriched Cu, Ni, and Co alloy produced is only about 5.5-6.0% of the input sea nodules weight, (b) concentrated leach solution of Cu, Ni and Co with reduced volume is generated by pressure leaching of alloy/matte and (c) with partial removal of iron in the alloy, the Mn rich slag generated in direct smelting of sea nodules is highly suitable for production of standard grade Fe-Si-Mn. Sridhar et al [3] reported prereduction roasting followed by smelting to produce Cu-Ni-Co alloy with about 85% metal recovery. Earlier, bench scale study on direct reduction smelting was reported by our group [4]. In the present paper, systematic syudy on direct reduction smelting of sea nodules was reported both on bench scale as well on pilot scale to recover Cu-Ni-Co in the form of alloy, which was further processed to recover individual metal.

Experimental

Sea nodules containing 1.15% Ni, 1.1% Cu, 0.076% Co, 22.3% Mn, 5.36% Fe,13.1% SiO_2, 3.19% Al_2O_3, 7.03% MgO, 0.54% CaO, 1.76% Na_2O+K_2O, 24.5% LOI, were crushed down to minus 20 – plus 10 mesh size fraction; commercial grade quartz & coke were crushed to minus 5 mesh size fraction. Factsage (6.1 version) was used for thermodynamic calculations. The authors have used FACT53, oxide and SGSL data bases of the software to predict equilibrium of slag, alloy and gaseous products. The smelting experiments were carried out with charge mix, by mixing the desired amount of reductant and fluxing material. Initial smelting experiments were carried out on 500 g scale using recrystallized alumina crucible, in a resistance furnace. Smelting on 20 kg scale was done using 50 kVA two phase AC arc furnace with double electrode in graphite brick-lined rectangular vessels. These electrodes were suspended through the top into the hearth of 50 kVA furnace. The electrodes were connected to the bus-bars via a water-cooled clamp connection. The electrodes and clamp formed part of a moveable electrode arm, which were used to control the current and voltage ratio by adjusting the arc length i.e. moving the arm up or down. Before the start of the sea nodules smelting the furnace was preheated for few minutes and charge materials were added gradually. After complete melting of the charge, it was hold for 10 minutes. Thereafter, the molten metal and slag was discharged through a tap hole to a pre-heated crucible. After complete cooling, the metal and slag were separated and weighed.

Both metal and slag samples were ground to fine powder and analyzed for their chemical composition.

Results and Discussion

The X-ray diffractogram of the bulk sample of sea nodules indicated most of the mineral phases were either amorphous in nature or very poorly crystalline. The identified peaks confirmed the presence of quartz only. However, X-ray diffraction on individual layer deposit of nodules and EDAX analysis revealed the presence of todorokite as the major phases of manganese. The only iron oxide phase indicated was goethite. The major components in the sea nodules were reported earlier also to be present in the form of complex/individual oxide [5, 6]. From the Ellingham Diagram [7] it is evident that in presence of carbon as reductant Cu, Ni and Co could be selectively reduced by controlling reduction temperature during smelting and carbon in the charge. Higher oxides of manganese and iron will reduce to their lower oxides and need higher temperature to get reduced to their metallic state. Theoretical calculations were done using the thermodynamic software Factsage, to determine the composition and temperature for selective reduction of Cu, Ni and Co. Table I represent theoretical alloy composition at different carbon composition during smelting of sea nodules at 1400 °C in presence of additional 4% quartz. Actual smelting result under similar experimental condition conducted in a recrystallise alumina crucible was found good co-relation between the theoretical and experimental alloy quantity and composition. Recrystalise alumina crucible used to avoid carbon pick-up and to optimise coke requirement during smelting of sea nodules.

Table I. Comparisons between experimental and theoretical alloy quantity and composition obtained for sea nodules smelting on 500 g scale at 1400 °C in recrystallise alumina crucible.

Constituents	4% Carbon		6% Carbon	
	Theoretical, %	Experimental %	Theoretical,%	Experimental,%
Cu	29.8	29.9	10.88	12.2
Ni	31.2	33.1	11.4	12.9
Co	2.06	2.01	0.75	0.89
Mn	0.35	1.24	21	17.1
Fe	36.6	29.7	52.9	55
Alloy wt.(g)	18.5	16.12	50.5	45.0

Inorder to optimise carbon, the coke containing 77.2% fixed carbon was varied between 2.5% to 15% in the charge and smelted at 1400 °C. Alloy formation was not observed for smelting experiment with 2.5% coke. However, alloy weight increased with further increase in the coke percentage. The alloy composition and metal recovery obtained with different coke percentage is incorporated in Table II. Smelting with 5% (w/w) coke produced a copper and nickel rich alloy containing very low iron (0.21%) and manganese (0.014%). Due to low iron recovery, the cobalt in the alloy was also found very low (0.09%). There was a sharp decrease in the concentration of copper and nickel with increase in coke percentage from 5% to 6.25%. The copper and nickel in the alloy decreased from 44.6 and 47.7% to 17.4 and 18.2% respectively with increase in coke from 5% to 6.25%. With the same increase in coke, the cobalt concentration in the alloy increased from 0.1% to 1.08%.

Table II. Metal recovery in the form of alloy obtained during smelting of sea nodules with varying coke addition.

% coke, (w/w)	Alloy wt. (g)	Alloy Composition, %					Metal Recovery, %				
		Cu	Ni	Co	Fe	Mn	Cu	Ni	Co	Fe	Mn
2.5	0	0	0	0	0	0	0	0	0	0	0
5.0	9.33	44.6	47.8	0.09	0.21	0.014	75.7	77.5	2.21	0.07	0
6.25	27.7	17.4	18.2	1.08	51.7	4.64	87.6	87.7	78.7	53.5	1.06
7.5	34.0	14.9	15.9	0.96	58.4	5.02	92.1	95.1	85.9	74.2	1.41
10	39.4	13.2	14.0	0.83	62.1	6.21	94.8	95.9	86.1	91.3	2.01
15	44.4	11.8	12.3	0.76	54.6	12.8	96.7	97.2	88.8	90.4	4.66

The iron concentration in the alloy increased from 0.2% to 51.7% with increase in coke percentage from 5% to 6.25% and there is a marginal increase in the iron in the alloy with further increase in coke up to 10% (Table II). With 15% coke, iron in the alloy decreased due to incorporation of more manganese in to alloy. The manganese concentration in the alloy increased from 0.01% with 5% coke to 12.7%with 15% coke.

It was observed that recovery of all metals increased with increase in coke during smelting (Table II). Copper and nickel recovery increased from 75.7% and 77.5% to 96.7% and 97.2% respectively with increase in coke from 5% to 15% in the charge. The cobalt, iron and manganese recovery was found almost nil at 5% coke, but increased to 88.8%, 90.4% and 4.66% respectively using 15% coke during smelting.

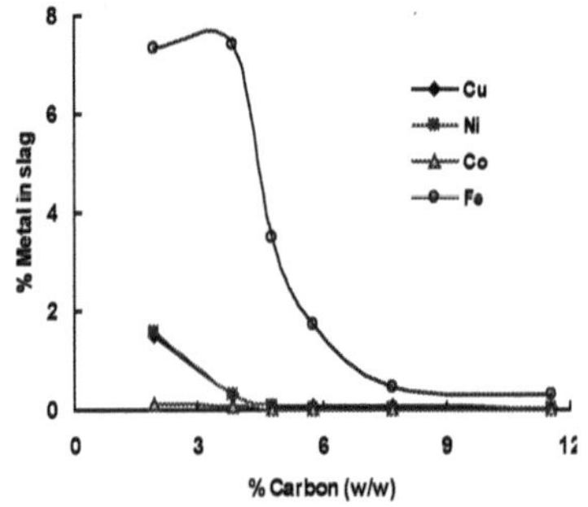

Fig. 1 : Metal distribution in slag obtained from smelting of sea nodules at different %ge of coke addition.

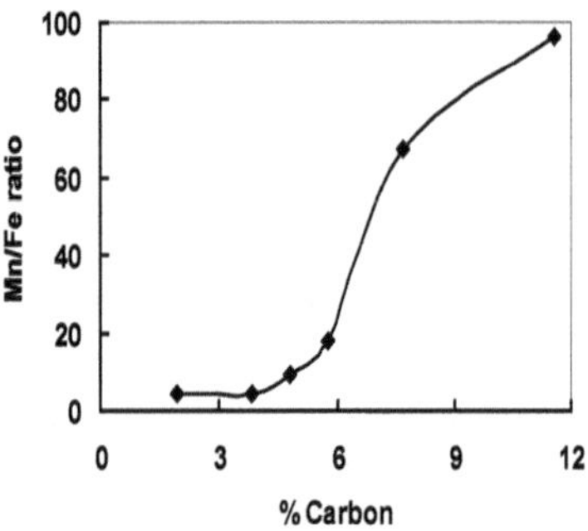

Fig. 2 : Mn/Fe in slag obtained during smelting of sea nodules at different %ge of coke addition.

The trend in metal content in the slag with increase in the coke during direct smelting of sea nodule in alumina crucible using muffle furnace is represented in Fig. 1. Percentage of metal in the slag gradually decreased with increase in the coke in the charge. There was a sharp decrease in the percentage iron from 7.4% to 3.5% with increase in the coke from 5% to 6.25%, thus removing almost 50% of iron from the system making the slag ideal for production of high grade Fe-Si-Mn. With further increase in coke, iron in the slag gradually decreased. With manganese composition remaining almost similar above 30%, the Mn/Fe ratio in the slag increased with increase in coke (Fig.2). The Mn/Fe ratio of 10 with 6.25% coke, increased to about 10 fold with 15% coke in the charge during direct smelting of sea nodules (Fig.2). The slag with such high ratio of Mn : Fe would be suitable for ferromanganese production.

Based on the data generated on 0.5 kg in recrystallise alumina crucible further scale-up experiments were carried out on 4 kg scale using graphite crucible and on 20 kg scale using a graphite brick lined 50 kVA electric arc furnace (EAF) with variation of coke and quartz. For optimisation of direct smelting of sea nodules on 20 kg scale, the coke percentage was varied from 6-8%. The alloy weight obtained for each experiment with variation of coke is incorporated in the Table III. The alloy quantity increased with increase in coke in the charge. With 6% coke and 20 minutes holding time, the alloy weight obtained was 980 g which increased to 1078 g with increase in coke to 8%. Experiment with 7% coke produced 1031 g alloy of composition: 18.2% Cu, 19.35% Ni, 1.28% Co, 53.9% Fe and 4.08% Mn (Table III) which corresponds to a metal recovery of 85.1% Cu, 86.7% Ni, 86.8% Co, 51.8% Fe and 0.94% Mn of the input metal present in sea nodule. However, metal recovery increased when holding time increased from 20 min. to 30 min. With 7% coke and 30 min. holding time, 90.5% Cu, 92.0% Ni and 86.6% Co recovery was achieved (Table-III). Under this condition iron and manganese in the alloy is 56.7% and 2.84%, which corresponds to recovery figure of 60.0% and 0.73% respectively. The smelting result was validated with three repeat experiments, which was found to yield almost similar recovery of Cu, Ni and Co. (Table III).

Table III. Copper, nickel and cobalt recovery in the alloy obtained during smelting of sea nodules on 20 kg scale in graphite lined 50 kVA arc furnace with varying coke and holding time in presence of 4% additional quartz.

% coke	Holding time	Alloy wt. (g)	Alloy Composition, %					Metal Recovery, %				
			Cu	Ni	Co	Fe	Mn	Cu	Ni	Co	Fe	Mn
6	20	980	18.1	19.4	1.24	54.3	3.72	80.4	82.5	80.0	49.7	0.82
7	20	1031	18.2	19.4	1.28	53.9	4.08	85.1	86.7	86.8	51.8	0.94
8	20	1078	17.4	18.6	1.17	56.9	3.42	85.6	87.3	83.0	57.2	0.83
7	30	1145	17.4	18.5	1.15	56.7	2.84	90.5	92.0	86.6	60.0	0.73
7	30	1177	16.5	17.9	1.19	56.2	3.45	88.5	91.6	84.3	53.7	0.83

Pilot Scale Smelting of Sea Nodules

Based on the optimization experiments on 20 kg scale, further scale-up experiment on 250 kg scale was carried out in a graphite brick lined 150 kVA electric arc furnace (Fig. 3). As optimized on 20 kg scale, 7% coke and 4% quartz were added during pilot scale smelting. With 7% coke the alloy and slag weight obtained with 20 minutes holding time were 9.45 and 137 kg respectively. However some unmelted / sintered mass found at the furnace bottom. The alloy of composition : 17.6% Cu, 19.3% Ni, 1.15% Co, 54.9% Fe and 2.4% Mn was obtained which corresponded to metal recovery of 60.5% Cu, 63.4% Ni, 57.2% Co, 36.8% Fe and 0.41% Mn. The melted slag collected through tap hole with composition: 0.65% CuO, 0.51% NiO, 0.049% CoO, 5.7% FeO and 41.7% MnO have percentage metal distribution of 21.4% Cu, 19.2% Ni, 26.7% Co, 43.3% Fe and 97.0% Mn. Incomplete reduction due to less holding of the melted charge resulted in low recovery of different metals. Therfore, further experiments were conducted with variation coke from 6-10% maintaining 1 h holding time.

(a) (b)

Fig. 3 : (a) 150 kVA electric arc furnace with (b) control pannel

Table IV. Copper, nickel and cobalt recovery in the alloy obtained during pilot scle smelting of sea nodules on 250 kg scale in graphite lined 150 kVA arc furnace with varying coke.

% Coke	Holding time, min	Slag wt. (kg)	Alloy wt. (kg)	Metal recovery, %				
				Cu	Ni	Co	Fe	Mn
6	60	149	10.5	73.3	75.6	74.6	39.5	0.54
7	60	143	14.6	87.1	89.9	86.1	59.6	0.97
8	60	145	15.3	89.8	90.9	87.8	61.6	1.31
10	60	142	19.1	90.8	93.2	89.5	72.7	5.04
8	60	142	14.9	90.4	91.7	87.8	58.8	1.18

Alloy quantity increased from 10.5 kg to 19.1 kg with increse in coke from 6% to 10% in the charge during smelting on 250 kg scale in presence 4% quarz (Table IV). With 6% coke, the alloy of composition : 19.2% Cu, 20.7% Ni, 1.35% Co, 52.9% Fe and 2.87% Mn was obtained which corresponded to metal recovery figure of 73.3% Cu, 75.6% Ni, 74.6% Co, 39.5% Fe and 0.54% Mn. With increase in coke to 8% and maintaining similar holding time (1 h) during smelting, the alloy (Fig. 4) quantity obtained was 15.3 kg of composition: 16.1% Cu, 17.0% Ni, 1.09% Co, 56.9% Fe and 4.77% Mn which corresponded to metal recovery of 89.8% Cu, 90.9% Ni, 87.8% Co, 61.6% Fe and 1.31% Mn (Table IV). There was a marginal increase in Cu (90.8%), Ni (93.2%) and Co (89.5%) recovery with further increase in coke to 10%. Iron and manganese recovery in the alloy was increased to 72.7% and 5.04% respetvively, which is not desiredrable. Therefore considering 8% coke is appropriate for getting optimum recovery; one repeat smelting campaign was conducted to validate the recovery result. Metal recovery of 90.4% Cu, 91.7% Ni, 87.8% Co, 58.8% Fe and 1.2% Mn was obtained with the repeat smelting trials using 8% coke (Table IV).

(a)

(b)

Fig. 4 : Slag tapping during smelting of sea nodules in 150 kVA arc furnace (b) Cu-Ni-Co rich alloy obtained from sea nodule smelting on 250 kg scale

Conclusion

As an alternate process on direct reduction smelting of sea nodules is developed at CSIR-NML to recover Cu, Ni and Co in the form of alloy with about 90% recovery, achieve good result. Using 8 % coke & 4 % quartz, 90.4% Cu, 91.7% Ni, & 87.8% Co recoveries were achieved on pilot scale. As a part of iron was recovered in the alloy, the process generate a slag rich in manganese with Mn/Fe ratio of about 10, from which standard grade Fe-Si-Mn was produced. The metal rich alloy was further processed to recover Cu, Co and Ni in pure form.

Acknowledgement

The authors wish to express their gratitude to Director, National Metallurgical Laboratory for granting permission to publish this paper. The authors owe their gratitude to Ministry of Earth Sciences, Govt. of India for financial support.

References

1. R.P. Das, and S. Anand, 1997, ''Aqueous reduction of polymetallic nodule for metal extraction.'' In: Proceedings of the Second Ocean Mining Symposium. International Society of Offshore and Polar Engineers (ISOPE), Seoul: Korea, 1997, 165–171.
2. D. Mishra, R.R. Srivastava, K.K. Sahu, T.B. Singh and R.K. Jana, "Leaching of roast-reduced manganese nodules in NH_3-$(NH_4)_2CO_3$ medium," *Hydrometallurgy*, 109(3-4). (2011), 215-220.
3. R. Sridhar, J.S. Warner, M.C.E. Bell, 1977, U.S Patent no 4049438.
4. S. Agarwal, K.K. Sahu, R.K. Jana, S.P. Mehrotra, 2009, Proceedings of the Eighth ISOPE Ocean Mining Symposium, (India), 2009, 131-136.
5. A.J. Monhemius, "The extractive metallurgy of deep sea manganese nodule, In Topics in non ferrous extractive metallurgy, R. Burkin Eds. Society of chemical industry, (London) 1980, 42.
6. F. Habashi, *Handbook of Extractive Metallurgy*, Volume 1, Part II, p. 421.
7. H.S. Ray, R. Sridhar, and K.P. Abraham, Extraction of Nonferrous Metals, Affiliated East-West press Pvt Ltd. (India), 1985, 19.

Ni-Co 2013
Edited by: Thomas Battle, Michael Moats, Violina Cocalia, Harald Oosterhof, Shafiq Alam, Antoine Allanore, Rodney Jones, Nathan Stubina, Corby Anderson, and Shijie Wang
TMS (The Minerals, Metals & Materials Society), 2013

ALTERNATIVE COOLANTS AND COOLING SYSTEM DESIGNS FOR SAFER FREEZE LINED FURNACE OPERATION

Mark William Kennedy[1,2], Per Nos[3], Mia Bratt[4], Mark Weaver[5]
[1]Department of Materials Science and Engineering, Norwegian University of Science and Technology, N-7491 Trondheim, Norway.
[2]ProVal Partners SA, 1004 Lausanne, Switzerland
[3]Termek Technology, Rudssletta 70, Rud N-1351, Norway
[4]Elkem AS Research, Fiskåveien 1, Kristiansand NO-4675, Norway
[5]Alcoa Technical Center, 15069, Pennsylvania, United States
Communicating author: mark.kennedy@material.ntnu.no, m.kennedy@provalp.com

Keywords: furnace, coolant, safety

Abstract

Water coming in contact with liquid metal is well understood in the metallurgical industry to be an extreme hazard and historically has been at the root cause of many fatal incidents world wide. A wide range of alternative heat transfer fluids exist, which can be used to replace water in both common and special high performance applications. These fluids have a range of physical characteristics leading to different process performances, which can be improved by changes in the configuration and design of the cooling system and furnace cooling elements. This article will examine a number of coolants including natural and synthetic organics, silicones, fluorinated oils, and lead-bismuth eutectic alloy. Peak transient heat fluxes (often where significant damage occurs), Critical Heat Flux and other key copper cooler design parameters will be discussed. Safety tests to determine the suitability of new fluids for use as furnace coolants are discussed and proposed.

Introduction

Water cooling is used extensively in many types of modern furnace designs to allow for higher operational intensity (i.e. increased throughput) and longer campaign lives (e.g. up to 10 or more years). Blast furnaces use water cooled plate and stave coolers, tap holes, and tuyeres for hot blast injection, amongst other equipment. Electric smelting furnaces use a number of cooled components as shown in Figure 1 [1], from electrode contact clamps, electrode seals, tap holes, feed chutes and cooled lining systems for use in the slag zone and more controversially in the metal or matte zones. Examples of advanced cooler designs are shown in Figure 2 [2]. Ferroalloy furnaces also use water cooled components, particularly in the furnace hoods, off-gas ducts, electrode clamps, slag zone (if slag is present) and around tap holes.

Nearly all commercial furnaces operating today use water as the cooling media due to its abundance, low cost and acceptable heat transfer properties. However, Elkem Solar operates a commercial oil cooled (Therminol® 66) furnace supplied by Tenova-Pyromet in their new facility in Kristiansand Norway and an oil cooled (Marlotherm® SH) Elkem Multi-Purpose Furnace (EMPF®) has been operational for more than 25 years in Iceland at Steinull H.F.

Recently Alcoa (in cooperation with Elkem) have been piloting high intensity (1 MW/m^2) slag resistance furnaces at temperatures exceeding 2000 °C, as part of the Carbothermic Aluminum project [3]. Fluorinated oil (Galden® HT 200) and currently Mono-Ethylene-Glycol (MEG) have been used by the project for deep cooling of slag/metal furnace panels. The use of coolants other than water has been found to reduce the peak pressure experienced during submerged leakages, avoid most equipment damage and most importantly, prevent injury to personnel.

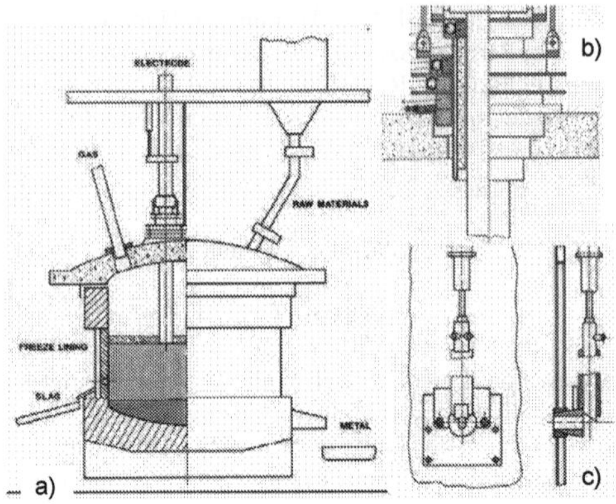

Figure 1. Overview of some cooled smelting furnace structures. a) Elkem Multi-Purpose Furnace®, b) electrode seal, and c) slag flow controller [4]. Cooled structures include: freeze-lined sidewalls, electrode clamps and collar, feed chutes, tap hole and slag flow controller.

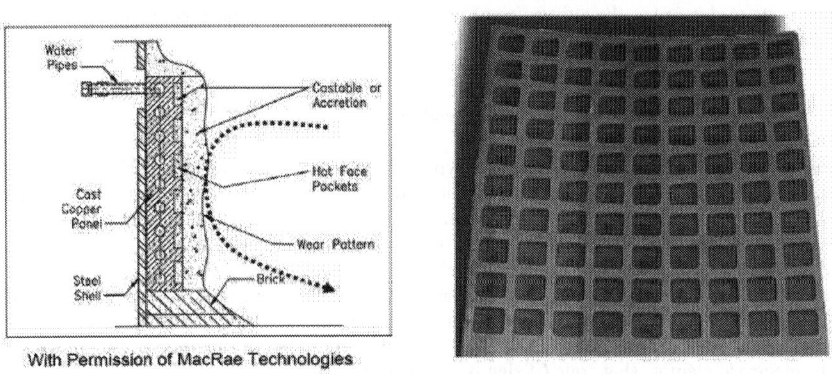

With Permission of MacRae Technologies

Figure 2. (a) Advanced copper cooler with cast-in-pipes and a patterned hot face. Freeze lining and 'natural' convective slag flow at the hot face are shown. (b) Photograph of such a cooler [2].

This report will examine some of the safety issues involved in the use of water as a furnace coolant, explore some of the alternative furnace coolants, and the design considerations required to use them successfully. This report will also suggest various active and passive strategies to achieve a higher level of safety for plant equipment and more importantly the personnel.

Safety Issues Associated with Water Cooled Furnace Components

Modern coolers such as pictured in Figure 2, have been installed on second and third generation smelting furnaces [1] and have brought the cooling media deep into the furnace, under the slag and more recently under the matte or metal levels. The use of such panels results in coolant being present only centimeters away from liquid materials with high temperatures and often high superheats.

The explosive potential of water in contact with molten smelting materials (metal, matte or slag) has been graphically displayed in many tragic and often fatal accidents in the metallurgical industry. One well documented incident occurred at Elkem Thamshavn and involved the leakage of large quantities of water from the electrode clamps into a silicon reduction furnace [5]. The resulting structural damage is documented in Figure 3. Similar incidents have occurred in ferroalloy [6] and platinum smelting [7], resulting in multiple fatalities, extensive equipment damage or both.

Figure 3. Photograph showing structural damage suffered by Elkem Thamshavn silicon furnace in 2006 water-metal explosion, which resulted in 1 fatality. Photograph by S. Fugelsøy [8].

Investigations have taken place primarily in the nuclear industry, to study the interaction between molten metal/oxide, e.g. liquefied fuel, fuel cladding or primary reactor coolant (e.g. liquid sodium), with water [9-11]. In some cases, metal and water are known to react to produce hydrogen gas, which adds to the explosion hazard [11]. This hazard was graphically illustrated by the recent events at Fukushima Japan beginning on March 11, 2011, where the build up of hydrogen gas lead to the explosion and failure of the reactor containment vessels and reactor buildings [12]. Similar effects have been reported in titanium and other reactive metal production [13, 14]. In vessels containing carbon and low oxygen, hydrogen can be produced by gasification from the water-gas reaction:

$$C_{(s)} + H_2O_{(g)} = CO_{(g)} + H_{2(g)} \quad (1)$$

Reaction (1) may have played a role in a multiple fatality resulting from melted water cooled components in a modern water cooled blast furnace. The root cause of this accident was operation without proper back up cooling systems in place and the subsequent reduction of cooling water flow during mechanical maintenance of the pumping system [15].

Few investigations have taken place in the metallurgical industry looking at the interaction of water even with common materials like steel and aluminum. One notable exception with high quality empirical data on the explosive potential of water falling onto liquid metal, has been published by Babaitsev and Kuznetsov [16]. Their data have been summarized in Tables I and II, showing that in many cases, water dropped onto liquid metal results in explosions with a force equal to 1 kg of TNT equivalent per kg of water, or higher.

Table I. TNT Equivalent of Explosion Resulting from the Specified Mass of Water Interacting with 1000 kg of a Given Molten Metal [16].

Metal	TNT equivalent of specified mass of water				
	1 kg	5 kg	10 kg	50 kg	100 kg
Steel 12Kh18N9T	0.82	4	7.8	33	57
Aluminum	3.17	15.9	32	176	200

Table II summarizes the resulting damage from an 80 kg equivalent explosion caused by 26 kg of water interacting with molten aluminum. Table II clearly indicates the inadvisability of placing control rooms next to furnaces, as graphically illustrated in the recent deaths of 6 people at a ferroalloy facility [6].

Table II. Effects due to 26 kg of Water Interacting with 1000 kg of Liquid Aluminum [16].

After-effects of explosion	Distance, m	Pressure, kPa
Breakage of glass	250	2
Collapse of roof	41	14
Destruction of interior walls	23	28

The physics of water thermal explosions, so called 'steam explosions' are complex and beyond the scope of this document. They often appear to act the same as Boiling Liquid Expanding Vapor Explosions (BLEVE's) [17, 18], but typically occur in open systems with no clear means to superheat the water prior to the phase change. They do however, involve a complex interaction of boiling heat transfer (e.g. film boiling), phase changes (vapor formation and solidification), contact area (droplet fragmentation), etc. Extremely rapid gas expansions are produced under the correct conditions of mutual phase dispersions (high interfacial area) and high metal superheats. Self-propagating (sonic) shock waves can then result [9, 10] with or without a triggering event. Due to the hazards associated with deep cooling using water, it behooves us to explore economically viable alternative cooling media, which have lower explosion potential in contact with liquid material and most particularly molten matte or metal.

Heat Transfer Theory and Design Considerations for Metallurgical Furnace Coolers

Before alternative coolants can be discussed, it is first necessary to have a firm grasp of the heat transfer theory and design considerations associated with metallurgical furnace coolers. The largest heat loss in a modern high intensity smelting furnace is from the bath-line coolers that are used to maintain furnace integrity below the slag level. These coolers also carry the greatest possibility of creating an explosion, due to the immediate proximity of the melt and the probability of releasing coolant below the upper surface of the molten material. Therefore bath-line cooler design (see Figure 2) will be discussed in detail. A review on modern furnace cooler technology is available by Verscheure *et al.* [19] and recommended for further information.

From an energy balance stand point, the most important design factor is the average cooler heat flux. The heat flux experienced by a given cooler can be calculated by:

$$\frac{Q_c}{A_c} = h\left(T_{bath} - T_{hotface}\right) \qquad [1]$$

where Q_c is the heat flow [W], A_c is the cooler area [m^2], h is the heat transfer coefficient between the melt and the cooler [W/m^2·K], T_{bath} is the temperature of the bath material arriving immediately adjacent to the 'stagnant' zone near the wall [K] and $T_{hotface}$ is the hot face temperature of the cooler [K]. Often $T_{hotface}$ is taken as the liquidus temperature, $T_{liquidus}$.

If the hot face is below the liquidus temperature, then h will contain both a convective component and an added heat flux due to the heat associated with solidification of the melt. Under dynamic conditions, such as exist during recovery from a slag spall (see Figure 4) or after a significant change (increase or decrease) in furnace power, the heat of solidification will add or subtract significantly to the convective heat flux component. Under steady state conditions it is conceivable to solidify some of the melt in the proximity to the wall, e.g. solidified onto suspended solid nuclei and convect such solid material back into the bulk liquid phase, where it is re-melted, i.e. it is possible to operate with a $T_{hotface}$ between the liquidus and the solidus of the bath material [20] creating a mushy zone. The mushy zone on the hot face of a copper cooler is a complex and dynamic system that is a topic for current study [21, 22].

For the purpose of discussion, it can however, be assumed that at steady state $T_{hotface} = T_{liquidus}$ in which case h becomes the heat transfer coefficient for a vertical plate undergoing natural convection heating. The magnitude of h is then of the order of 425 [W/m^2·K] for a cooler immersed in only 'silicate' slag. More precise estimates accounting for different material properties can be made using [20, 23, 24]:

$$Nu_{m-c} = 0.32\left(Gr_{m-c}Pr_m\right)^{0.3} \qquad [2]$$

where Nu_{m-c} is the Nusselt number at the melt-cooler interface [unitless], Gr_{m-c} is the Grashof number [unitless], and Pr_m is the Prandtl number [unitless].

$$Nu_{m-c} = \frac{h_{m-c}L}{k_m},\ [3] \qquad Gr_{m-c} = \frac{g\rho_m^2\beta\left(T_{bath} - T_{hotface}\right)L^3}{\mu_m^2},\ [4] \qquad Pr_m = \frac{C_p\mu_m}{k_m},\ [5]$$

where h_{m-c} is the natural convection heat transfer coefficient at the melt-cooler interface [W/m^2·K], L is a characteristic length, i.e. the height along the cooler [m], k_m is the thermal conductivity of the molten material [W/m·K], g is the acceleration due to gravity, 9.81 [m/s^2], ρ_m is the melt density [kg/m^3], β is the melt volumetric expansion coefficient [m^3/m^3·K], μ_m is the viscosity [Pa·s] and C_p is the heat capacity of the liquid melt [J/kg·K].

Equation [1] implies that the sidewall heat flux will increase with bath superheat, and superheat increases significantly at higher smelting rates in many smelting processes [25]. With typical slag superheats in the range of 50-250 K, a heat transfer coefficient of 425 [23] [W/m^2·K] implies a Q_c/A_c of about 21-106 kW/m^2. These fluxes are in the range typically experienced by slag coolers, which are not in contact with metal or matte or experiencing forced convection, due to poor furnace design or operating practices. Molten metal and matte typically produce about one order of magnitude higher heat fluxes and require refractory to be placed in front of the coolers, i.e. direct contact between matte/metal and coolers should be avoided.

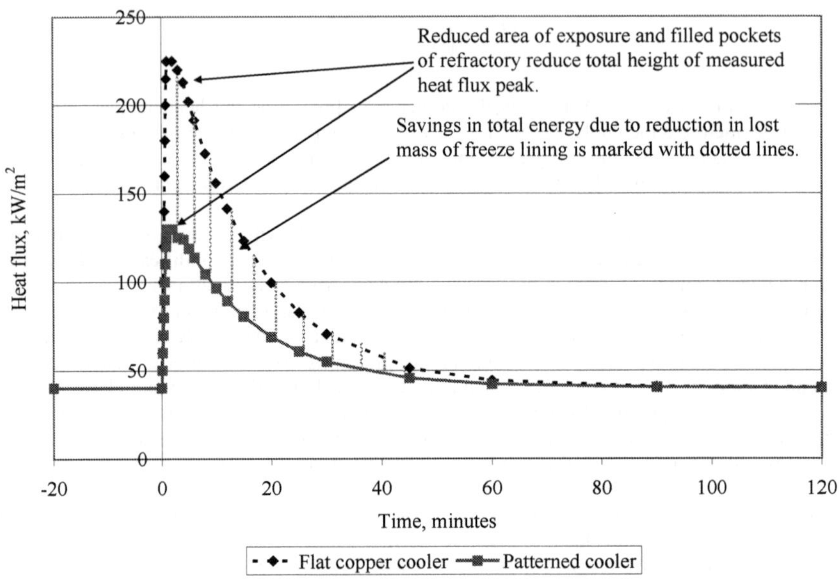

Figure 4. Illustrative example of the dynamic responses of a flat copper cooler and copper cooler with a hot face pattern, due to a slag spall occurring at time zero. A reduced peak heat flux and lower total energy loss are indicated for a copper cooler with a hot face pattern.

Safety Considerations in Advanced Furnace Cooler Design

In designing a forced convection cooled furnace component, the most critical factor is the expected peak heat load or heat flux that it is expected to withstand under emergency conditions, as visualized for a slag spall in Figure 4. Peak heat flux is often difficult to estimate and is scenario dependent. Not all emergency scenarios can be identified a priori. If previous operational data for similar furnaces are available (for at least one entire smelting campaign), then the design or peak heat flux, should be specified as being equal to the expected average plus 4.5 σ (standard deviations), as this should result in a failure rate of <1 in the life of a typical smelting plant due to excessive heat flux.

Bath-line coolers can experience several common types of emergency situations:

1) Contact with matte or metal at the hot face (e.g. for coolers designed for slag operation),
2) Transition from natural convection to forced convection heat transfer at the hot face,
3) Corrosion of the material of construction (typically copper),
4) Spontaneous and catastrophic loss of hot face protective layer (either refractory or frozen bath material, i.e. a spall as shown previously in Figure 4),
5) Damage due to the conduction of electric current (steel panels are particularly susceptible), and
6) Loss or reduction of coolant flow.

Furnace level control is critical to preventing metal or matte from rising onto a slag cooler and leading to extreme heat flux. Failure under abnormal heat flux conditions can also be encountered due to transition to forced convection, e.g. due to a slag 'boil', combined with highly superheated liquid [26]. Corrosion of copper coolers by sulfur [27-29] resulting in water leaks and explosions have also been reported [27, 28, 30, 31]. Monitoring systems are available to detect corrosion, leaking water passages and extreme local heat fluxes in copper coolers and particularly in tap hole blocks, which see especially severe service [32-38].

Refractory failures sometimes occur, particularly when the thickness of the residual refractory or frozen slag is thin (e.g. < 25 mm) and it becomes mechanically unstable. The use of a high conductivity refractory lining will result in a thicker more mechanically sound hot face protection for a given heat flux boundary condition and can provide added protection against chemical corrosion, due to reduced mass-transport by diffusion [28]. A thick highly conductive refractory also provides added thermal mass in a critical near 'burn-out' heat flux condition, by absorbing more total energy in the process of melting.

Rapid changes in furnace power can result in spalling of the freeze lining from the hot face of the coolers due to mechanical stresses induced by thermal expansion. Such spalls result in a characteristic spike, followed by a first order exponential decay on a panel's heat flux with time. Refractory pockets as shown in Figure 2, reduce both the height of the spike (the peak emergency heat flux) and the area under the response curve (total heat loss), due to the fact that less slag must be refrozen and less heat of fusion must be withdrawn as illustrated in Figure 4. A hot face pattern also improves the mechanical adhesion of slag to a panel and reduces both the probability and severity (total area) of a spall. Panels with no hot face pattern can therefore not be recommended for use in a modern furnace design. If "dove-tail" grooves are used in the hot face, these can also be used as a means of securing refractory to the cooler, for added integrity in the lining system.

The steady state heat flux that a cooler must withstand is often known with some accuracy, either due to calculation (e.g. from Equations [1-5]), piloting or previous commercial operation. The peak heat flux can be considered a known-unknown [39], i.e. it is known that it will occur, but the magnitude is uncertain. Estimations can be made using heat transfer modeling for certain dynamic scenarios, such as a simulated spall or partial refractory failure. These values can be compared to the mean heat flux and a suitable safety factor applied to the nominal value for design purposes. It is recommended that peak design heat flux, should be at least 4 times higher than the nominal heat flux over the full area of the cooler [40]. Where higher ratios can be achieved within conventional design limits, they should be used.

A simple 2D simulation has been performed using COMSOL® 4.2a, to model a cast cooler as shown in Figure 5. Copper with 80% of the theoretical copper thermal conductivity, i.e. $k = 320$ W/m·K, has been used to simulated phosphorus deoxidized copper. The copper cooler shown in Figure 5 is equipped with 'dove-tail' grooves containing refractory with a $k = 1.8$ W/m·K. A heat flux of 1 MW/m^2 has been applied to the hot face, by manipulating the hot face temperature (set to 1100 °C). It can be seen that the sharp tips of the copper at the hot face are at incipient melting and are well above the recommended 400 °C maximum copper hot face temperature for continuous cooler operation without oxidation [41].

Figure 5 clearly indicates that if embedded thermocouples or other sensors are to be used to monitor copper cooler temperature or estimate local heat fluxes, that they must extend past the plane of the cooling passages. Outside this plane, the copper has essentially the temperature of the cooling media; 120 °C in the example shown. The copper cold face has been modeled as undergoing combined natural convection and radiative heat loss (~5.6 kW/m^2), while the other sides of the cooler are perfectly insulated to simulate the presence of similar coolers. The modeling of the cooling channels has been simplified to be a constant 120 °C (simulating high velocity forced convective cooling).

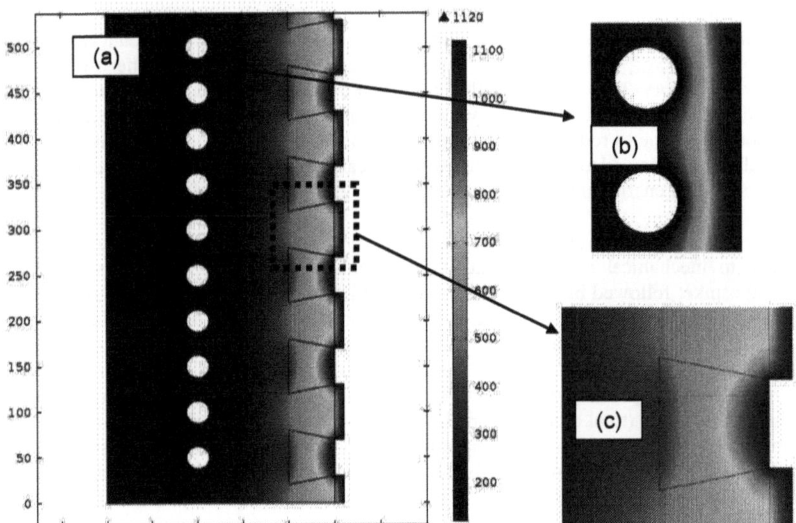

Figure 5. (a) Simplified 2D FEM model of a furnace cooler (half section shown) at steady state, simulating a critical 'emergency' heat flux of 1 MW/m^2, 1100 °C hot face temperature, 10 mm of frozen slag on the copper, and exposed refractory lined dove-tail grooves on the hot face. (b) Close up of a cooler, where the color scale has been modified such that red indicates 200 °C and dark blue 120 °C. (c) Close up of one refractory lined pocket showing incipient melting of the copper in the corners at the hot face.

Critical Heat Flux (CHF) and Cooler 'Burn-Out'

It is highly recommended to design such that the peak heat flux can be transferred into the cooling media without boiling. The concern is that if boiling takes place that vapor can obstruct the flow of coolant or that the heat flux will exceed the Critical Heat Flux (CHF). At CHF, nucleate boiling transitions to film boiling, where the heat transfer surface becomes blanketed with gas, insulating the surface and reducing heat transfer, as indicated in Figure 6. Past the CHF the furnace cooler temperatures increase rapidly and 'burn-out' quickly ensues.

The cooling channel heat transfer coefficient and total heat flow can increase by orders of magnitude before the transition to film boiling occurs as indicated in Figure 6 [42], due to the enhancement of heat transfer by vapor bubble agitation. Figure 6 shows a CHF of 1 MW/m^2 for water under stagnant or 'pool boiling' conditions. Using forced convection cooling, the coolant velocity can be used to increase CHF by one or more orders of magnitude, by stripping away vapor before it can insulate the heat transfer surface. Methods for predicting and enhancing CHF have been extensively studied by those designing modern nuclear reactors [43].

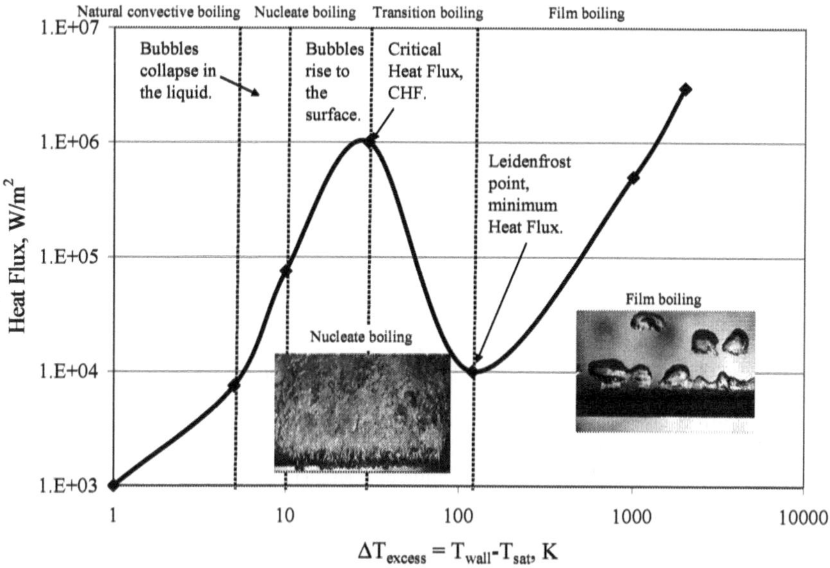

Figure 6. 'Pool' boiling heat transfer for water as a function of excess temperature T_{wall} - T_{sat}, showing improved heat transfer (and heat transfer coefficients) up to the Critical Heat Flux (CHF), [42] often referred to as the "Burnout Heat Flux" in furnace literature. Past the CHF, film boiling blankets the inner heat transfer surface as shown in the inserted photo, causing the furnace cooler to rapidly heat and melt.

If care is taken in the panel design to prevent low velocity regions near the tops of panels, which can become easily blocked by vapor, and if a sufficiently high coolant velocity is used (>1 m/s and preferably between 2 and 3 m/s or even higher for extreme conditions), then film boiling and 'vapor lock' can be prevented until heat fluxes are so high, that the copper approaches melting temperatures as indicated in Figure 5. It must be understood that such a 'boiling' case represents a critical emergency and not a normal 'design case'. Permanent loss of copper due to melting is the 'true' limiting factor for well designed copper cooling elements. Maximizing the Copper Melting Heat Flux and not the Burn-Out Heat Flux is the most reasonable target in modern cooler design.

For the transition to film boiling, the location of the critical heat flux is not at the panel hot face, but at the internal cooling channels. Therefore to design robust panels, the maximum internal to external surface area should be created. The panel picture in Figure 5 uses 25 mm bored passages, on 50 mm centers, resulting in a ratio of internal/external heat transfer areas of 1.5. So hot face heat flux of 1 MW/m^2 is reduced to an average of 667 kW/m^2 over the area of the internal cooling channels. Figure 5 b) shows that it is unlikely that the side of the cooling channels opposite the cooler hot face will fully participate in cooling, emphasizing the importance of a very large total area, particularly facing the hot face (alternate shapes e.g. triangular, square or oval channels could conceivably be used if the panel were of an extruded design).

Internal Cooling Channel Heat Transfer Coefficient Estimation

For fully developed forced convective cooling in smooth internal passages, at turbulent Reynolds numbers, the Colburn equation [44] can be used to estimate the Nusselt number:

$$Nu_{w-c} = 0.023\, Re_i^{0.8}\, Pr_c^{1/3} \qquad [6]$$

where Nu_{w-c} is now the internal wall-coolant Nusselt number, Re_i is the cooling channel Reynolds number [unitless] and the Pr_c is the coolant Prandtl number, all evaluated using the average bulk fluid properties. Average properties are evaluated at $(T_{inlet}+T_{outlet})/2$. Equation [6] is sometimes multiplied by the Sieder-Tate viscosity relation $\left(\frac{\mu_c}{\mu_w}\right)^{0.14}$ [45], when large temperature differences are experienced between the wall and bulk of the fluid. μ_c is the average bulk coolant viscosity and μ_w is the coolant viscosity [Pa·s] evaluated at the average wall temperature. With use of the Sieder-Tate relation the calculation becomes iterative in nature.

$$Nu_{w-c} = \frac{h_{w-c} D_h}{k_c},\ [7] \quad D_h = 4\frac{Channel\ area}{Channel\ perimeter},\ [8] \quad Pr_c = \frac{C_p \mu_c}{k_c},\ [9] \quad Re_i = \frac{\rho_c v_i D_h}{\mu_c},\ [10]$$

where h_{w-c} is the average turbulent forced convection heat transfer coefficient between the channel wall and the bulk coolant [W/m^2·K], D_h is the hydraulic diameter of the channel [m], k_c is the average thermal conductivity of the coolant [W/m·K], C_p is now the mean heat capacity of the coolant [J/kg·K], ρ_c is the average bulk coolant density [kg/m^3], and v_i is the average velocity in the cooling channel [m/s].

Comparison between the Cooling Effectiveness of Water and Alternative Coolants

A number of heat transfer fluids can be considered to replace water as a cooling media. Natural and synthetic organic oils (one or two phase) or other organic compounds (e.g. Mono-Ethylene-Glycol), silicones, fluorinated oils and even liquid metals (such as Pb-Bi) can be considered. A 2-phase heat transfer fluid is a substance, which boils before decomposition takes place.

A number of fluid options are listed in Table III. Heat transfer coefficients have been evaluated using Equations [6] through [10] at 3 m/s velocity, using 25 mm channel diameters and 1.5 surface area ratio, as shown previously in Figure 5. A peak emergency heat flux of 4 times the highest slag heat flux of 106 kW/m^2, has been selected for discussion purposes. Two phase fluids are highlighted in blue. For reference the copper cooler hot face temperature is ~200 °C higher than the indicated cooling channel surface temperature for the given heat flux.

Table III. Comparison of Different Coolant's Physical and Heat Transfer Properties, Evaluated for 3 m/s, 25 mm Channel Diameter and at a 424 kW/m^2 'Peak' Design Heat Flux [46-52]

Cooling fluid	Approximate Range of Operation Min/Max °C	Temp for Ref. Data, °C	Density, kg/m^3	Heat Capacity, J/kg·K	Thermal Conductivity, W/m·K	Viscosity, mPa·s	Prandtl Number	Reynolds Number at 3 m/s	Nusselt Number at 3 m/s	25 mm Tube Forced Convection Heat Transfer Coefficient, W/m^2·K	Cooling Channel Surface Temperature, °C for 424 kW/m^2 heat flux	Approximate Copper Hot Face Temperature, °C
Water	0/100	20	1000	4183	0.604	1.0	6.9	75000	348	8414	54	252
MEG	-13/240	60	1087	2343	0.261	5.2	47	15682	188	1964	204	403
Dowtherm A	15/400	60	1028	1687	0.132	1.7	22	44818	339	1790	218	417
Galden HT 200	-85/300	60	1715	1080	0.063	2.1	36	61475	514	1295	278	477
Texatherm HT 22	-50-320	60	885	1936	0.126	7.8	119	8506	158	798	414	613
Marlotherm SH	-5/350	60	1016	1700	0.125	8.2	112	9259	165	826	402	601
Therminol 66	-32/345	60	982	1699	0.116	11.5	169	6388	141	653	493	691
Xceltherm 500	-62/260	60	766	2327	0.134	2.8	49	20487	236	1269	283	482
Syltherm 800	-40/400	60	899	1677	0.128	5.1	67	13181	185	943	360	558
44.5%Pb-55.5%Bi (123.5 °C)	200/600	225	10150	146	14.2	1.4	0.015	535714	215	122304	227	426

The following can be observed from Table III at the selected conditions:

1) Water at 3 m/s is well within its design limits, and has a safe heat flux limit of ~1 MW/m^2 under these conditions.
2) Eutectic Pb-Bi is more than an order of magnitude improvement over water as a heat transfer fluid.
3) Texatherm® HT 22, Marlotherm® SH and Therminol® 66 (marked in yellow) will not maintain the bulk of the copper below the recommended maximum of 400°C, which substantively eliminates long term oxidation [41]; however, as this represents a critical emergency and not a steady state operating condition, this is not a serious design issue.
4) MEG, Dowtherm® A, and Galden® HT 200 (marked in blue) may experience boiling at the copper interface depending on the system operating pressure and local peak heat fluxes, as these are 2-phase fluids.
5) Syltherm® 800 a silicone based oil, and Xceltherm® 500 an organic one phase oil, both show improved performance over the traditional organic based one phase oils.

The combination of high internal surface area (1.5 times larger than the hot face area) and high fluid circulation rate of 3 m/s, has permitted organic and silicone based fluids to perform within thermal limits at 400% of the maximum expected steady state slag only heat flux, thus achieving a good standard of safety. At 3 m/s all of the coolants shown in Table III are capable of protecting the hot face of the copper from melting at heat fluxes up to ~600 kW/m^2. The reduced emergency peak heat flux achievable by the use of dove-tail grooves as shown in Figure 4, further improves safety and the technical viability of organic coolants.

While seemingly exotic, Pb-Bi eutectic alloy has been investigated in the laboratory as a cooling media, since 1951 and has previously been used to cool Russian nuclear submarine reactors [52]. 4th generation reactor research is currently investigating the use of Pb-Bi coolants and a great volume of design information is available upon which the metallurgical industry could build. At 0.1 m/s circulating velocity, Pb-Bi could satisfactorily maintain the temperature of a copper cooler exposed to a heat flux of up to 4.5 MW/m^2, while using a fraction of the pumping cost of water (even after considering the high density of the alloy). Pb-Bi eutectic alloy having a 'pour' point of 125 °C, does not represent a significant engineering challenge. At high velocity, e.g. 1 m/s, Pb-Bi would allow metallurgical engineers to construct processes that could not be conceived of otherwise.

In the event of a Pb-Bi leak into or out of the furnace, neither fire nor explosion will occur. Pb-Bi has the advantage of being non-flammable and virtually non-chemically reactive in contact with liquid water, unlike sodium or potassium. The environmental implications of a lead 'leak' into a high temperature smelting vessel require evaluation before such a coolant could be applied in the metallurgical industry.

Some coolants such as MEG, Dowtherm® A and Galden® HT 200, are 2-phase fluids like water. Under extreme heat transfer conditions, 2-phase fluids will boil, enhancing heat transfer as shown previously in Figure 6. Boiling is preferable to the alternative, which for organic oils is permanent thermal degradation. The most critical emergency that a cooler can experience is a complete loss of coolant circulation. In such an emergency, pool boiling will ensue for 2-phase fluids. The cooled structures will remain at the boiling point of the fluid, until such time as the coolant contained in the equipment and its associated piping has 'boiled dry'. This provides a substantial time for system recovery or plant evacuation. The piping can be designed to ensure that gravity supply of coolant is maximized, by placing reservoirs high within the system and ensuring that the resulting vapor can be vented safely from the system on the return-side headers.

Special Safety Issues Associated with Alternative Coolants

The use of alternative heat transfer fluid to water is often discounted based on assumptions, rather than proper technical analysis. As shown in the previous section, most of the commonly available heat transfer fluids can safely be used for most furnace applications, when combined with advanced cooler designs such as those pictured in Figures 2 and 5, and high fluid circulating velocities. Some coolants, like liquid metals, significantly improve on the performance of water as a heat transfer media, even at greatly reduced circulating velocities.

It is often assumed that oils and other fluids will create BLEVE-like explosions when released under superheated liquid metal in a manner similar to water. Controlled testing by Elkem and Alcoa, have shown that the peak pressure produced by fluids other than water is significantly reduced. Elkem performed relatively small scale experiments (sub-kg scale) using liquid filled metal ampoules plunged into superheated silicon and aluminum. Alcoa has performed larger scale testing (including flammability) with liquid aluminum. Anecdotal experience from leakages in commercial and large scale pilot operation have corroborated the laboratory testing, indicating that organic heat transfer fluids represent a safer alternative based on explosion risk.

The precise mechanism for the reduced peak pressures from the alternative fluids is not understood, but could be due to the: energy absorbed by cracking of the large molecules, reduced vapor pressures at the metal temperature, or the lower thermal conductivity of the fluids.

Fire risk is an issue associated with organic oils, which is not present with water. The heats of combustion of all of the organics listed in Table III are high and of the same magnitude as 'gasoline'; however, the limited empirical evidence is that they are difficult to combust in open air in the liquid state. Testing should be conducted to determine the energy released per unit area and time [W/m^2] for these fluids at different initial ignition temperatures. This would simulate the liquid release of hot coolant, followed by subsequent ignition and provide more practical information useful for the selection of furnace coolants. Standardized tests exist to guide metallurgical engineers in the development of targeted methods for such flammability testing [53]. Research has also been published on the use of halogen additives in hydraulic oils for military tanks to reduce flammability [54]. Research into additives for furnace cooling oils (and hydraulic systems) is therefore recommended to reduce the intensity of fires and hopefully produce self extinguishing cooling oils.

Galden® is a fluorinated oil and is completely non-flammable. The decomposition products of Galden® at extreme temperature and in the absence of oxygen are primarily COF_2 and CF_4. While low levels of HF have been detected during explosion testing, toxic levels of HF have not been observed in practice, although HF can theoretically be produced by the reaction of COF_2 and water vapor. Concerns about HF formation have lead to the replacement of Galden with MEG by Alcoa for the Carbothermic Aluminum project.

Conclusions and Recommendations

Water cooled components have allowed metallurgical vessels to simultaneously achieve greater productivity and longer campaigns; however, the use of water has exposed plant workers to significant risk in the form of 'steam explosions'.

Viable technical alternatives to the use of water exist, in the form of organic and fluorinated fluids and liquid metal. It is recommended that the metallurgical industry take action to explore these alternative heat transfer fluids.

Improved and standardized testing methods for explosivity and flammability should be identified, developed if necessary and adopted to allow systematic analysis of alternatives. Oil additives should be explored that might allow for the creation of self extinguishing cooling (and hydraulic) oils.

Pb-Bi eutectic alloy is perhaps the best alternative to water for the metallurgical industry, offering massively improved cooling capability, lower pumping costs and the elimination of BLEVE-like explosions. The issue of environmental contamination in the event of a leak into a smelting vessel needs to be addressed, and active development is recommended.

References

1. M.W. Kennedy, H. Haaland, J.A. Aune, and C. Headquarters, "High Intensity Slag Resistance Furnace Design," *Proceedings of the Conference of Metallurgists, Nickel-Cobalt 2009*, CIM, Sudbury, Ontario, (2009) 101-110.
2. G. Slaven, A. MacRae, and L. Valentas, "The Implementation of Ultralife™ Copper Casting Technology in the EAF," *AISE*, Pittsburgh, Pennsylvania, (2003), 1-9.
3. C.V. White, Ø. Mikkelsen, and D. Roha, "Status of the Alcoa Carbothermic Aluminum Project," *Light Metals*, (2012), 81-88.
4. G. Ellefsen and Ø. Hallquist, "The Elkem Multi-Purpose Furnace® Used for Slag Production, Technical Description with Plant Features," *presented at the Elkem seminar*, Bhubaneswar, India, (1989), 1-17.
5. H. Tveit, M. Garcia, H. Delbeck, A.T. Haug, B. Saugestad, and I.J. Eikeland, "Water Leakages in Ferroalloy and Silicon Reduction Furnaces - Experience Gained from a Severe Accident in 2006," *presented at the Silicon for the Chemical and Solar Industry IX*, Oslo, Norway, (2006), 1-16.
6. "Assmang Plant Operations Ban Lifted," *in South African Government News Agency*, SAnews.gov.za, (2008).
7. J. Robinson, "Lonmin Plc - Update on Smelter Incident," *Lonmin Press Release*, (2004).
8. C. Solbakk Johnsen, "Kritisk Etter Thamshavn-Ulykken," *in Adresseavisen*, Trondheim, (2006).
9. M. Rizk, "A Review of Large Scale and Small Scale Underwater Thermal Explosions," *DTIC Document*, (1990), 1-118.
10. P.E. Shick, T.M. Grace, and I. o. P. Chemistry, "Review of Smelt-Water Explosions" *Institute of Paper Chemistry*, (1982), 1-69.
11. D. Cho, D. Armstrong, and R. Anderson, "Combined Vapor and Chemical Explosions of Metals and Water," *Nuclear Engineering and Design*, vol. 155, (1995), 405-412.
12. D. Biello, "Partial Meltdowns Led to Hydrogen Explosions at Fukushima Nuclear Power Plant " *in Scientific American*, (2011).
13. E. Poulsen, "Safety-Related Problems in the Titanium Industry in the Last 50 Years," *JOM Journal of the Minerals, Metals and Materials Society*, vol. 52, (2000), 13-17.
14. S.C. Evans and D.F. McLaughlin, "A Thermodynamic Investigation into Reactive-Metal Melting-Furnace Explosions," *JOM Journal of the Minerals, Metals and Materials Society*, vol. 57, (2005), 49-52.
15. "The Explosion of No. 5 Blast Furnace, Corus Uk Ltd, Port Talbot, 8 November 2001," *Health and Safety Executive*, UK, (2008), 1-60.
16. I. Babaitsev and O. Kuznetsov, "Energy of Explosions Occurring When Water Falls onto a Layer of Molten Metal," *Metallurgist*, vol. 45, (2001), 185-188.
17. T. Abbasi and S. Abbasi, "The Boiling Liquid Expanding Vapour Explosion (BLEVE): Mechanism, Consequence Assessment, Management," *Journal of Hazardous Materials*, vol. 141, (2007), 489-519.
18. J. Venart, G. Rutledge, K. Sumathipala, and K. Sollows, "To BLEVE or Not to BLEVE: Anatomy of a Boiling Liquid Expanding Vapor Explosion," *Process Safety Progress*, vol. 12, (1993), 67-70.
19. K. Verscheure, A.K. Kyllo, A. Filzwieser, B. Blanpain, and P. Wollants, "Furnace Cooling Technology in Pyrometallurgical Processes," *Sohn International Symposium, Advanced Processing of Metals and Materials* Vol 2., Eds. F. Kongoli and R.G. Reddy, TMS, San Diego, California, USA, 27-31 August, (2006), 139-154.

20. H. Joubert, "Designing for Slag,"Freeze Linings" on Furnace Sidewalls – an Engineering Perspective," *Molten Slags, Fluxes and Salts*, Stockholm, Eds. S. Seetharaman and DU Sichen, Sweden-Helsinki, (2000), 12-17.
21. F.J. Guevara and G.A. Irons, "Simulation of Slag Freeze Layer Formation: Part I. Experimental Study," *Metallurgical and Materials Transactions B*, (2011), 652-663.
22. F.J. Guevara and G.A. Irons, "Simulation of Slag Freeze Layer Formation: Part II: Numerical Model," *Metallurgical and Materials Transactions B*, (2011), 664-676.
23. D. Robertson and S. Kang, "Model Studies of Heat Transfer and Flow in Slag-Cleaning Furnaces," *Fluid Flow Phenomena in Metals Processing*, (1999), 157-168.
24. S. Kang, "A Model Study of Heat Transfer and Fluid Flow in Slag-Cleaning Furnaces," *Dissertation Abstracts International (USA)*, vol. 52, (1992), 1-182.
25. J. Merry, J. Sarvinis, and N. Voermann, "Designing Modern Furnace Cooling Systems," *JOM Journal of the Minerals, Metals and Materials Society*, vol. 52, (2000), 62-64.
26. L. Nelson, R. Sullivan, P. Jacobs, E. Munnik, P. Lewarne, E. Roos, M. Uys, B. Salt, M. de Vries, and K. McKenna, "Application of a Highintensity Cooling System to Dc-Arc Furnace Production of Ferrocobalt at Chambishi," *SAIMM Journal*, vol. 104, (2004), 551-561.
27. B.M. Thethwayo and A. Garbers-Craig, "Corrosion of Copper Coolers in PGM Smelters," *The 4th International Platinum Conference, Platinum in transition 'Boom or Bust'*, SAIMM, (2010), 187-194.
28. B.M. Thethwayo, "Sulphidation of Copper Coolers in PGM Smelters," *Masters Thesis*, University of Pretoria, (2010).
29. B.M. Thethwayo and A. Garbers-Craig, "Laboratory Scale Investigation into the Corrosion of Copper in a Sulphur-Containing Environment," Corrosion Science, (2011), 3068–3074.
30. T. Raymond, "Polokwane Smelter Shutdown Update," *Anglo American Platinum Press Release*, (2005).
31. "Lonmin Shuts Number One Furnace for a Week," *in miningmx*, (2009).
32. A. Sadri, P. Gebski, H. Ghorbani, G. McGarrie, and T. de Vries, "Monitoring Deterioration of Waffle Cooler Thickness at Polokwane Smelter," *JOM Journal of the Minerals, Metals and Materials Society*, vol. 61, (2009), 69-73.
33. M. Hopf and E. Rossouw, "New Opportunities-Exhaustive Monitored Copper Coolers for Submerged Arc Furnaces," *SAIMM Journal*, vol. 106, (2006), 373-378.
34. J. Eksteen, B. Van Beek, and G. Bezuidenhout, "Cracking a Hard Nut: An Overview of Lonmin's Operations Directed at Smelting of UG2-Rich Concentrate Blends," *SAIMM Journal*, vol. 111, (2011), 681-690.
35. R. MacRosty, S. Nitschke, T. Gerritsen, and C. Crowe, "Advances in Furnace Monitoring: Instrumentation," *Proceedings of the Sixth International Copper-Cobre Conference*, vol VII, August 25–30, Toronto, Canada, (2007), 203–215.
36. T. Plikas, L. Gunnewiek, T. Gerritsen, M. Brothers, and A. Karges, "The Predictive Control of Furnace Tapblock Operation Using CFD and PCA Modeling," *JOM Journal of the Minerals, Metals and Materials Society*, vol. 57, (2005), 37-43.
37. A. Sadri, P. Gebski, and D. George-Kennedy, "Development of the Taphole Acoustic Monitoring (TAM) System for Water-Cooled Copper Tapblocks," *47th Conference of Metallurgists*, Winnipeg, (2008), 7-19.
38. P. Tracy, R. MacRosty, Q. Zhao, L. Gunnewiek, and T. Gerritsen, "Tapblock Diagnostic System," *Proceedings of the Sixth International Copper-Cobre Conference*, vol. III, August 25–30, Toronto, Canada, (2007), 453–464.
39. W.P. Imrie, "New Technology Developments in the Production of Non-Ferrous Metals," *ERZMETALL*, vol. 59, (2006), 9-18.

40. F. Stober, T. Miraza, A. T. Hidyat, I. Jauhari, K. Belanger, D. Fowler, T. Gerritsen, A. Matyas, C. Nichols, and N. Voermann, "Furnace Upgrade with Hatch Technology at PT Antam Feni-II in Pomalaa, Indonesia," *INFACON XI*, New Delhi, India, (2007), 638-653.
41. U. Aniekwe and T. Utigard, "High-Temperature Oxidation of Nickel-Plated Copper vs. Pure Copper," *Canadian Metallurgical Quarterly*, vol. 38, (1999), 277-281.
42. Y.A. Çengel and A.J. Ghajar, *Heat and Mass Transfer: Fundamentals & Applications*: McGraw-Hill, (2011).
43. S.H. Chang and W.P. Baek, "Understanding, Predicting, and Enhancing Critical Heat Flux," *10th International Topical Meeting on Nuclear Reactor Thermal Hydraulics (NURETH-10)*, Seoul, Korea, October 5-9, (2003), 1-20.
44. A.P. Colburn, "A Method of Correlating Forced Convection Heat Transfer Data and a Comparison with Fluid Friction," *Transactions of the American Institute of Chemical Engineers*, vol. 29, (1933), 174-210.
45. E. Sieder and G.E. Tate, "Heat Transfer and Pressure Drop of Liquids in Tubes," *Industrial & Engineering Chemistry*, vol. 28, (1936), 1429-1435.
46. J.P. Holman, *Heat Transfer*: McGraw-Hill, New York, (1986).
47. Monoethylene Glycol (MEG). Available: http://www.chem-group.com/services/meg.tpl, accesed, September 3, 2012.
48. Dow Chemical Co. Supplied Computer Software: "Fluidfile [the Dow Chemical Co.]," (1995).
49. Sasol Product Information Sheet, "Marlotherm SH Heat Transfer Fluid," (2004).
50. Solvay Solexis Product Information Sheet, "Galden PFPE: Heat Transfer Fluids," www.solvaysolexis.com, (2004).
51. SOLUTIA Product Information Sheet, "Therminol 66 Heat Transfer Fluids by SOLUTIA," (2004).
52. P.N. Martynov, A.V. Gulevich, Y.I. Orlov, and V.A. Gulevsky, "Water and Hydrogen in Heavy Liquid Metal Coolant Technology," *State Scientific Center of the Russian Federation – Institute for Physics and Power Engineering*, (2004), 1-20.
53. L. Deleanu, D. Buzoianu, M. Rîpă, Ş. Crăciunoiu, and A. Drug, "Flammability Tests on Hot Surfaces for Industrial Fluids," *The Annals of University "Dunărea de Jos "of Galaţi, Fascicle VIII, Tribology*, (2007), 22-31.
54. B. Wright, R. Mowery, and M. LePera, "Survivability? It Could Be a Matter of Hydraulic Fluid," *Proc. 11th Annual US Army Ground Vehicle Survivability Symposium*, 27-30 March (2000), 1-14.

Ni-Co 2013
Edited by: Thomas Battle, Michael Moats, Violina Cocalia, Harald Oosterhof, Shafiq Alam, Antoine Allanore, Rodney Jones, Nathan Stubina, Corby Anderson, and Shijie Wang
TMS (The Minerals, Metals & Materials Society), 2013

Outotec's Ausmelt Top Submerged Lance (TSL) technology for the Nickel Industry

Ross Andrews[1], Robert Matusewicz[1], Lauri Aspola[1], Stephen Hughes[1]

[1]Outotec, Level 2, 2 Brandon Park Drive, Wheelers Hill, VIC 3150, Australia

Keywords: Nickel, TSL, Ausmelt, Outotec

Abstract

Beginning with early lab studies in the 1970's examining recovery of nickel from slags through to currently operating furnaces treating over 1 million tonnes annually of nickel concentrate, Outotec's Ausmelt Top Submerged Lance technology has developed to become a viable and proven method of treating nickel bearing materials. This paper covers the technology over the two decades since the first implementation of Outotec's Ausmelt TSL technology on a commercial furnace. Over this time period, the scale and capabilities of the technology have been greatly expanded. With an increasing focus on the sustainable use of metals with particular interest focusing on utilizing secondary sources, including metal bearing residues, this efficient and flexible technology is ideally suited to increasingly complex and changing feed materials.

Introduction

Nickel demand has slowly been increasing from 1.1 million tonnes to 1.57 million tonnes in 2011 [1]. Whilst the past 10 years has seen a steady growth in demand, between 2007 and 2010 the price and demand have shown large peaks and troughs. A key driving force in the growing demand can be attributed to China, particularly in the area of stainless steels.

Of the nickel produced about 59% comes from smelting nickel sulphide concentrate, another 30% from smelting nickel laterites, with the remainder coming from hydrometallurgical processing of nickel laterites [2]. Of the nickel sulphide concentrate that is smelted, in 2004 69% of these concentrates was smelted using flash furnace technology, while the remaining 31% was smelted using electric furnace technology [2]. Since 2004 Ausmelt technology has become a commercially accepted alternative to flash furnace and electric furnace smelting of nickel sulphides.

It has been estimated that about 60% of the nickel reserves are in the form of oxides (laterites) with the remainder sulphides [3]. An increasing proportion of nickel production is from laterites and predictions have been made that by 2012 more nickel will be produced from laterites than from nickel sulphides [4]. With the increasing importance of processing nickel laterites, focus will begin to move toward improving the processing techniques. Current smelting techniques involve electric furnace smelting of the nickel laterites following drying calcining/reduction [5]. Electric furnace smelting of nickel laterites is energy intensive requiring high energy inputs to achieve the high temperatures required. With the rising costs of energy, alternative technologies stand to take an increased proportion of the workload.

The development of Outotec's Ausmelt Top Submerged Lance (TSL) technology was derived from the original Sirosmelt research undertaken at the CSIRO in Australia. The use of a submerged lance to deliver fuel and air/oxygen was quickly seen as advantageous to bath smelting processes. Some of the initial investigations [6] examined recovery of metals such as tin, copper and nickel from slags. It is directly from this research work that Outotec's Ausmelt TSL process was born, via the desire to commercialize and improve metals extraction technology.

Outotec® Ausmelt Top Submerged Lancing (TSL) technology utilizes a lance to inject fuel and oxygen enriched air into the furnace bath. This results in intense mixing and thus promotes rapid reactions within the bath. The Ausmelt TSL technology is used in many commercial furnaces across a wide range of scales and processing a wide range of materials such as copper, lead, nickel, tin, zinc, platinum group metals (PGM), ferrous and various industrial waste materials. Figure 1 shows a typical configuration of an Ausmelt furnace.

Figure 1: Schematic of a typical Outotec Ausmelt furnace

With over 50 Outotec Ausmelt furnaces started up since 1992, the technology has matured and been accepted by the market as providing reliable, cost effective, environmentally friendly and highly flexible pyrometallurgical processes.

Initial Commercialization of the Outotec Ausmelt technology for Nickel

Two industrial Outotec Ausmelt furnaces were commissioned in the 1990's dedicated to treating nickel leach residues; one in Eiffel Flats, Zimbabwe and the other in Bindura, Zimbabwe. These initial steps represented the growing commercial application of the technology and the growth of expertise within nickel process applications. Table I provides a list of the commercial operations using the Outotec Ausmelt for treatment of nickel bearing materials.

Table I: Outotec Ausmelt smelting and converting operations for Ni-containing materials

Client	Location	Year	Feed	Design	Product
Jilin	China	2009	Concentrate	275,000	LG Matte
Jinchuan	China	2008	Concentrate	1,100,000	LG Matte
Anglo Platinum	South Africa	2002	EF Matte	213,000	HG Matte
Bindura Nickel	Zimbabwe	1995	Residues	10,000	Blister Cu
RTZ	Zimbabwe	1992	Residues	7,700	HG Matte

Throughout this period pilot plant test work continued. Nickel processes investigated ranged from smelting of concentrates to produce low grade matte, to the converting of low grade (LG) matte into high grade (HG) matte. This work lead to the next significant nickel process for Outotec Ausmelt technology.

Traditionally nickel matte from a smelting furnace has been treated in Peirce-Smith converters to produce a matte low in iron (known as a Bessemer matte). This material is then refined hydrometallurgically into nickel metal.

Operating difficulties inherent in Peirce-Smith converters are well known [7]. Peirce-Smith converters operate in a batch-wise manner, and hence the offgas from these converters varies in composition and volumetric flowrate. These variations increase the difficulty of offgas treatment. The converters are poorly sealed leading to fugitive SO_2 and fume emissions causing hygiene and environmental issues.

One key problem area with Peirce-Smith converters is their limited ability to control temperature. Hot matte transfer is a requirement for the converters and they are limited in their ability to process cold revert material. The management of movements of ladles and charging boats within the converter aisle becomes vital to efficient operations. Additional problems also relate to blockage of tuyeres and poor control of the end point for the converter blow.

An attractive alternative to Peirce-Smith converting is to perform the matte converting in an Ausmelt furnace. This is done commercially at Anglo Platinium's Rustenburg site in South Africa, where an Ausmelt furnace is used to convert a low grade Ni/Cu/PGM electric furnace matte [8]. Converting of a conventional nickel matte derived from flash smelting has also been successfully carried out on a pilot plant scale [9]. Pilot plant work has even studied direct production of high grade matte from concentrates. A flowsheet of the process at Anglo Platinum, as well as a flowsheet for basic nickel matte converting, is shown in Figure 2.

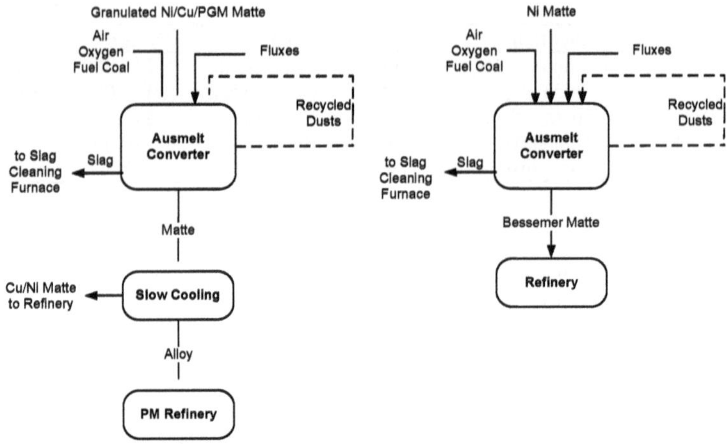

Figure 2: Flowsheet of Anglo Platinum's matte converting (left) and basic matte converting (right)

The process for converting nickel matte is achieved by feeding low grade nickel matte with fluxes to achieve a target slag chemistry. The Ausmelt lance provides fuel, air and oxygen to manage the process. Dust is collected in the offgas system and recycled back to either the smelter or the converter to maximize recovery of nickel.

The oxygen enriched air and fuel that are injected into the converter bath via the Ausmelt lance serve to control the temperature of the matte and reduce the iron levels in the matte by oxidising the FeS as shown in Equation 1. Oxidation of the bath will also lead to magnetite formation in the slag and oxidation of the Ni_3S_2 from the matte to NiO which reports to the slag, Equations 2 and 3. The iron level in the final matte may be customised to suit downstream requirements. A downstream requirement for low iron in matte must be balanced against nickel losses to the slag, for the lower the iron level achieved in the final matte, the higher the resulting nickel level will be in the discard slag.

$$FeS_{(matte)} + O_{2(gas)} \rightarrow FeO_{(slag)} + SO_{2(gas)} \quad (1)$$
$$3FeO_{(slag)} + \tfrac{1}{2}O_{2(gas)} \rightarrow Fe_3O_{4(slag)} \quad (2)$$
$$Ni_3S_{2(matte)} + 3\tfrac{1}{2}O_{2(gas)} \rightarrow 3NiO_{(slag)} + 2SO_{2(gas)} \quad (3)$$

Nickel matte may be converted continuously in an Ausmelt furnace. This can result in benefits such as: improved control of bath chemistry and temperature, and consistent offgas composition and flowrate. This makes treatment less difficult and the continuous operation allows an increased furnace throughput to be achieved [7]. A single Ausmelt furnace may be capable of treating the same tonnage of matte as three Peirce-Smith converters combined. Economic decisions may result in utilizing two Ausmelt converting furnaces to provide the same level of redundancy that comes with a series of Peirce-Smith converters.

A significant amount of pilot plant scale work has been conducted processing nickel laterites in an Ausmelt furnace [10]. The Ausmelt furnace was shown to be able to treat a full range of ores, and has the ability to utilise lower grade ores and indigenous fuels (e.g. low value coals). With the increasing significance of processing nickel laterites it is becoming more likely that nickel laterites may be processed in an Ausmelt furnace on a commercial scale.

The desulphurisation of nickel mattes to sulphur contents of <5% is required to provide a feed suitable for Sherritt-Gordon refining. The copper content of the matte has a significant influence on how readily the matte may be desulphurised. A nickel matte with a high copper content may be desulphurised to a significant extent in a Peirce-Smith converter. This cannot be done in a Peirce-Smith converter with a nickel matte low in copper.

The desulphurisation of nickel matte with low copper content has been successfully achieved in an Ausmelt furnace on a pilot plant scale [12]. This area shows promise for full scale developments in the future.

Continued Growth of Nickel Processing in an Outotec Ausmelt Furnace

The use of an Ausmelt furnace as an alternative to either a flash furnace or an electric furnace to treat nickel sulphide concentrates has gained commercial acceptance. A flowsheet incorporating an Ausmelt furnace to smelt nickel sulphide concentrate is shown in Figure 3. In the same manner as matte converting, nickel sulphide concentrates are fed into the Ausmelt furnace along with fluxes to control slag chemistry. Any dusts from the Ausmelt furnace and downstream furnaces are recycled to maximize recovery of nickel. Fuel and enriched air are injected through the Ausmelt lance into the bath to maintain the bath at the target temperature and react the nickel sulphide with oxygen to form a matte rich in Ni_3S_2. A discard slag is generated along with an offgas rich in SO_2, see Equations 4 to 7. The gas injected through the lance into the bath also provides agitation promoting rapid reaction.

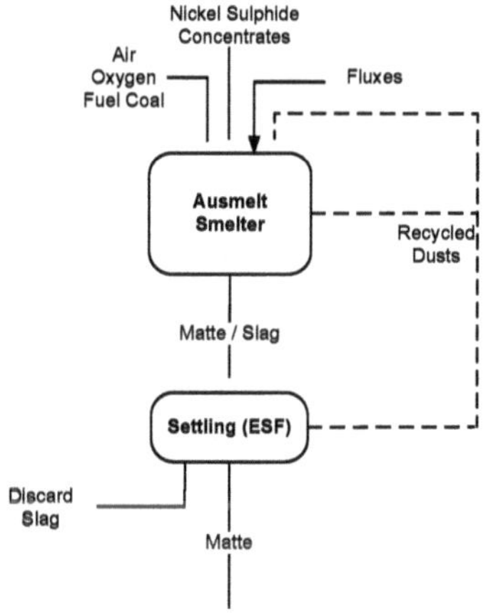

Figure 3: Flowsheet for smelting nickel sulphide concentrate in an Ausmelt furnace

$$3(FeS)_2.NiS_{(concentrate)} + O_{2(gas)} \rightarrow Ni_3S_{2(matte)} + 6FeS_{(matte)} + SO_{2(gas)} \quad (4)$$
$$Fe_7S_{6(concentrate)} + 9\tfrac{1}{2}O_{2(gas)} \rightarrow 7FeO_{(slag)} + 6SO_{2(gas)} \quad (5)$$
$$FeS_{(concentrate)} + 1\tfrac{1}{2}O_{2(gas)} \rightarrow FeO_{(slag)} + SO_{2(gas)} \quad (6)$$
$$2CuFeS_{2(concentrate)} + 4O_{2(gas)} \rightarrow Cu_2S_{(matte)} + 2FeO_{(slag)} + 3SO_{2(gas)} \quad (7)$$

The choice of slag chemistry will be optimised for the specific chemical composition of the feed, with appropriate fluxes added with the feed to achieve the desired slag chemistry. An MgO-SiO_2-FeO phase diagram constructed using Factsage [11] is shown in Figure 4. The smelt slag target chemistry will generally be located in the shaded region, dependant on the feed concentrate composition. It can be seen that as MgO content increases so too does the liquidus temperature and hence the required operating temperature. As higher MgO levels are associated with many nickel concentrates, it may not be possible to process feeds with higher MgO contents in a flash furnace due to process temperature limitations. However both an Ausmelt furnace and an electric smelting furnace are capable of tolerating higher operating temperatures.

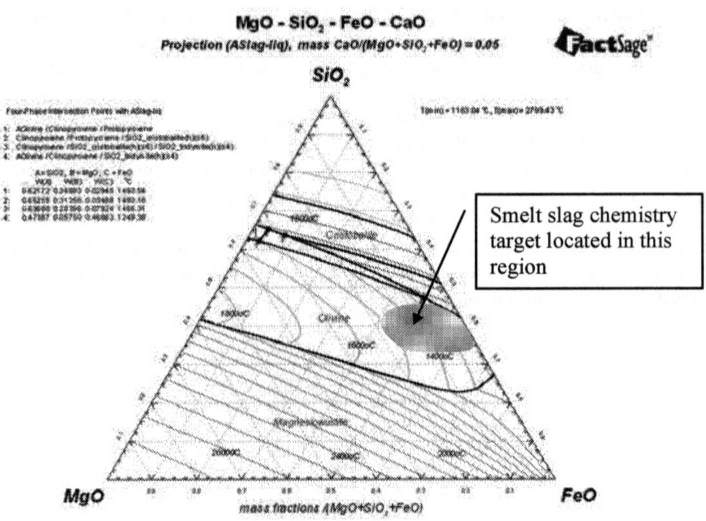

Figure 4: MgO-SiO_2-FeO ternary phase diagram generated using FACTSAGE [13]

The continuous operation of the Ausmelt furnace allows a greatly increased furnace throughput. However the continuous operation of the intensely mixed Ausmelt furnace does not give opportunity for separation of the slag and matte phases. This requires settling and phase separation to occur in a downstream settling furnace. This furnace will generate a slag for discard and a low grade matte from the settling furnace is sent for converting while the slag is discarded.

In late 2008 the Ausmelt furnace at Jinchuan China was commissioned. A cooperative relationship was formed between JNMC, Ausmelt Limited and China Nonferrous Engineering and Research Institute (ENFI) to take knowledge from these parties and jointly develop a smelting technology called JAE Nickel Smelting Technology [14]. This furnace is rated to treat more than 1,000,000 t/y of nickel sulphide concentrate, making it of a scale comparable to the largest nickel flash furnaces and exceeding the scale of the largest nickel electric furnaces.

A second Ausmelt furnace at Jilin China was commissioned in late 2009. This furnace is rated to treat 250,000 t/y of nickel sulphide concentrate, as well as a small amount of secondary nickel material. These commercial examples show that Ausmelt is an industrially accepted alternative to flash furnace or electric furnace treatment of nickel sulphide concentrate.

The application of Ausmelt technology in the nickel industry has established commercial acceptance both in the smelting of nickel sulphide concentrate and in the converting of matte.

Solutions

With the addition of the Ausmelt technology to Outotec's suite of solutions, Outotec can now provide tailored solutions in nickel processing, from hydrometallurgical processing to pyrometallurgical processes, or a combination of technologies to provide a final saleable end product. Options include:

- Primary smelting in either a Flash Furnace or an Ausmelt TSL
- Direct smelting in a Flash Furnace
- Slag cleaning in an Electric Furnace
- Nickel matte refining (leaching process, solvent extraction and electrowinning/reduction)

Additional relevant Outotec solutions centre around traditional smelter solutions and include:

- Offgas Handling
- Sulphuric Acid Plant

An idea of the total plant solution available is seen in Figure 5. This range of solutions allow Outotec to tailor a well integrated process to specific client needs, ensuring cost efficiency and maintaining high environmental standards.

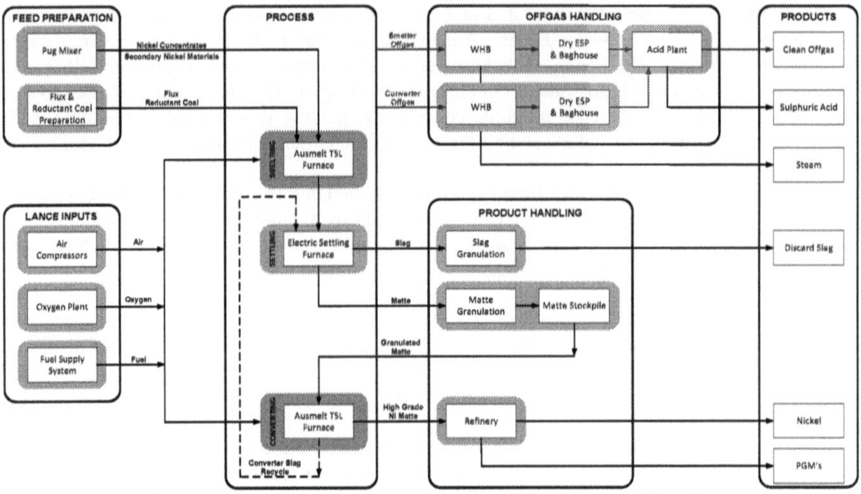

Figure 5: Overall Outotec Ausmelt TSL Nickel Process Flowsheet

Conclusions

From a concept examined in the 1970's, extensive work and effort has brought the Outotec Ausmelt TSL technology for nickel from its first small commercial scale implementation to the current large scale facilities being used today. This has resulted in the efficient and effective processing of a range of different nickel feed materials including some previously difficult to treat feeds.

References

1. International Nickel Study Group, About Nickel: Production, usage and prices, accessed 1 October 2012, URL: http://www.insg.org.

2. A.E.M. Warner, C.M. Diaz, A.D. Dalvi, P.J. Mackey, A.V. Tarasov and R.T. Jones, "JOM World Non-ferrous Smelter Survey, Part IV- Nickel-Sulphides", Journal of Metals, Vol. 59, No. 4, 2007, 58-72.

3. C.M. Diaz, C.A. Landolt, A. Vahed, A.E.M Warner and J.C. Taylor, "A Review of Nickel Pyrometallurgical Operations", Journal of Metals, Vol. 48, No. 9, 1988, 28-33.

4. A.D. Dalvi, W.G. Bacon, and R.C. Osborne, "Past and the Future of Nickel Laterite Projects," Laterite Nickel 2004, TMS, W.P. Imrie and D.M. Lane, Eds., Warrendale, USA, 2004, 1-27.

5. A.E.M. Warner, C.M. Diaz, A.D. Dalvi, P.J. Mackey and A.V. Tarasov, "JOM World Non-ferrous Smelter Survey, Part III- Nickel-Laterites", Journal of Metals, Vol. 58, No. 4, 2006, 11-20.

6. J. M. Floyd, D. S. Conochie, "Nickel Recovery from Slags". Paper to the Extractive Metallurgy Symposium, Melbourne, 1975, 9 pp

7. J. Wood, R. Matusewicz and M.A. Reuter, "Ausmelt C3 Converting", The International Peirce-Smith Converting Centennial Symposium, J. Kapusta and T. Warner Eds., TMS, Warrendale, USA, 2009, 397-406.

8. M. Jacobs, "Process Description and Abbreviated History of Anglo Platinum's Waterval Smelter", Southern African Pyrometallurgy 2006, R.T Jones Ed., South African Institute of Mining and Metallurgy, Johannesburg, South Africa, 2006, 17-28.

9. K.R. Robilliard; R.W. Matusewicz and B.R. Baldock, "The Ausmelt Process for Smelting Nickel Concentrates and Further Treatment of the Matte Produced" CIM 33rd Conference of Metallurgists '94, Recent Developments in Non-Ferrous Pyrometallurgy Symposium, Toronto, Canada, August 21-25, 1994, (Unpaged).

10. J. M. Floyd; B. W. Lightfoot and W. E. Short, "Application of the Ausmelt Process to Treatment of Nickel Laterite Ores", Nickel/Cobalt Laterites Seminar, Alta Metallurgical Services, Melbourne, Australia, 4-5 May, 1995, (Unpaged).

11. Factsage [1976-2009], Version 6.0, Thermfact and GTT-Technologies. URL: http://www.factsage.com.

12. B.R. Baldock; I.L. Chard; J.M. Floyd; K.R. Robilliard; B.J. Elliot and G.M. Willis, "Desulphurisation of High-grade Nickel Matte from Kalgoorlie Nickel Smelter using Ausmelt Technology". Paul E. Queneau International Symposium, Extractive metallurgy of copper, nickel, and cobalt, R.G Reddy and R.N. Weizenbach Eds, Denver, USA. Vol. 1, 1993, 1081-1100.

13. J.A. Coveney, B.R. Baldock, S. Hughes and M.A. Reuter, "Ausmelt Technology For Nickel & PGM's Processing", Pyrometallurgy of Nickel and Cobalt 2009, J. Liu et al, Eds, Sudbury, Canada, August 23-26, 2009, 169-180

14. M. Zhou, A. Wan, G. Li, B.R. Baldock and H. Li, "Industrial Operations of JAE Nickel Smelting Technology at Jinchuan Nickel Smelter", International Symposium on High-Temperature Metallurgical Processing: Smelting and Reduction Processes, TMS, Seattle, USA, February 14-18, 2010

Ni-Co 2013
Edited by: Thomas Battle, Michael Moats, Violina Cocalia, Harald Oosterhof, Shafiq Alam, Antoine Allanore, Rodney Jones, Nathan Stubina, Corby Anderson, and Shijie Wang
TMS (The Minerals, Metals & Materials Society), 2013

Processing of PGM containing Ni/Cu bulk concentrates in a sustainable way by Outotec Direct Nickel Flash Smelting process

Satu Jyrkonen[1], Kaarlo Haavanlammi[2], Matti Luomala[1], Janne Karonen[1], Paivi Suikkanen[3]

[1]Outotec (Finland), Kuparitie 10; FIN-28330 Pori, Finland
[2]Outotec (Finland), Puolikkotie 8; FIN-02201 Espoo, Finland
[3]Boliden Harjavalta, Teollisuuskatu 1; FIN-29200 Harjavalta, Finland

Keywords: Direct Nickel Flash Smelting, bulk concentrates, PGM

Abstract

Large deposits of low grade mixed Ni/Cu sulphide ores with important PGM values exist in North America and elsewhere. The challenge is producing saleable concentrates at acceptable recovery degrees. Producing bulk concentrates and treating them by Outotec Direct Nickel Flash Smelting (DON) to matte and then hydrometallurgical refining, offers a reliable route from concentrate to product. In the DON process high grade matte with low Fe-content is produced in the Flash smelting furnace (FSF) and metal values in slag are recovered in an electric furnace (EF). Matte refining is based on atmospheric pre-leaching and total pressure leaching. Nickel, cobalt and iron are leached leaving a leaching residue containing small amount of gangue minerals and PGM's. Iron is removed from the leach solution in an autoclave to produce high quality hematite as by-product. Cobalt and nickel will be recovered with standard methods. Overall > 95% metal recovery can be expected with low energy consumption and low environmental impact.

Introduction

Since laterite projects are energy and capital intensive exploiting low-grade sulphide deposits have become much more attractive alternative. Especially low grade Ni/Cu ores have not been fully exploited to date due to metallurgical challenges. Separate, saleable Ni and Cu concentrates would naturally be ideal but flotation usually does not produce very good recoveries when producing separate concentrates. Thus, bulk concentrates are the best reasonable product from mill. Recovering both base metals and PGM's efficiently from above-mentioned bulk concentrates is however, possible with Outotec's tailored process solutions (combined pyro and hydro). This paper discusses the applicability of Outotec Direct Nickel Flash Smelting process and subsequent hydrometallurgical processes on processing Ni - Cu - PGM sulphide concentrates into end products. Some experiences of Cu containing Ni concentrate from Boliden Harjavalta's DON smelter are also presented.

Outotec Direct Nickel Flash Smelting Process

50% of the world annual primary Ni-production from sulfide raw materials is processed by Outotec Flash Smelting technology. The original concept of Ni flash smelting was in operation in Harjavalta but since 1995 (Boliden) Harjavalta has utilized the DON technology. The same technology was applied to Fortaleza, Brazil, in 1998. As shown in the flow sheet in Fig. 1, the concentrate and flux are dried before feeding to the flash smelting furnace. Then the concentrate is fed to a flash smelting furnace (FSF) where it is smelted and oxidized to produce high-grade matte, slag and SO_2 containing off-gas. Oxygen enrichment is used in the process gas and thus no additional fuel is needed in the reaction shaft. Matte is tapped from the furnace and granulated and thereafter it is refined by hydrometallurgical methods to recover copper, nickel, cobalt and PGMs.

SO_2 containing process gas from the FSF is cooled and de-dusted by a heat recovery boiler and a electrostatic precipitator (ESP). After that the gas goes to a sulphuric acid plant. Flue dust is re-circulated to FSF.

Because the oxidation degree in the FSF is high, some nickel, copper and cobalt oxidize and end up to the slag phase. Therefore the FSF slag is laundered to an electric furnace, where the metal values are recovered from the slag by coke reduction. To adjust the liquidus temperature of EF matte (increase S-level of matte), some concentrate is injected into the EF. The EF process produces metallized matte, which is taken to granulation directly from the furnace. Also EF matte is treated by hydrometallurgical methods. The slag coming from the EF is granulated, and it is suitable to be used e.g. for sandblasting or clean enough to be discarded.

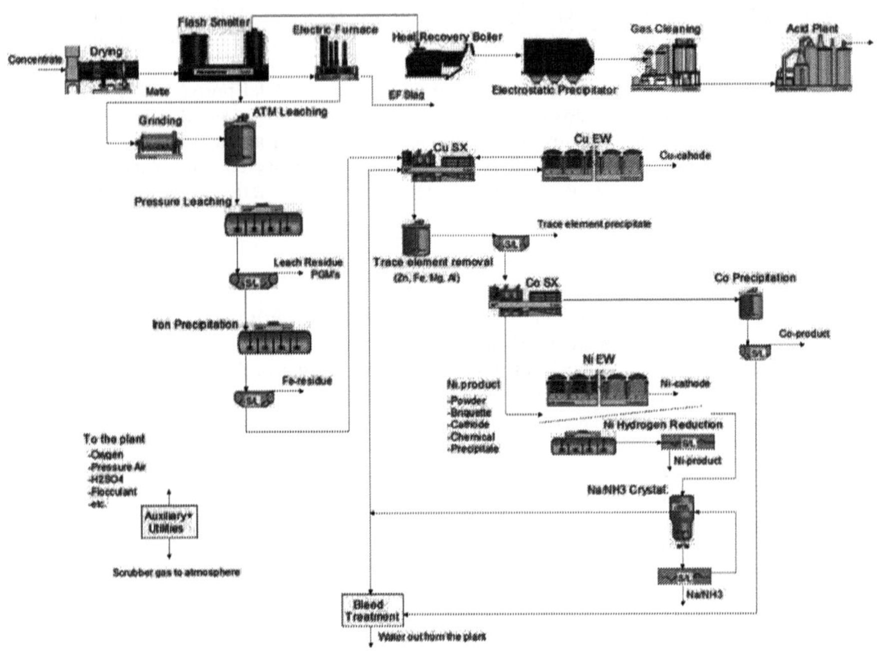

Figure 1. Smelter and refinery flowsheet.

Features of DON Process

Raw Material Flexibility

Nickel content in sulphidic nickel concentrates ranges normally from 5 to 12% Ni. Concentrates from Western Australia have higher nickel concentrations typically 20% and even higher, up to 25%. Copper contents in these concentrates are very low, even below 0.1% Cu while in other concentrates copper contents are relatively high, 1-6%. An essential constituent in nickel concentrate is often cobalt, which is more valuable than nickel. Concentrations of cobalt vary in range of 0.2-1%. High recovery of cobalt is of great importance to the profitability of a smelter.

A wide variety of Ni concentrates ranging from low-copper raw materials to concentrates with high Cu as well as from MgO-lean feed mixtures to those with low Fe/MgO can be successfully smelted in the DON process.

Flexibility and Variability of the DON Process

The DON technology allows optimization of the smelting and refining process conditions according to the requirements of raw materials and the products, as well as other boundary conditions. Oxidation degree in the flash smelting furnace can be selected based on the feed composition and required products specifications. For example in Fortaleza, where nickel content of the concentrate is relatively low, the oxidation is less than in Harjavalta and the FSF matte contains more iron. Conditions of the subsequent slag cleaning furnace can be adjusted independently from the flash smelting. In the EF area, in addition to the industrial scale plant trials, an extensive CFD study has been conducted on the fluid flows and metal droplet settling in order to ensure sufficient metal value recoveries and to develop tools for modeling of the slag cleaning process. Also for the sulphidation of EF matte several methods are available. Harjavalta utilizes concentrate injection carried out batch-wise using a consumable steel lance immersed into the slag layer, while e.g. liquid flash smelting matte has been tested in Fortaleza. Concentrate injection has proved efficient and reliable for modification of the EF-matte composition.

The DON process can be operated also with a full EF matte circulation to the flash furnace. Thus, only a low iron FSF matte is produced which simplifies the refinery flow sheet as a result of decreased iron removal. This concept is applicable to feed materials, which do not contain such elements in significant concentrations, which would accumulate in the circulation between the furnaces, e.g. cobalt. On the other hand, the concentrates containing PGM's may benefit from this concept. Matte circulation has been successfully tested on a pilot scale. In addition, industrial tests have been carried out at Fortaleza.

Environmental Aspects

The best available technique, both environmentally and economically, is accepted on the BAT reference list of the EU. The flash smelting method has been classified on the BAT reference list, because it reduces emissions and energy consumption in metal production.

The biggest problem area in sulfur-dioxide emission control at smelters arises from converting which utilizes rotary (PS-) converters and the transfer of molten material in ladles, both open processes which release off-gases, including sulfur-dioxide into the working atmosphere. One major advantage of the DON technology is that it produces only one continuous, concentrated SO_2-gas stream, when the fluctuating PS converter gas flow is eliminated. Thus, investment costs especially in the gas cleaning section but also in the other areas of the sulphuric acid plant are lower. The DON smelting is continuous and no ladle transportation is needed but matte as well as slag are transferred via covered and hooded launders. For these reasons fugitive emissions are avoided and the working conditions inside the plant can be kept clean and hygienic.

Energy consumption

The Flash Smelting process is based on utilization of the feed material's internal energy for smelting. Sulfide compounds of the feed ignite, oxidize and release heat, acting as a fuel for the process and no, or small amount of external energy is needed for smelting. Moreover, the consumption of fossil fuel is low in the DON process because only two hot units are needed, thus the direct greenhouse gas emissions of the smelter are small compared to conventional processes where fossil fuel is used for smelting and for keeping standby converters hot.

Heat recovered from process off-gas as high-pressure steam can be used for drying the feed material or for electricity generation, reducing the demand for electricity from power plants and minimizing green house gas emissions.

Recovery of PGM's

In the DON smelter PGM's are effectively collected directly into high grade matte formed in the flash smelting furnace, and the amount of PGM's going into slag cleaning furnace is almost negligible. In the refinery, the PGM's containing leach residue can be further processed to concentrate the PGM's.

Experiences of High Copper Feed Material in Boliden Harjavalta

Normally, Boliden Harjavalta's nickel smelter operates feed mixture containing about 10-12% Ni and 1-2% Cu in single-burner Outotec FSF. Due to the change in concentrates available for smelting, nickel and copper contents in feed mixture changed quite drastically in 2010. Consequently, ratio of Ni/Cu in the feed mixture of FSF reduced from about 8 to as low as 2.

The smelter has managed the rise of Cu content in the feed mixture very well. At first, some problems were encountered e.g. with increased build-up formation in the uptake shaft and WHB throat area. The role of copper in the build-up problem is somewhat unclear – apparently the prevailing balance in the dust chemistry was altered and the flue dust became stickier. Problems were overcome mostly by introducing extra flow of pressurized air to the settler/uptake shaft area to cool down the temperature of dust particles below their "liquidus" temperature.

Above-mentioned change does not affect on the control of smelting process, i.e. on the oxidation degree, although short periods of lower degree of oxidation have also been practised. According to the earlier experiences from risen Cu contents it was expected that some bottom build-up might occur. It hasn't, however, at any stage, posed any real threat to the smooth operation of FSF. Bottom build-ups are traditionally managed by feeding cast iron bars or by controlling bottom cooling when possible.

The flowability of Ni(Cu) matte has not reduced significantly despite the increased Cu content. Equally, the granulation of matte, by Outotec Matte Granulation, has remained trouble-free.

Furthermore, it seems that the total recoveries (combined FSF and EF) of nickel and copper have decreased. Decreased recoveries result partially from insufficient reduction capacity in EF but mainly because of decreased metal contents in concentrates.

Hydrometallurgical process concept

The ground DON matte is leached in atmospheric leaching reactors followed by total pressure leaching in autoclave. The solution from the leaching (PLS) is pumped to the iron removal and then to the metal recovery steps. Copper is extracted from the solution and recovered in copper electrowinning. Raffinate, now containing nickel and cobalt is forwarded to cobalt solvent extraction where cobalt is extracted and afterwards precipitated with hydrogen sulfide. Raffinate from cobalt extraction is pumped to nickel electrowinning or hydrogen reduction in autoclave. In both of these options, nickel is reduced to metal nickel. After the reduction water balance and neutralizing agent level is controlled with crystallization process.

Matte Grinding

The purpose of the grinding is to decrease concentrate solids particle size. This will be done in ball mill as wet grinding. The classification is done with hydro cyclones classification. Matte is fed with belt conveyor to the ball mill together with process water from the grinding thickener overflow. The target particle size is D80 <75 μm (200 mesh) and D50 <37 μm (400 mesh). Ground slurry will be pumped to the hydro cyclone classification circuit. Underflow of the cyclones returns to the ball mill as circulating load and the cyclone overflow continues to the grinding thickener. Flocculant is added to the thickener feed. The thickener underflow, grinded matte is then pumped to the leaching area. The grinding thickener underflow solid content can vary from 60-80 w-%, depending of the thickener type used. The thickener overflow water is circulated back to the ball mill as previously explained. Considering the water balance, the grinding circuit is closed circuit. Make up water has to be taken in the same amount that exits the grinding circuit as grinding thickener underflow moisture to the leaching circuit

Atmospheric Leaching

The matte slurry from the grinding circuit, is pulped with water, possible anolyte from the nickel electrowining, possible raffinate recycle from the copper solvent extraction and possibly with fresh sulphuric acid. The pulping and leaching is done in agitated atmospheric reactors. The main purpose of quick atmospheric leaching step is safety: The matte may contain some metallic nickel, cobalt, copper etc. elements. If the metallic elements react with the acid in non-aerated, non-agitated conditions, hydrogen gas may be produced as per formula (1) and create risk of explosion. It is necessary to dissolve these metallic elements as far as possible before the pressure leaching in autoclaves.

$$Me_{(S)} + H_2SO_{4(aq)} \rightarrow MeSO_{4(aq)} + H_{2(g)}, Me = Ni, Cu, Co, Fe\ etc. \quad (1)$$

The hydrogen formation risk is eliminated in the well agitated atmospheric leaching reactors. These reactors also have good ventilation system. Ventilation gases are washed in the gas scrubber.

The metallic compounds in the matte are leached as per the equation (2).

$$Me_{(s)} + H_2SO_{4(aq)} + 0{,}5O_{2(g)} \rightarrow MeSO_{4(aq)} + H_2O, Me = Ni, Cu, Co, Fe\ etc \quad (2)$$

Also some of the nickel containing minerals in the matte start to dissolve in the reactors (equation 3).

$$Ni_3S_{2(s)} + H_2SO_{4(aq)} + 0{,}5O_{2(g)} \rightarrow NiSO_{4(aq)} + NiS_{(s)} + H_2O \quad (3)$$

Depending of the case, the slurry from the atmospheric reactors may be pumped straight to the autoclave feed tank. If there is need to adjust the autoclave feed solid content, then the slurry from the reactors may first go to thickener. The thickener underflow is pumped to the pressure leaching feed tanks. Thickener overflow is directed straight to the iron removal circuit.

Pressure Leaching

In this step the remaining copper, nickel and cobalt are dissolved in an autoclave with the help of oxygen and sulfuric acid in process solution. Slurry from the atmospheric leaching step is pumped to autoclave feed tanks where solid content and acid concentration are adjusted by addition of recycle process solutions.

Slurry from the feed tank is fed to the autoclave and all components except PGM's are dissolved. The dissolved metals form aqueous sulphate solution as per the reaction presented below.

$$MeS + 2O_2 = MeSO_{4(aq)}, Me = Ni, Cu, Co, Fe\ etc \quad (4)$$

The leaching reactions are exothermal. Temperature in the autoclave is kept at about 140-160 °C with cooling coils (closed cooling water circuit) and the pressure is 15-16 bar. The autoclave vent gases are directed to the gas scrubbing system.

The pressure leaching autoclave is continuously operating autoclave. The autoclave is divided in several compartments and the slurry flows inside the autoclave from compartment to compartment as overflow. Every compartment is equipped with an agitator. Oxygen is fed into the slurry, under the agitator. The oxygen overpressure in the autoclave is about 7 bar. The results from existing operations and laboratory test work show total matte to final product recoveries of over 98%.

The slurry from the autoclave flows by the pressure difference to a flash tank. In the flash tank the slurry temperature decreases from 140-160 °C to about 95 °C. The evaporation gas from the flash tank flows to the gas scrubbing system.

From the flash tanks the slurry flows by gravity to the solids and liquid separation phase, first to the leach thickener. Thickener underflow slurry is pumped to the filter feed tank. The filter cake is washed to remove the acidic and PLS solution in the cake. The filtrate is pumped back to the main process. The further treatment of the leach residue depends of the amount of precious metals in the matte. If the precious metals concentration is high in the matte, filter cake containing insoluble precious metal components (such as iron silicates etc.) may be transported to the PGM cake treatment area. If the precious metals do not play important part, the filter cake is transported to the disposal pond.

Iron Removal in Autoclave

Solution from the atmospheric leaching thickener overflow and leach filtrate tank is pumped to the iron autoclave feed tank. From the autoclave feed tank, the solution is pumped through iron removal autoclave. The optimum operating temperature in the autoclave is 85-120 °C. The temperature is controlled with steam (steam coils) depending on situation. The oxygen overpressure in the autoclave is 1-2 bar.

The iron precipitation autoclave is also continuously operating and divided to compartments. Each compartment is equipped with an agitator. Soda or lime milk may be added to the autoclave feed tank to control pH. Iron precipitates either in the form of goethite (5) or hematite (6). The reactions are presented below.

$$2FeSO_{4(aq)} + 3H_2O + 0{,}5O_{2(g)} \rightarrow 2FeOOH_{(s)} + 2H_2SO_4 \quad \text{(Goethite)} \qquad (5)$$

$$2FeSO_{4(aq)} + 2H_2O + 0{,}5O_{2(g)} \rightarrow Fe_2O_{3(s)} + H_2SO_4 \quad \text{(Hematite)} \qquad (6)$$

The pH adjustment is done in this stage to keep the pH high enough to prevent the iron compound back dissolving. Slurry from the autoclave flows by pressure difference to a flash tank. The ventilation gas from the flash tank flows to the gas scrubbing system. From the flash tank, the cooled slurry flows to a pH adjust tank where the slurry pH is adjusted by $Ca(OH)_2$) to the level of 2-3. The pH adjustment is done in this stage to keep the pH high enough to prevent the iron compound back dissolving.

Iron residue slurry is pumped to a thickener where the underflow is pumped through a filter feed tank to residue filtration. After the filtration and the cake washing, the cake is collected into a container or dumper. Iron residue is transported to the lined disposal pond. Overflow of the thickener is pumped to the solvent extraction circuit.

If gypsum level in the whole plant solution circulation has to be controlled, the iron thickener overflow can also be treated in a cooling tower. Cooling tower is used to cool the process solution down to 40 °C. During the cooling some water is evaporated and more solid gypsum is formed to the surface of seed material received from the gypsum thickener underflow. The gypsum thickener is located after the cooling tower. If the gypsum removal circuit is used, the gypsum thickener overflow is the PLS that is pumped further in the process, to the copper solvent extraction circuit.

Copper and cobalt solvent extraction

Pregnant leach solution containing copper, cobalt and nickel, is pumped from the PLS storage tank through heat exchangers to the extraction trains. The PLS minimum temperature is 15 °C and maximum 50 °C when entering into solvent extraction. The extraction train contains DOP unit, spirok unit and mixer-settlers. After the last mixer settler, the aqueous phase is directed to raffinate after settler. From the after settler the solution is pumped to raffinate filter and to cobalt solvent extraction.

Organic phase from extraction train flows to loaded organic tank and then to a stripping stage. Some of the lean electrolyte from copper electrowinning is also fed to this stage. Organic phase from stripping stage goes back to extraction stages. Copper rich electrolyte is fed to copper electrowinning. After the CuSX the PLS is pumped to the cobalt solvent extraction. From there the $CoSO_4$(aq)-solution is transported to cobalt precipitation. The cobalt end product is cobalt sulphide CoS.

The basic chemistry in Cu- and CoSX is similar. In the different solvent extraction steps the pH of the solution is also controlled with ammonia solution (NH_3) or lye ($Na(OH)_2$).

Organic reagent removes impurities/metals from the pregnant leach solution according to reaction 7. RH is aldoxime or ketoxime (or a mixture).

$$2RH + Impurity^{2+} + SO_4^{2-} \rightarrow R_2Impurity + 2H^+ + SO_4^{2-} \quad (7)$$

Stripping reaction 8 is the reverse of reaction 7.

$$R_2Impurity + 2H^+ + SO_4^{2-} \rightarrow 2RH + Impurity^{2+} + SO_4^{2-} \quad (8)$$

Copper electrowinning

Copper is recovered from the electrolyte solution by connecting direct current between the lead anodes and the stainless steel cathodes. The electric current is fed to the electrolytic cells by one transformer-rectifier. The copper electrowinning products are copper metal cathodes, oxygen gas at the anode and regenerated sulphuric acid in the electrolyte solution. Reactions are shown in equations 9-11.

Cathode reaction:

$$Cu^{2+} + 2e^- \rightarrow Cu^o \qquad E^o = +0.34\ V \quad (9)$$

Anode reaction:

$$2H_2O \rightarrow O_2 + 4H^+ + 4e^- \qquad E^o = +1.23\ V \quad (10)$$

The net electrowinning reaction including sulphate ions:

$$Cu^{2+} + SO_4^{2-} + H_2O \rightarrow Cu^o + 0{,}5O_2 + 2H^+ + SO_4^{2-} \quad (11)$$

The standard normal potential (against SHE) needed for copper electrowinning is approximately 2 V and is made up of:

Theoretical voltage for reactions	~ 0.9 V
Oxygen deposition over voltage on lead	~ 0.5 V
Copper deposition over voltage	~ 0.05 V
Electrical resistance of solution	~ 0.5 V
Electrodes contact losses	~ 0.05 V

The electrowinning feed electrolyte is pumped from the electrolyte circulation tank to the electrolytic cells. The electrolyte temperature is maintained at 45-50 °C. The change of the copper concentration in the electrolyte is about 2-3 g/l. The electrolyte from the cells flows by gravity back to the circulation tank.

The cathodes are allowed to grow 7 days after which they are harvested, washed and transferred to the receiving conveyor of the stripping station and stripped in semi automatic stripping machine. Stripped blanks are lifted from the discharge rack of the stripping station and transferred back to the cells. The average current efficiency is in the range of 92%.

The acid mist created in the electrolysis process by the anodic oxygen is effectively removed by the acid mist capture system. Electrolytic cells are covered with cell hoods to prevent the acid mist spreading to the tank house atmosphere.

Nickel electrowinning or hydrogen reduction

After the copper and cobalt removal the PLS contains now almost purely $NiSO_4(aq)$. The nickel can be recovered as cathodes from electrowinning or nickel powder from the hydrogen reduction In the electrowinning process, nickel is precipitated from a nickel sulphate solution into metallic form by electric current. The cathodic reaction is the reduction of nickel ions on the cathode and the anodic reaction is the oxidation of water on the surface of the lead anode.

Cathode reaction: $Ni^{2+} + 2e^- \rightarrow Ni^o$

Anode reaction: $H_2O \rightarrow 2H^+ + ½\, O_2 + 2e^-$

Cell reaction: $NiSO_4 + H_2O \rightarrow Ni^o + H_2SO_4 + ½\, O_2$ (12)

Nickel is less noble than hydrogen in the electrochemical series and the reduction over potential for nickel is higher than for hydrogen. Hydrogen evolution is a preferential reaction and part of the electric current is consumed in the hydrogen evolution reaction $2H^+ + 2e^- \rightarrow H_2$. To minimize the hydrogen evolution, the pH of the solution should be sufficiently high to have a good current efficiency for nickel. Current efficiency decreases drastically if the electrolyte pH is 2.0 or less. The Outotec anode back technology guarantees lowest possible nickel emissions to the cell house air and to environment.

The reduction is done in continuous working autoclave with the low temperature and low pressure method, LTP. Nickel sulfate solution is fed to solution feed tanks where some ammonia solution and ferrous sulfate is mixed with the solution. Ferrous sulfate acts as a catalyst in the hydrogen reduction reaction. In the autoclave nickel is precipitated with the presence of ammonia and hydrogen as shown in reactions below. Hydrogen gas is fed directly to the autoclaves.

$$Ni^{2+} + 2NH_3 \rightarrow Ni(NH_3)_x{}^{2+} + (2-x)NH_3 \quad (13)$$

$$Ni(NH_3)_x{}^{2+} + H_2 \rightarrow Ni^o + 2NH_4{}^+ \quad (14)$$

$$2NH_4^+ + SO_4^{2-} \rightarrow (NH_4)_2SO_4 \quad (15)$$

After autoclaves the slurry is flashed in flash tanks. Vent gases from autoclaves and flash tanks are fed to a scrubber. Slurry from the flash tanks flows to a screw thickener. Nickel powder containing slurry from the screw thickener is fed to a nickel powder filter. After the filter the nickel powder is dried, cooled, briquetted and packed for the customer.

Conclusions

Outotec provides complete tailored process solutions for customers' needs. Decades of experience in the supply of Flash Smelting plants have resulted in extensive metallurgical know-how that ensures the most profitable process choice and its optimal performance. Outotec's life-cycle technology partnership with the customer spans from the early stages of the process solution design all through the supply and commissioning of the entire process chain with integrated process control and holistic service and support concept. Continuous research and development ensure that the customer is provided with the latest advancements in process and equipment solutions.

References

1. T. Ahokainen and T. Mäkinen, "50 Years of Nickel Flash Smelting - Still Going Strong" COM 2009 August 23-26, 2009, Sudbury, ON, Canada.

2. T. Makinen and P. Taskinen, "State of the art in nickel smelting: direct Outokumpu nickel technology," Mineral Processing Extractive Metallurgy: Transactions of the Institution of Mining and Metallurgy Section C 117 (2) (2008), 86-94.

3. T. Makinen, K. Fagerlund, Y. Anjala, and L. Rosenback, "Outokumpu's technologies for efficient and environmentally sound nickel production," COM 2005: Nickel and Cobalt, August 21-24, 2005, Galgary, Alberta, Canada

Ni-Co 2013
Edited by: Thomas Battle, Michael Moats, Violina Cocalia, Harald Oosterhof, Shafiq Alam, Antoine Allanore, Rodney Jones, Nathan Stubina, Corby Anderson, and Shijie Wang
TMS (The Minerals, Metals & Materials Society), 2013

NICKEL-CHROMIUM-BORON ALLOYS PRODUCTION BY ALUMINOTHERMIC PROCESSES

Özge Caglar Yilmaz, Murat Alkan, Onuralp Yücel

Istanbul Technical University, Faculty of Chemical & Metallurgical Engineering, Metallurgical and Materials Engineering Dep., ITU Ayazaga Campus, Maslak, 34469, Istanbul, Turkey

Keywords: Intermetallics, Ni-based alloys, Cr-based alloys, Metallothermic process

Abstract

In this study, Ni-Cr-B alloys were synthesized using a mixture of NiO-Cr_2O_3-B_2O_3 powders reducing with Al powders by a metallothermic process which is a high energy efficient, fast and low-cost production technique for synthesis of cast alloys. Detailed thermodynamical studies were simulated by using FactSage 6.2 Thermochemical software in order to calculate adiabatic temperature of the reactions and effects of initial molar composition on the final products. In the SHS experiments, effects of initial mixture ratios ($NiO:Cr_2O_3:B_2O_3$) on the metal recovery and compositions of final alloys were carried out. Final products were characterized by using AAS, XRD, EPMA and microhardness techniques.

Introduction

Self Propagating High Temperature Synthesis (SHS) has significant advantages compared to the high technology furnace applications, such as very short processing time, simple operation, low energy requirement and low cost [1, 2]. In a SHS process, the ignition starts the combustion and propagates throughout the reactant mixture yielding the desired product [3, 4]. However, the disadvantages of the process such as inhomogeneous or unreacted products due to undesirable reaction rates needs to be overcome by changing some parameters such as ignition temperature, particle size, additive, atmosphere etc. The earliest studies on Ni-Cr-B alloys were started with the boron additives into the Cr-Ni brazing alloys which led to decrease the melting temperature of these alloys [5]. In this study, effect of initial mixture compositions on SHS process, metal recovery and alloy compositions were examined.

Experimental Procedure

In the SHS experiments, NiO, Cr_2O_3, B_2O_3 and Al powders were used in order to produce Ni-Cr-B alloy. The raw materials have over 96 % purity and 200 μm average grain sizes. The initial mixtures were calculated and prepared to produce alloy containing 80-15% Ni, 15-80% Cr, and 5% B by weight. The advanced thermochemical simulations of the reactions were investigated in detail including different ratios of initial mixtures, and their effects on metal recoveries, metal distributions and alloy compositions were carried out.

The reaction mixtures were mixed thoroughly 15 minutes in a turbula mixer and powder mixtures (approximately 150 g) were charged into Cu crucible and compacted. Cr-Ni wire was placed at the top of copper crucible and the reaction realized by passing current through the wire.

After initiation, a highly exothermic reaction became self-sustaining and propagated throughout the SHS mixture. The obtained SHS products were discharged from the crucible after cooling.

The phase compositions of the SHS products were characterized by using XRD, XRF and EPMA techniques. The morphologies of the products were characterized by electron probe microanalysis (EPMA, CAMECA SX-100) with using three WDS (Wavelength Dispersive Spectrometer) units. X-ray analyses of obtained alloys and slags were performed with Thermo Scientific Niton XL3t XRF device and PANalytical X'Pert Pro PW3040/60 XRD device.

Results and Discussions

Before the SHS experiments, a thermochemical simulation was performed to estimate the effect of precursor compositions on the SHS process. Effect of Al addition to the mixture of 3 mole NiO, 1 mole Cr_2O_3 and 1 mole B_2O_3 the final products under 1atm pressure at the adiabatic temperature was given in Figure 1. Cr, B solubility in the liquid metal and the adiabatic temperature of the reactions could be increased with increasing in Al addition in the green mixtures. Due to the unreduced formation of CrO and B_2O_3 into the slag, Al solubility in the alloys could be obtained at higher degrees.

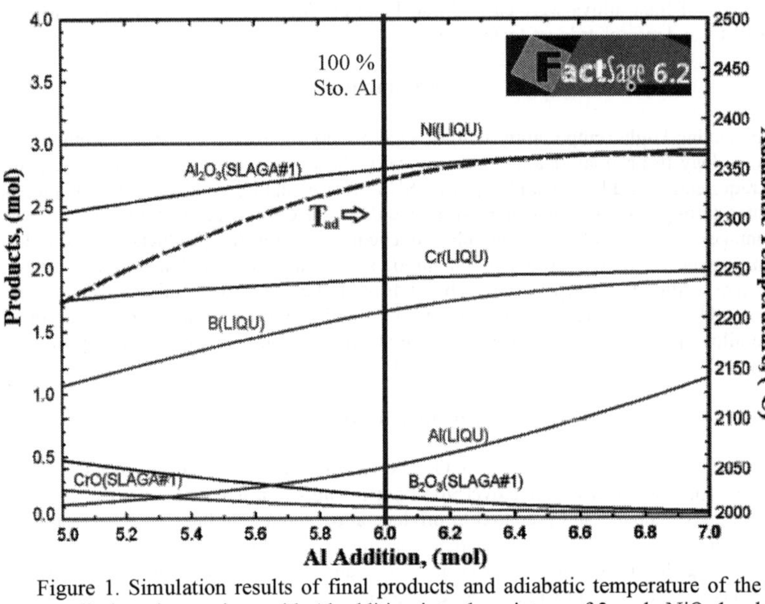

Figure 1. Simulation results of final products and adiabatic temperature of the metallothermic reactions with Al addition into the mixture of 3 mole NiO, 1mol Cr_2O_3 and 1 mole B_2O_3 [6].

Simulation result for cooling of SHS alloys obtained by using a mixture of 3 mole NiO, 1 mole Cr_2O_3, 1 mole B_2O_3 and 6 mole Al was given in Figure 2. The adiabatic temperature of this mixture was calculated as 2340 °C. In Figure 2, it could be seen that, the first crystallized phase NiB started to solidify at 2040 °C and totally dissolved at 410 °C. Cr_2B phase started to crystallized at 1405 °C and it had been stable during the cooling process. The melting/solidification temperature was measured as 1295 °C.

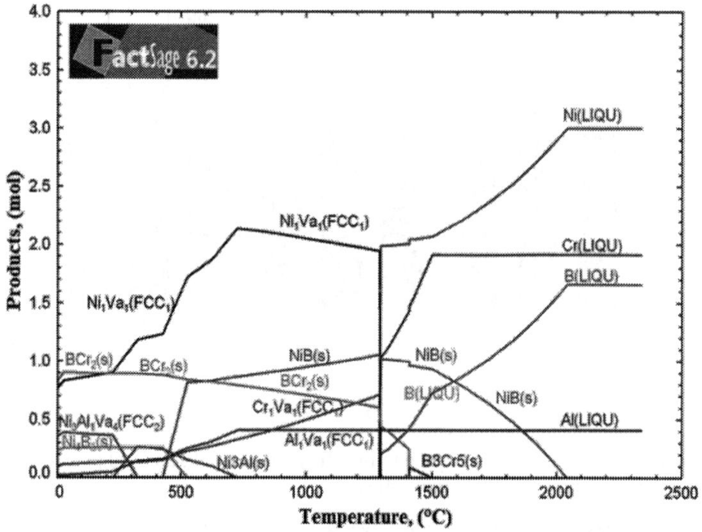

Figure 2. Simulation results of the cooling of SHS alloys obtained by using a mixture of 3 mole NiO, 1 mole Cr_2O_3, 1 mole B_2O_3 and 6 mole Al [5].

Metallic distributions into alloy, slag and scattered parts obtained by different initial mixtures were given in Figure 3. The compositions of the alloys and slags produced with different initial mixtures were given in Table I. With the increasing in Cr_2O_3 ratio in the green mixtures, Cr recoveries were increased despite that B recoveries were decreased. B and Al contents were also decreased with the increasing in Cr_2O_3 ratio. Due to the metallic leaks into the slag during rapid cooling, Ni and Cr contents in the slag were measured at higher degrees.

Table I. The Compositions of Alloys and Slags.

Estimated Alloy Compositions, wt.%			T_{ad}, °C	Composition of SHS Alloys (AAS), wt.%						Composition of Slag, wt.%		
Ni	Cr	B		Ni	Cr	B	Al	Fe	Si	Ni	Cr	B
75.0	20.0	5.0	2416	70.62	17.30	4.13	5.83	1.34	0.75	23.25	5.96	2.10
47.5	47.5	5.0	2289	48.14	48.73	2.15	0.40	0.22	0.13	11.98	12.87	1.85
20.0	75.0	5.0	2112	16.73	81.24	1.10	0.26	0.90	0.00	6.02	18.99	1.69

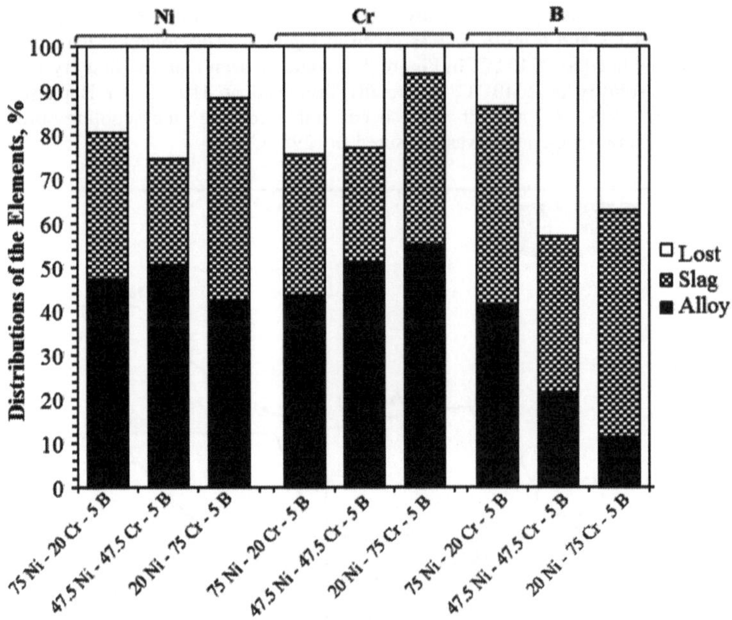

Figure 3. Metallic distributions into the alloys, slags and scattered parts obtained with the different initial mixture ratios.

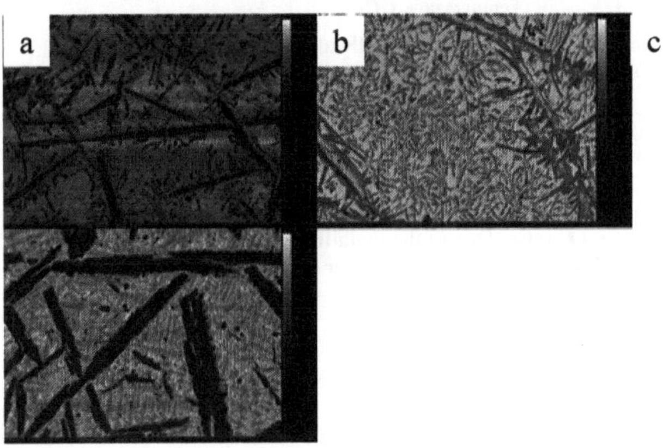

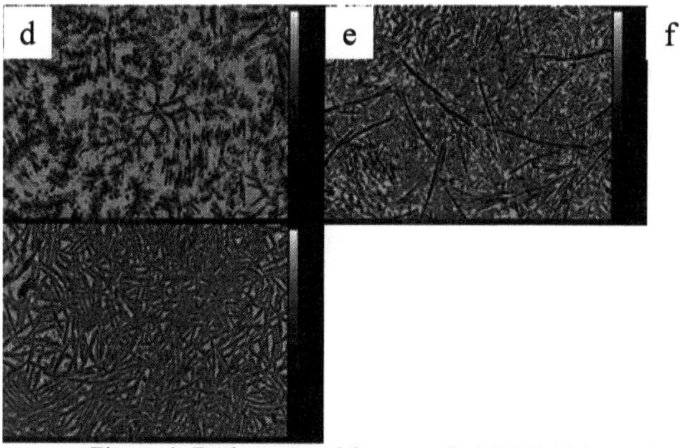

Figure 4. Back scattered images of a) 75Ni-20Cr-5B b) 47.5Ni-47.5Cr-5B, c) 20Ni-75Cr-5B metallothermic alloys, d) 75Ni-20Cr-5B e) 47.5Ni-47.5Cr-5B, f) 20Ni-75Cr-5B refined alloys.

The comparisons of the microstructures of the alloys obtained after metallothermic processes and after refining processes were given in Figure 4. The sizes of the Cr_2B phases were grown with the increasing in Cr content in the final alloy. After refining process, alloys became more homogeneous.

Conclusions

In this study, production parameters of Ni-Cr-B alloys via self propagating high temperature synthesis were investigated. The adiabatic temperatures of the reactions are increased with Al addition into the green mixtures. With the increasing in Cr_2O_3 ratio in the green mixtures, Cr recoveries were increased despite that B recoveries were decreased. It was found that, the matrix of the alloys was found as (Ni-Cr-Al) solid solution and there was also a secondary phase Cr_2B, with the eutectic formation includes Cr_3B_4 and (Ni) solid solution phases.

More studies are currently under way to examine the effects of flux addition into the green mixtures and to mechanical properties with microhardness equipments.

Acknowledgement

The authors are pleased to acknowledge the financial support for this research from ITU-BAP (Istanbul Technical University, Office of Scientific Researches Projects).

References

1. O. Yücel, F.C. Sahin, A. Tekin, "The Preparation of Ferroboron and Ferrovanadium by Aluminathermic Reduction," *High Temperature materials and Processes*, 15 (1-2) (1996) 103–106.

2. A.G. Merzhanov, "Self-propagating High-temperature Synthesis (SHS)," (ISMAN, Russia, 2002).
3. I.P. Borovinskaya, "Chemical Classes of the SHS Processes and Materials," *Pure and Applied Chemistry*, 64 (1992) 919-940.
4. V.I. Yukhvid, "Modifications of SHS Processes," *Pure and Applied Chemistry*, 64 (1992) 977-988.
5. A. Bondar, 2006. Boron–Chromium–Nickel, in G. Effenberg, ve S. Ilyenko (Eds.), *Ternary Alloy System Phase Diagrams, Crystallographic and Thermodynamic Data Vol.11D1*, 320–343.
6. FactSage 6.2 Thermochemical Software for WindowsTM. Thermfact and GTT-Technologies (2010).

Ni-Co 2013

Applications & Recycling

Session Chairs

Xingbo Liu
Jon Groh

Ni-Co 2013
Edited by: Thomas Battle, Michael Moats, Violina Cocalia, Harald Oosterhof, Shafiq Alam, Antoine Allanore, Rodney Jones, Nathan Stubina, Corby Anderson, and Shijie Wang
TMS (The Minerals, Metals & Materials Society), 2013

NUMERICAL SIMULATION OF TEMPERATURE FIELD IN DIRECTIONAL SOLIDIFICATION OF TURBINE BLADE BY LIQUID METAL COOLING METHOD

TANG Ning, XU Qingyan, LIU Baicheng

Key Laboratory for Advanced Materials Processing Technology, Ministry Of Education, Department of Mechanical Engineering, Tsinghua University, Beijing 100084 China

Keywords: IGT blade; Directional solidification; parallel conputing; numerical simulation

Abstract

Blades with excellent high temperature performance are required for industrial gas turbines (IGT). However, defects such as stray grains are almost inevitable, which sharply decrease the properties. Liquid metal cooling (LMC) is used as a new process in directional solidification of large blade castings in recent years, and it still needs to be improved. In order to optimize the casting process, the mathematical models of solidification during investment casting of IGT blade were developed. In this model, the convection between the shell and the cooling liquid metal as well as the influence of the height of liquid metal surface was taken into account. Due to large size of IGT blade, the massive data in the model consumes very much memory and CPU time. A modified FD method is established to reduce memory usage; parallel computing is developed to speed up the calculation. Based on simulation, the temperature and mushy zone evolution could be studied. Validation experiments were carried out. The cooling curves either from experiment or simulation corresponded well with each other.

Introduction

The properties of turbine increase with the temperature and pressure of the burning gas. The Hotter the Engine, the Better [1]. Thus, the mechanical property, such as fatigue life, creep resistance, etc, at high temperature of key parts, especially the blades, is very important. Directional solidified (DS) turbine blades are widely used in advanced gas turbines. A very strict process control is needed to prevent from defects occurring. However, defects easily occur during solidification.

In the past 10 years, Gas turbine blade casting is mainly manufactured by Bridgman directional solidification technology [2]. However, during Bridgman process, the mushy zone is sometimes very wide, and usually concave, which will easily result in stray grain. Some times Bridgman process cannot provide enough temperature gradient, especially for the blades with large transverse cross-section. A case in point is the heavy-duty gas turbine blade. In a word, Bridgman method is not appropriate to IGT blades.

In recent years, liquid metal cooling (LMC) is used as a new process in large blade casting[3], but it still needs to be improved. Fig. 1 is the schematic of a LMC furnace. In this process, low melting-point metal is used as cooling liquid metal. Instead of radiation, heat convection happens between the cooling liquid metal and the shell below the liquid surface. A new way is needed to find more optimized process by LMC without too many experiments.

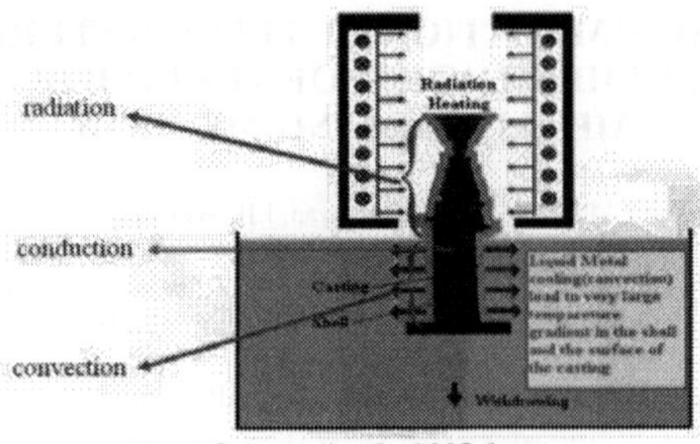

Fig. 1 Schematic of a LMC furnace

With the development of solidification simulation technology, Numerical methods provide effective ways to simulate the directional solidification process, and then to optimize the process[4]. studying the directional process and predicting the formation of casting defects are helpful to optimize technical parameters, avoid casting defects and reduce the development cycle and cost. Finite element (FE) and finite difference (FD) are most widely used discrimination methods in computer aided engineering. FE relatively reduces the memory cost. In the other side, FD can solve the microstructure coupled with temperature field, without the complex FE shape function. Computational efficiency is core problem of numerical simulation. With the popularity of computer aided engineering, the numerical models become larger and larger, consuming more and more memory and CPU time. Due to large size of IGT blade, the numerical model of temperature and grain growth of the casting is particularly huge; and it is difficult to be solved by ordinary serial program with tranditional FD method. Recently, with the constant advancement in computer manufacture and technology, computers are developing towards multi-core and multi-processor. Together with multi-core and multi-processor computers, parallel computation can greatly fasten the calculation.

Some experimental researches has been done by Giamei A F[3], Elliott A J[5] , Liu C [6]and Zhang J[7] , revealing that higher thermo gradient can be obtained by LMC, which get finer dendrite and precipitation as well as less defect. Haipeng J [8] and Dong H B [9] simulated the directional solidification of aero turbine blade by Bridgman process. They calculated temperature by FE and microstructure by FD. Their Microstructure calculation is limited in the start block and the grain selector because of too massive data, let alone the IGT blade has much larger volume than aero turbine blade. Simulation of temperature in LMC process is attempted only by Elliott A J [10], Kermanpur A [11]. Simulation of real IGT blade higher than 200mm is not reported.

In this paper, a model of casting, shell and furnace is established. The convection between the cooling liquid metal and the shell as well as the influence of the height of liquid metal surface is taken into account. To cope with the large model and to increase the computational efficiency, a quasi FE method is proposed and applied, as well as multithreading. The temperature evolution and solidification is calculated and the temperature result is validated by experiment.

Modeling

2.1 Mathematical model

Heat transferring equation: The non-linear conduction equation is described as follows:

$$\rho c \frac{\partial T}{\partial t} = \lambda \left(\frac{\partial^2 T}{\partial x^2} + \frac{\partial^2 T}{\partial y^2} + \frac{\partial^2 T}{\partial z^2} \right) + \rho L \frac{\partial f_S}{\partial t} + Q_R \tag{1}$$

Where T is the temperature, t is the time, ρ is the density, L is the latent heat, c is the specific heat, λ is the heat conductivity; x, y and z are the coordinates; f_S is the mass fraction of solid phase, Q_R is the heat radiation exchange volume.

The type of heat transfer at the surface is determined by the value of z. Suppose the height of the cooling liquid metal surface is h, when $z>h$, the surface element radiates with the furnace wall, a Monte Carlo algorithm[12] is applied to calculate the view factor varying with withdrawal distance. When $z<h$, the following equation is used to calculate the heat transfer:

$$\rho c \frac{\partial T}{\partial t} = \alpha \left(T - T_l \right) \tag{2}$$

Where α is the convection coefficient, T_l is the temperature of liquid metal.

h is updated by:

$$\frac{dh}{dt} = \frac{dh_c}{dt} \frac{S_c}{S_c - S_p} \tag{3}$$

Where h_c is the withdrawal distance, S_c the area of the horizontal section of mould and casting at the liquid surface level, S_p is the area of the horizontal section of the liquid metal container.

2.2 Quasi FE – modified FD method

FD method is applied to solve the temperature field. In traditional ordinary FD method, the 3D model is discretized into a 3D matrix, whose size depends on the smallest circumscribed box of all the geometry. Fig. 2 schematically shows that the vacuum elements in white occupy memory as well, but they are never used in the solution.

Fig. 2 Geometry discretization in traditional FD

To reduce memory, the box method is abandoned. Similar to that of FE, only the solid elements are recorded, ignoring the vacuum elements. The topology relationship is stored in the variable of each element. This algorithm is not used only in temperature calculation; it will be even more helpful in microstructure calculation in further works.

2.3 Parallel computing

In spite of the reduced memory, the element number is still too large to calculate by single processor, it takes months to solve a casting, which delays the research period. Multi thread program with parallel algorism is constructed to save time (Fig. 3). Except for the temperature, microstructure calculation in future works will also be based upon parallel algorithm.

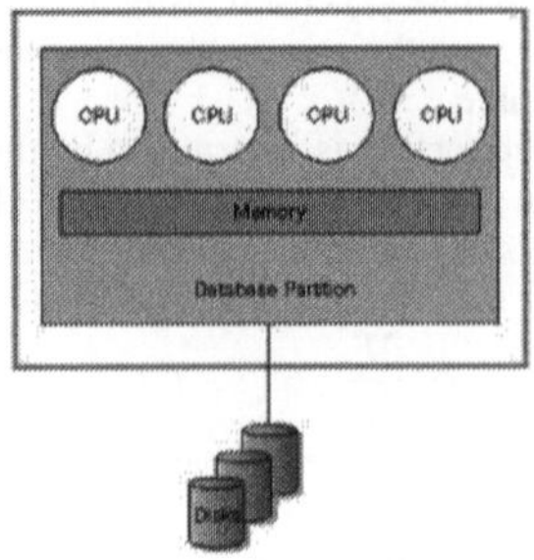

Fig. 3 Schematic of Parallel computation

The acceleration efficiency is defined as:

$$R_s = T_s / (T_m \cdot m)$$

Where T_s is the calculation time by single thread, T_m is the calculation time by multi thread , m is the thread number.

In parallel program with ideal efficiency, this ratio equals to 1, e.g. the speed is strictly in direct proportion to the number of cores. In practical parallel program, it will be less than 1 due to efficiency losses.

Experimental

Practical process is done in a LMC furnace (Fig. 12). W-Re thermocouples are installed in the casting, protected by quartz tube with 1mm thickness. Slurry is smeared on the thermocouple wires outside the thermocouple hole (Fig. 5). The geometry of the solidified slurry is also measured and meshed in the FD model. Four holes are prepared for thermocouples.

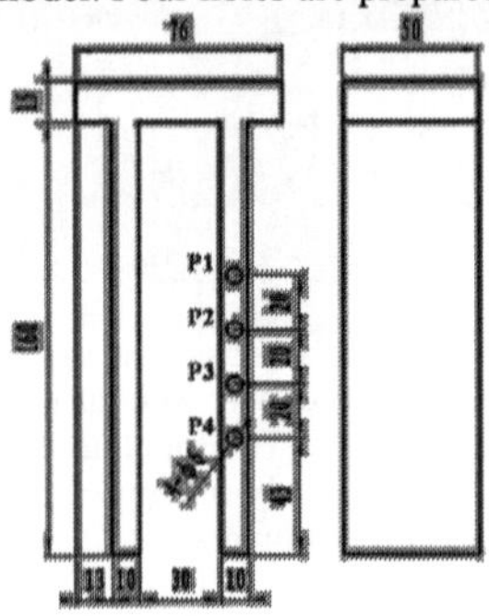

Fig. 4 Practical sample

Fig. 5 shell ready to be filled in the experiment

Results and discussion

Modified FD sharply reduced the memory consumption (Tab. 1).

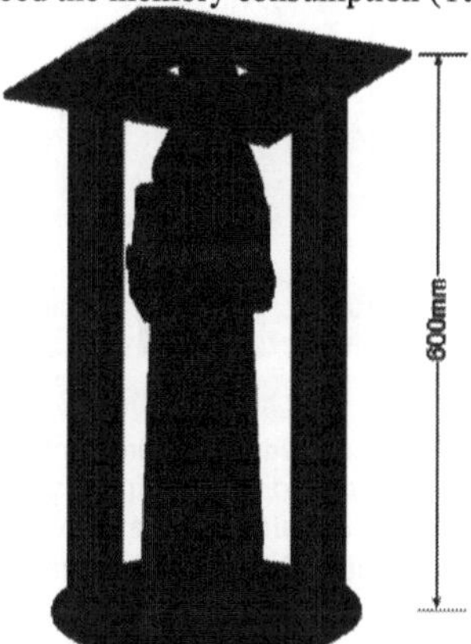

Fig. 6 Casting and shell used for memory test

Tab. 1 Memory consumption of two methods by the model in Fig. 6

Method	Solid elements/10^4	Elements stored/10^4	Memory used/Mb	Memory reducement
Traditional	393.25	3174	1289.48	76.5%
Modified	393.25	393.25	303.26	

By parallel computing, The speedup of a blade with 12.8×10^4 elements is show in Fig. 7. The calculating speed is defined as the iterative times per millisecond in calculation.

It is revealed that the acceleration efficiency is better with 8 or 4 threads except for single thread. 16 threads got the highest speed, but the acceleration efficiency is lowered by communication between threads.

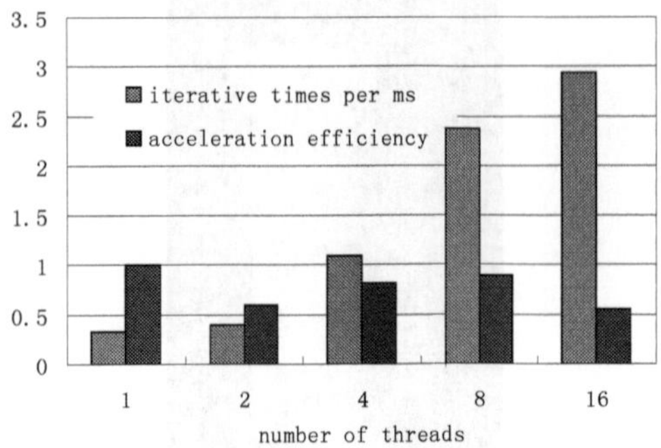

Fig. 7 Speedup of a blade

The speedup of blades in different sizes is shown in Fig. 8. The heights of them vary from 200mm to 600mm. For comparison, the calculation time is normalized as time in hour divided by element number. The speed up ratio deceases sharply with the increase of element number. It has something to do with the Memory access optimization.

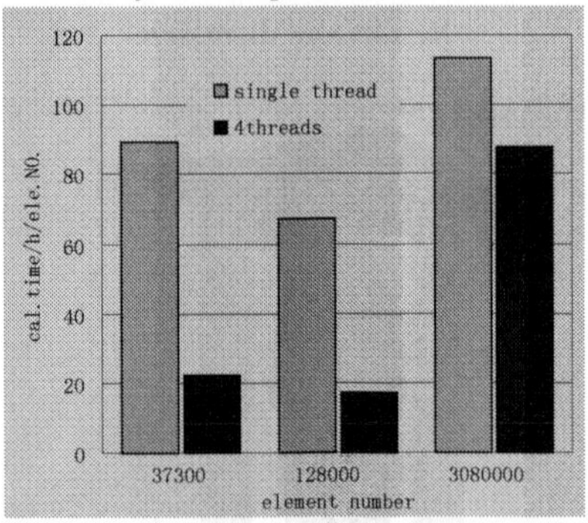

Fig. 8 Speedup of blades in different mesh sizes

As same as speedup, accuracy is also a core problem in calculation. Speedup often leads to Accuracy losses. The results by single and multi thread are compared, indicating the results are nearly identical (Fig. 9). That's to say, multi thread cause no more error than single thread.

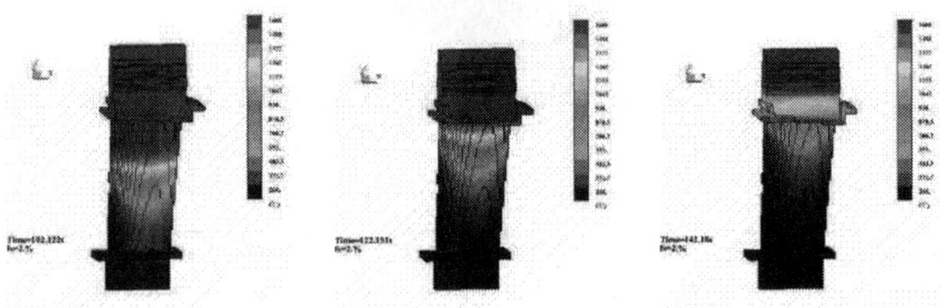

a) solidify 102s, single thread b) solidify 122s, single thread c) solidify 142s, single thread

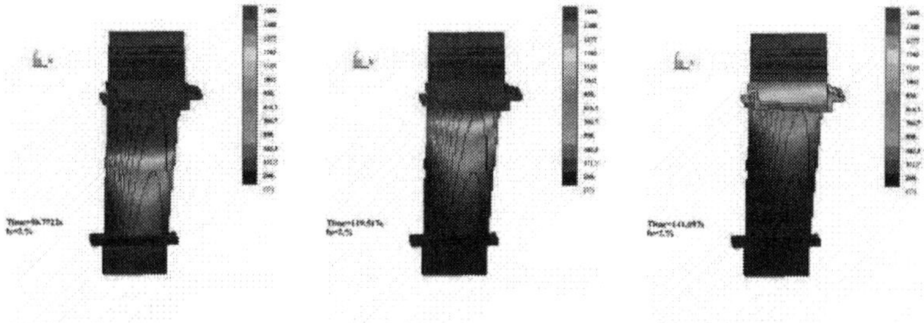

d) solidify 98s,multithread e) solidify 109s,multithread f) solidify 141s,multithread

Fig. 9 Comparison of results by single and multi thread

A blade in real practical process is calculated (Fig. 10,Fig. 11). With the withdrawing mould, the mushy zone is convex at first. It becomes flat in the middle level, and get concave finally in the upper end.

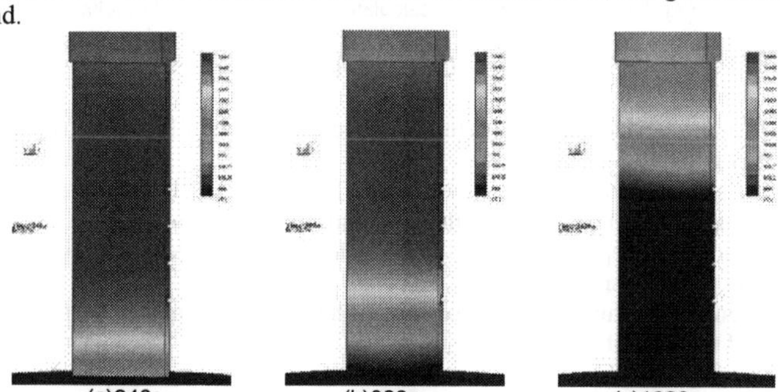

(a)240s (b)900s (c)1620s

Fig. 10 Calculated result of temperature field

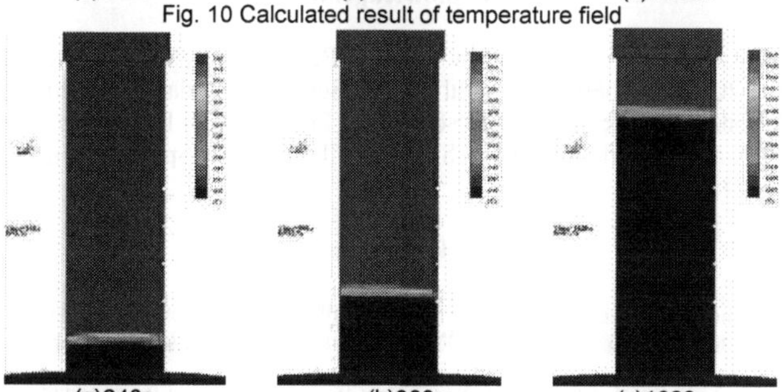

(a)240s (b)900s (c)1620s

Fig. 11 Calculated result of mushy zone

The calculated temperature curves agree well with the experimental , shown in Fig. 12. The max error is approximately 6%.

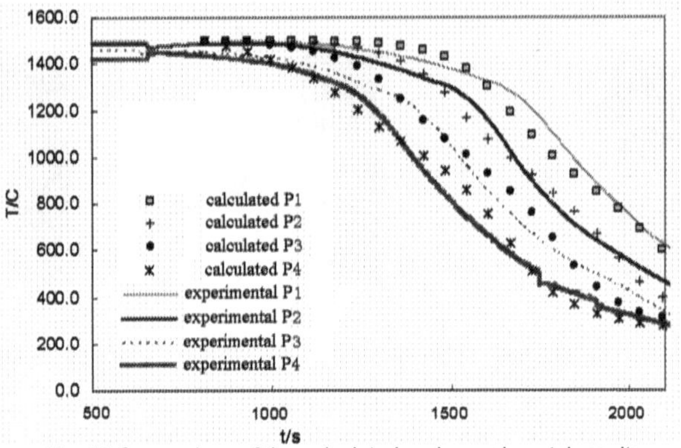

Fig. 12 Comparison of the calculated and experimental results

Conclusions

1) Numerical simulation can be used in directorial solidification of IGT by LMC to predict the temperature Field.
2) In FD modeling, data store and access by quasi FE method largely decreased the cost of computer memory and thus increased the utilization rate of memory.
3) Parallel computing is applied in this calculation; the solution speed is multiplied without significant loss of accuracy.
4) The calculated temperature curves agree well with the experimental .The simulation is demonstrated to be reliable.

Acknowledgements

The work is financially supported by National Basic Research Program of China (No. 2005CB724105, 2011CB706801), National Natural Science Foundation of China (No. 10477010), National High Technology Research and Development Program of China (No. 2007AA04Z141), and Important National Science & Technology Specific Projects (2009ZX04006-041-04).

References

[1] Perepezko J H. The hotter the engine, the better [jet engines][J]. Science. 2009, 326(5956): 1068-1069.

[2] 潘冬，许庆彦，柳百成. 考虑炉壁温度变化的高温合金叶片定向凝固过程模拟[J]. 金属学报. 2010(3): 294-303.

[3] Giamei A F, Tschinkel J G. LIQUID METAL COOLING: A NEW SOLIDIFICATION TECHNIQUE.[J]. Metallurgical Transactions A (Physical Metallurgy and Materials Science). 1976, 7 A(9): 1427-1434.

[4] Stanford N, Djakovic A, Shollock B A, et al. Seeding of single crystal superalloys - - role of seed melt-back on casting defects[J]. Scripta Materialia. 2004, 50(1): 159-163.

[5] Elliott A J, Tin S, King W T, et al. Directional solidification of large superalloy castings with radiation and liquid-metal cooling A comparative assessment[J]. METALLURGICAL AND MATERIALS TRANSACTIONS. 2004, 35A: 3221-3231.

[6] Liu C, Shen J, Zhang J, et al. Effect of Withdrawal Rates on Microstructure and Creep Strength of a Single Crystal Superalloy Processed by LMC[J]. Journal of Materials Science & Technology. 2010, 26(4): 306-310.

[7] Zhang J, Lou L. Directional solidification assisted by liquid metal cooling[J]. Journal of Materials Science and

Technology. 2007, 23(3): 289-300.
[8] Haipeng J, Jiarong L, Jing Y, et al. Study of Heat Transfer Coefficient Used in the Unidirectional Solidification Simulation Based on Orthogonal Design[J]. Rare Metal Materials and Engineering. 2010, 39(5): 767-770.
[9] Dai H J, Dong H B, D'Souza N, et al. Grain Selection in Spiral Selectors During Investment Casting of Single-Crystal Components: Part II. Numerical Modeling[J]. 2011: 1-8.
[10] Elliott A J, Pollock T M. Thermal analysis of the bridgman and liquid-metal-cooled directional solidification investment casting processes[C]. 101 Philip Drive, Assinippi Park, Norwell, MA 02061, United States: Springer Boston, 2007.
[11] Kermanpur A, Varahram N, Davami P, et al. Thermal and grain-structure simulation in a land-based turbine blade directionally solidified with the liquid metal cooling process[J]. Metallurgical and Materials Transactions B: Process Metallurgy and Materials Processing Science. 2000, 31(6): 1293-1304.
[12] Yu J, Xu Q, Cui K, Liu B, Kimatsuka A, Kuroki Y, Hirata A. Modeling of directional solidification process of superalloy turbine blade castings. vol. 2. Opio, France: Minerals, Metals and Materials Society, Warrendale, PA 15086, United States, 2006. p.1011.[J].

Ni-Co 2013
Edited by: Thomas Battle, Michael Moats, Violina Cocalia, Harald Oosterhof, Shafiq Alam, Antoine Allanore, Rodney Jones, Nathan Stubina, Corby Anderson, and Shijie Wang
TMS (The Minerals, Metals & Materials Society), 2013

INFLUENCE OF THERMOMECHANICAL TREATMENT ON STRUCTURE AND PROPERTIES OF A COBALT BASED SUPERALLOY

P Sarkar[1], S Narahari Prasad[1], M Chatterjee[1], M Narayana Rao[1]

[1]Mishra Dhatu Nigam Limited, P.O. Kanchanbagh, Hyderabad – 500058, India

Keywords: Cobalt Based Superalloy, Thermo-Mechanical Treatment, Grain Size, Tensile Properties.

Abstract

L605 (equivalent to KC20WN), is a single phase cobalt based superalloy, finding extensive applications in liquid propulsion engines. For optimal performances in service, a uniform fine grained structure is desirable. Controlled thermomechanical treatment was imparted to the material while manufacturing hot forged components. The present paper highlights the influence of the thermo-mechanical treatment on structure and properties of a cobalt based superalloy.

Introduction

Cobalt based superalloy, by virtue of their excellent high temperature strength and oxidation resistance coupled with good workability as well as weldability, find extensive applications as gas turbine components in liquid propulsion engines. L605 (equivalent to KC20WN), with a nominal composition of 10Ni-20Cr-15W-1.5Mn- balance cobalt, is one such single phase alloy deriving its strength through solid solution hardening and M_6C carbides. L605 can be work hardened and remains non-magnetic in this condition. It can be readily machined, hot worked and cold formed. The alloy can also be welded by various arc and resistance welding processes [1].

For optimum combination of strength, ductility and creep properties as well as good ultrasonic inspectability, a uniform fine grained structure is desirable. Carbide morphology and its distribution play an important role in determining the final properties of the alloy. Carbides stringers, if present, adversely affect the processing as well as achievement of satisfactory mechanical properties. These requirements call for optimization of hot working schedule and heat treatment parameters. Inadequate hot working, specially in the finishing stage, high finishing temperatures (> 1000°C) and use of high solution treatment temperature (~ 1175°C), results in coarse grained or mixed grain structure, with unacceptable levels of tensile strength and ductility.

Experimental Procedure

In the present study, the effect of varying the hot working and solution treatment parameters on the microstructure and mechanical properties of cobalt base superalloy L605 (equivalent KC20WN) has been investigated during the course of manufacturing ring forgings for the Indian Space Program. The alloy was manufactured in vacuum induction melting furnace and further remelted through ESR, in order to attain a clean alloy with high degree of homogeneity.

Material Challenges

One of the biggest challenges has been to control the grain size within a narrow range of 45 to 90 microns, in order to meet the conflicting requirements of ultrasonic inspectability and acceptable tensile strength and ductility on one hand and realize the required creep / stress rupture properties, on the other hand, for which coarser grain structure is desirable.

For certain applications involving welding, fine grained structure is desirable, to improve the weld quality. Finer grains ensure that the carbides have more sites for precipitating and lessen the detrimental effect of HAZ. The normal solution annealing temperature specified for this alloy is in the range of 1177 - 1232°C. Solution annealing at such a high temperature assures that majority of the carbides are taken into solution but, results in coarsening of grains to ASTM No. 3 and coarser. Below this temperature the precipitates remain and may reduce both the strength and elongation.

Initial Approach

To overcome this problem, initially the approach taken was to ensure proper distribution of alloying elements across the matrix and lowering the solution annealing temperature. In this approach the material was homogenized at suitable semi stage and forged with high finishing temperature. High finishing temperatures promotes dynamic recrystallization resulting in coarse grained structure of ASTM No. 2 (figure 1a). Moreover due to inadequate reduction in the final stages coarse carbides and stringers of carbides along the prior austenite grain boundary were also observed (figure 1b). Owing to very high finishing temperature of around 1100°C, subsequent solution annealing at lower temperature of 1070°C did not produce any microstructural refinement. Consequently, lower yield strength as well as unacceptably low tensile ductility was observed in the product. The following factors were considered responsible for the unacceptable quality of product –

a) Higher finishing temperature resulting in coarse grain.
b) Inadequate reduction coupled with high finishing temperature resulting in coarse grains as well as unacceptable carbide distribution

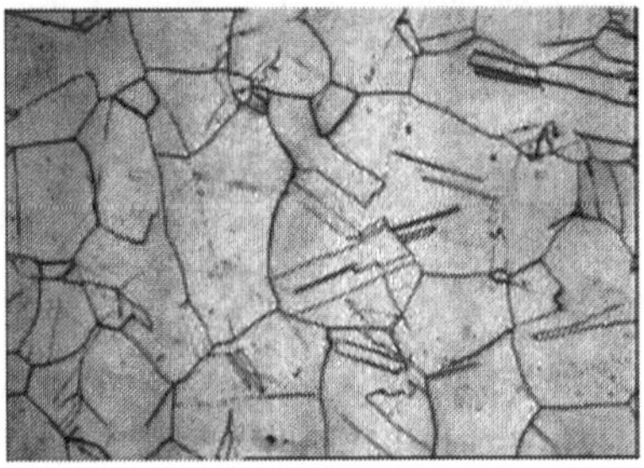

a

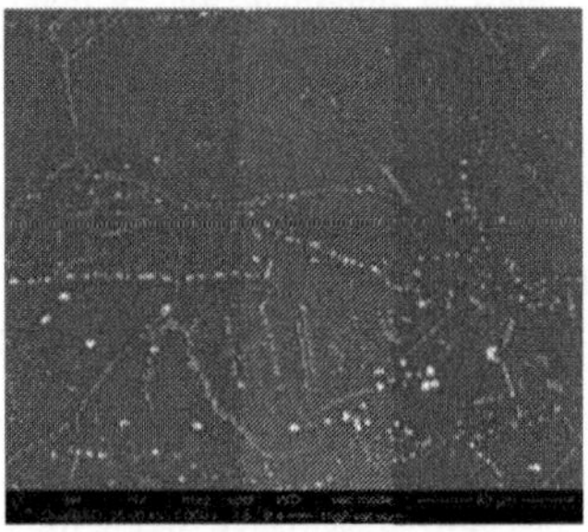

b

Figure 1: a) Optical micrograph showing coarse grain size of ASTM No. 2, 100X;
b) SEM micrograph showing carbides stringers and coarse carbides.

Modified Approach

In the modified approach a thermo-mechanical treatment was devised, where the final reheating temperature was lowered to around 1100°C (to ensure very low finishing temperature) and adequate final reduction of 25%, so as to derive adequate driving force for recrystallization during subsequent solution treatment. Figure 2 shows the flow chart for manufacture of ring forgings.

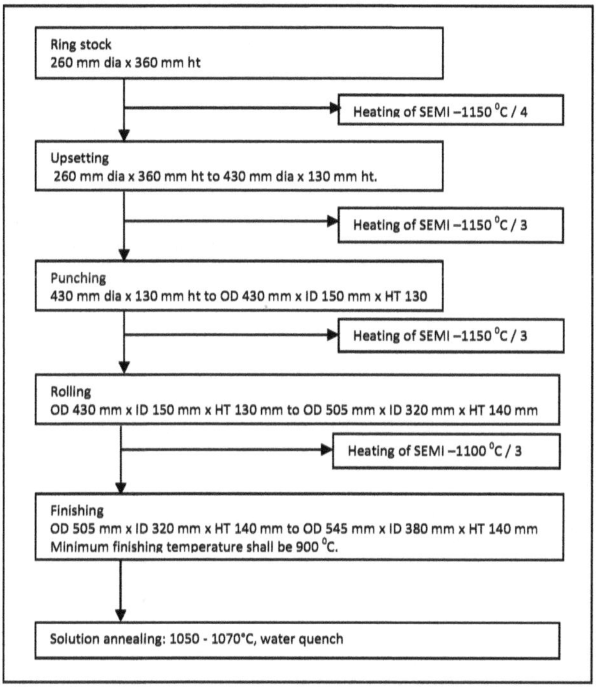

Figure 2: Flow chart for forging of rings

Results and Discussions

By lowering the final reheating temperature and increasing the final reduction to 25%, lower hot working finish temperatures of around 900°C were realized. Dynamic recrystallization being a diffusion controlled process is greatly retarded with lowering of forging temperature and also with onset of carbide precipitation during the cooling of the work piece. Hence, as a result of lower finish temperature the microstructure of the as forged components comprised of uncrystallized grains. However, the introduction of extensive working at lower temperature introduces sufficient strain energy in the material, which provides the driving force for static recrystallization during the solution annealing treatment. It may also be mentioned that due to sufficient strain energy available in the material, even lower solution annealing temperature of about 1050°C is adequate to bring about recrystallization. The resultant microstructure emanating from the thermomechanical treatment is fine grained recrystallized structure with uniform distribution of carbides (figure 3 a&b). The microstructure refinement brought about by the modified treatment resulted in enhancement of yield strength and tensile ductility, as seen from Table 1 (which also compares the mechanical properties of the alloy with conventional as well as modified treatments). At the same time finer and uniform carbide distribution, which is responsible effective grain boundary pinning enhances the stress rupture properties.

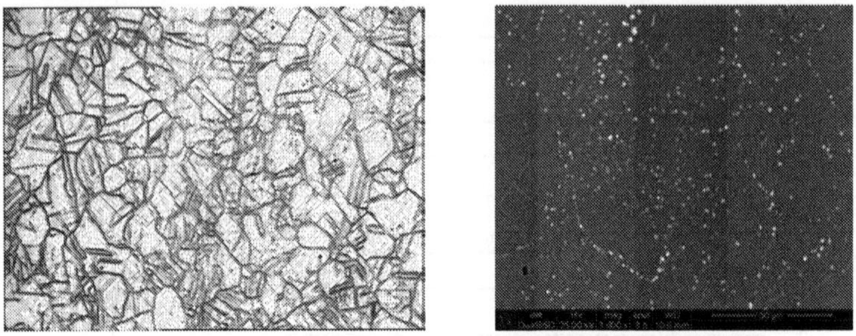

a b

Figure 3: a) Optical micrograph showing fine recrystallized grains of ASTM No. 5.
b) SEM microstructure showing fine uniform carbides distribution

Table 1: Mechanical Properties of Forgings After Conventional and Modified Processes

Condition	Direction	0.2% PS, MPa	UTS, MPa	E, %	Stress Rupture	
					Time (Hrs)	%EL
Specified	T/R	340 - 540	880 - 1080	≥ 30	15	8
Conventional process	T	470	865	28	47	27
	R	485	860	26		
Modified process	T	500	1040	33	57	46
	R	510	1070	39		

Conclusions

The microstructure refinement from the modified thermomechanical treatment is responsible for enhancement of tensile strength and ductility by 18% and 25% respectively. Simultaneously, the stress rupture life as well as ductility is enhanced by virtue of controlled grain size and uniform carbide distribution brought about by the thermo-mechanical treatment.

References

1) UNS R30605 (L605ALLOY) DATA SHEET, Fine Tubes, 2005

Ni-Co 2013
Edited by: Thomas Battle, Michael Moats, Violina Cocalia, Harald Oosterhof, Shafiq Alam, Antoine Allanore, Rodney Jones, Nathan Stubina, Corby Anderson, and Shijie Wang
TMS (The Minerals, Metals & Materials Society), 2013

EFFECT OF PROCESSING CONDITIONS ON STRUCTURE, PROPERTIES AND PERFORMANCE OF A NICKEL BASE CAST SUPERALLOY FOR HIGH TEMPERATURE APPLICATIONS

M. Chatterjee[1], A. Pani Kishore[1], P. Sarkar[1] M.Narayana Rao[1]

[1]Mishra Dhatu Nigam Limited, Kanchanbagh, Hyderabad – 500 058, India

Keywords: Strength, Creep, Fatigue, Equiaxed, Directional Solidification, Ni-base Superalloy

Abstract

Nickel base cast superalloys exhibits excellent combination of high temperature strength, creep, fatigue as well as oxidation and corrosion resistance and find wide applications in aerospace gas turbine components. Depending upon the severity of environmental conditions in gas turbine, components are cast in equiaxed (EX) or directionally solidified (DS) modes and subsequently heat treated to achieve a wide variety of structure and properties. Accordingly, a study was undertaken to evaluate the performance of a cast Ni-base superalloy and an attempt has been made to correlate processing conditions, structure and mechanical properties. Under similar heat treatment conditions, the directionally solidified material exhibited 15% to 20% superior high temperature strength and creep properties as compared to the equiaxed product.The superior performance of DS material is attributable to more effective solutionizing of γ/γ^1 eutectic, absence of interdendritic microporosity and fineness of microstructural features, which can be correlated to the progressive solidification under steep thermal gradient during vacuum investment casting.

The present paper highlights the structure property relations in a Ni- base cast superalloy, which was processed in equiaxed and directionally solidified castings, through investment casting route.

Introduction

Materials development for gas turbine engines has been largely driven by ever increasing demand for higher efficiency propulsion systems. Requirements of higher engine thrust, thrust to weight ratio and fuel efficiency have pushed engine operating temperatures and stresses progressively higher. Nickel base cast superalloys, due to their excellent combination of high temperature strength, resistance to creep, oxidation, hot corrosion, fatigue and thermal cycling, have been extensively used for manufacture of a wide variety of aerospace gas turbine components. While vacuum melting processes have opened the door to higher levels of alloying (thereby exploiting full potential of γ^1 strength), the precipitation of detrimental phases and the subsequent loss of properties during service, revealed the need for stringent composition control based on phase stability, during manufacturing process. Advances made in the area of vacuum investment casting technology has helped development of complex aerofoil components, cast in equiaxed or directionally solidified modes, depending upon the severity of operating conditions. In the present study, a frontline polycrystalline nickel base cast superalloy bar stock manufactured with precise chemistry control, has been further processed to produce equiaxed and directionally solidified castings. The present paper provides a comparative assessment of structure and mechanical properties of the two types of castings.

Manufacture of Ni-base Superalloy Bar Stocks

A specially designed 600Kg vacuum induction melting furnace having a monolithic $MgO-Al_2O_3$ spinel refractory lining was used for the manufacture of the required Ni-base superalloy bar stocks. The magnesia based spinel lining was specifically chosen to ensure inertness and avoid reactions of reactive metals like Ti and Hf present in the liquid metal with the refractory lining of the crucible. A specific charging sequence was used to ensure smooth melting of the constituents and avoid bridge formation during melting. The molten metal was subjected to extensive refining under high vacuum and the progress of refining was monitored by checking the hot leak rate and dissolved oxygen level using solid electrolyte based oxygen sensors and assessment of the atmosphere above the liquid metal using mass spectrometer. Residual oxygen and nitrogen levels of less than 5ppm could be consistently ensured with this technique. The chemistry of the alloy was tailored to satisfy the desired electron vacancy number ($N_v3B \leq 2.15$). Final composition adjustments were made based on PHACOMP analysis using the formula [1].

$$N_v3B = \sum^n{}_{i=1} \ (X^*{}_{ir})\,(N_v)s$$

Where $X^*{}_{ir}$ – residual atomic fraction of each element.
N_v – The electron vacancy number of respective element.

The liquid metal was cast into 80 mm dia remelt stock by teeming through a tundish fitted with dams, weirs and ceramic filters. Table: 1 provides typical analysis of remelt bar stocks.

Table 1. Typical Composition of Ni-Base Superalloy Bar Stock

Elements	C	Cr	Mo	Co	Ti	Al	Hf	W	Ta	B	Zr	Ni
Wt (%)	0.074	8.33	0.49	9.25	0.8	5.69	1.54	9.48	3.12	0.019	0.014	Bal

Manufacture of Equiaxed and Directionally Solidified Components

Equiaxed Casting Process

Dedicated 20Kg vacuum induction melting furnaces having pre-sintered zirconia crucible was used for further processing of re-melt bar stocks to EX and DS components. Cleaned and dressed re-melt bar stocks were melted in induction melting crucible and super heated to around 1580°C.The molten metal was poured in hot ceramic mould manufactured by investment casting processes and pre heated to around 1060°C. To obtained fine grain structure, the mold surface was coated with suitable inoculants. Further reduction in the grain size in the bulk for the equiaxed cast component (integral rotor castings), could be achieved by agitating the mold through rotary oscillating motion, to break the growing dendrites into fine fragments due to shear forces produced during oscillation.Figure1(a) shows schematic representation of mold oscillation during vacuum investment casting processes[2] and figure.1(b) shows the macrostructure of an equiaxed integral rotor casting without mold agitation, while Figure.1(c) shows the influence of mold agitation on the macrostructure of an equiaxed integral rotor casting[3].

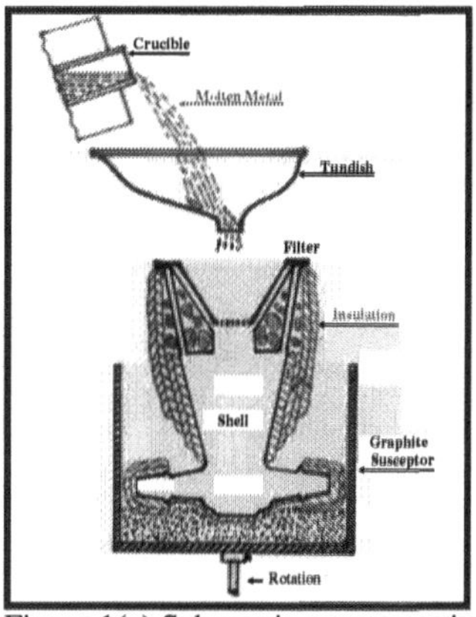

Figure 1(a) Schematic representation of mold oscillation.

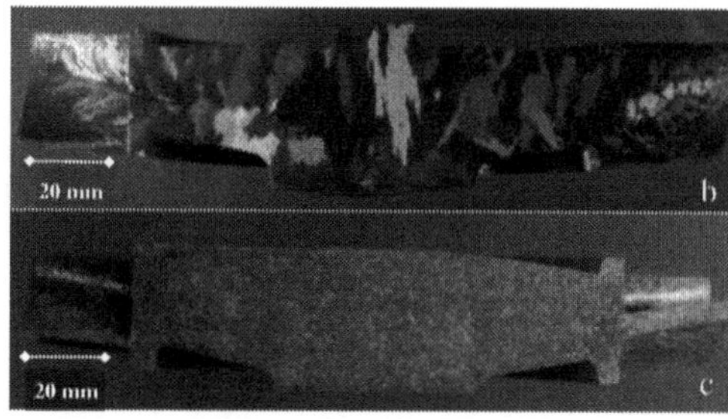

Figure1 (b). Macrostructure without mold agitation.
(c). Macrostructure with mold agitation.

Directional Solidification

For directional solidification, the molten remelt stock superheated to around 1500°C was teemed into investment cast mold preheated to around 1400°C. The mold was withdrawn out of the hot zone across a steep temperature gradient at a controlled withdrawal rate, until the entire component length was solidified in a directional manner. Figure.2 shows the macrostructure of a turbine blade and vane, which were directionally solidified.

Figure 2. Macrostructure of the directionally solidified turbine components.

Comparison of Mechanical Properties of Directionally Solidified and Equiaxed Materials

Mechanical property evaluation for directionally solidified and equiaxed materials have been carried out on 20mm. diameter cylindrical test coupons. Samples for mechanical property evaluation were drawn from the centre in the longitudinal direction.

Tensile Properties

The variation of strength and ductility with testing temperature for directionally solidified and equiaxed materials are shown in figure 3 a & b.

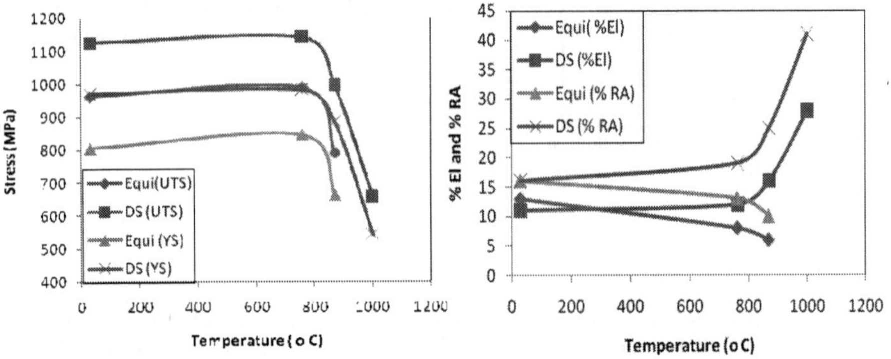

Figure 3. a) Variation of strength with temperature for DS and equiaxed castings.
b) Variation of ductility with temperature for DS and equiaxed castings.

Both directionally solidified as well as equiaxed materials retain their strength up to around 750 °C. This kind of behavior is typical in Ni-base superalloys strengthened predominantly by ordered and coherent γ^1 precipitates, where, formation of dislocation locks is responsible for retention of strength up to fairly high temperature. Beyond 870 °C, the strength rapidly falls in both the materials because of γ^1 coarsening and their shearing by dislocation partials [4].However, the UTS and YS of directionally solidified material was higher by about 25% as compared to equiaxed material. The superior strength of directionally solidified material is on account of higher level of γ^1 solutionizing during heat treatment, as borne out from figure 4(a) , which shows lesser volume fraction of undissolved γ/γ^1 eutectic after heat treatment in the directionally solidified material [undissolved $\gamma/\gamma^1 < 7$ % in directionally solidified material as compared to around 12% in equiaxed castings figure 4(b)]. In addition, the finer size of γ^1 in the directionally solidified material (< 0.5μ) as compared to around 1μ in the equiaxed material, is also responsible for its superior strength. In spite of its higher strength, the directionally solidified material shows significantly superior ductility than its counterpart as seen from figure 3b. The superior ductility can be attributed to absence of micro-porosity and finer scale of microstructure observed in the directionally solidified materials (figure-5a & 5b).

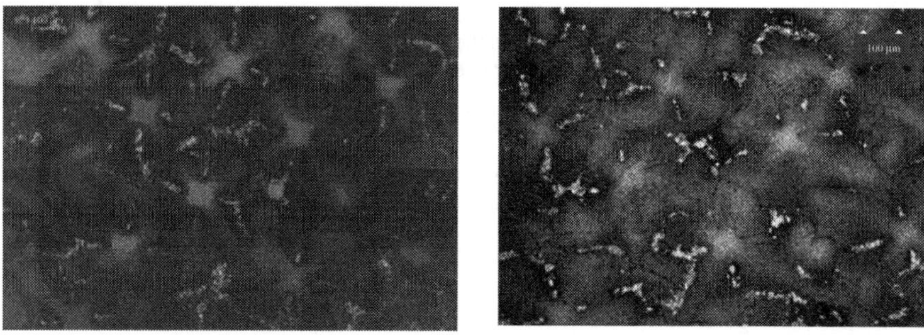

(a) (b)

Figure 4. a) Undissolved eutectic in DS material after solution treatment (100X).
b) Undissolved eutectic in equiaxed material after solution treatment (100X).

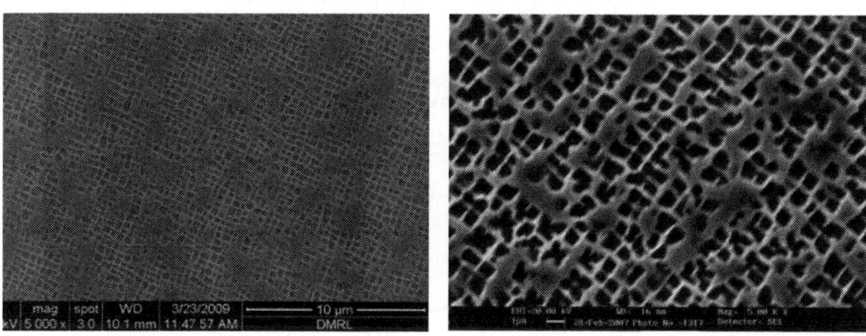

(a) (b)

Figure 5. SEM, Secondary electron images showing:
a) γ^1 structure in DS material after heat treatment.
b) γ^1 structure in equiaxed material after heat treatment.

Stress Rupture

Under different combinations of stress and temperatures, the DS material had far superior stress rupture life and ductility. This is evident from the variation of Larsen-Miller parameters (LMP) with stress as shown in figure 6. The superior performance of DS material is again attributable to its higher strength and finer γ^1 structure.

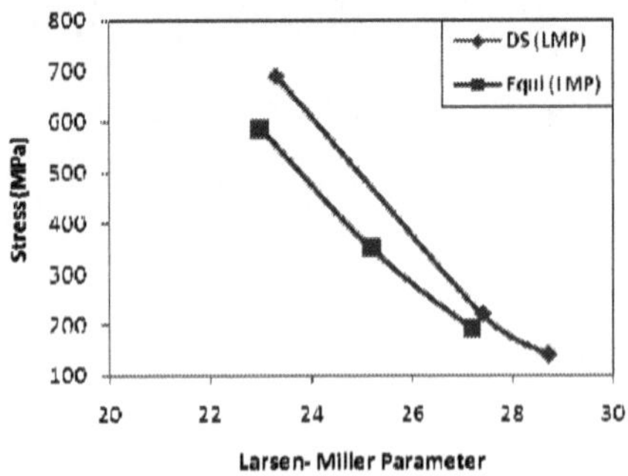

$$LMP = [(T\ (^{o}C) + 273)\ (20 + Log_{10}\ (t))]\ 10^{-3}$$

Figure 6. LM Plots for DS and equiaxed materials.

Fatigue Properties

The S-N curves for equiaxed and DS test coupons generated at 760 °C are compared in figure 7.

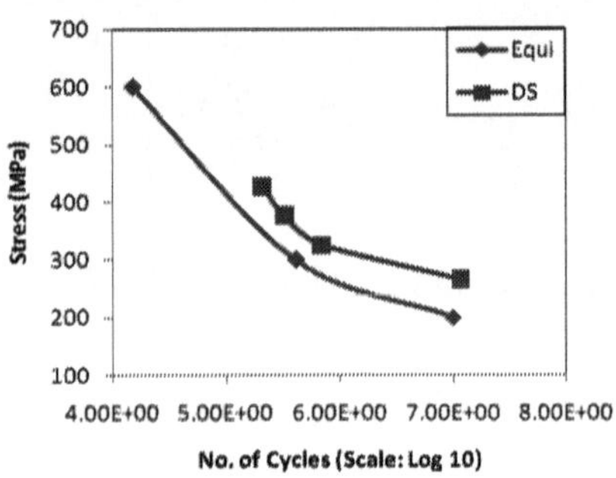

Figure 7. S- N curves for DS and equiaxed materials at 760 °C.

At different stress levels the DS materials showed superior fatigue life due to its higher strength, superior ductility, finer γ^1 structure and absence of micro-porosity. In the case of the equiaxed component(integral rotor castings, which were manufactured both with and without mould agitation),it was observed that mold oscillation during solidification resulted in a finer macrostructure (figure 1b &c) with enhanced low cycle fatigue life evaluated at 650 °C (life =

3100 cycles without mold oscillation as compared to 7800 cycles with mold oscillation), both the tests being carried out under identical conditions in the radial direction[3].

Discussion

The superior structure and properties of the directionally solidified material arises from the manner in which the liquid metal is processed during manufacture of component and test coupons. Progressive nature of solidification under steep thermal gradient results in significantly lower local solidification times and consequently much lower levels of micro-segregation in DS materials. This is borne from the fact that the average secondary dendrite arm spacing in DS test coupons was found to be around 125 microns as compared to about 230 microns in the equiaxed material. This kind of structure responds adequately to high temperature solution treatment resulting in better γ^1 solutionization without any danger of incipient melting. Moreover, the presence of a liquid metal head above the solidification front during directional solidification ensures proper feeding and elimination of micro-porosity in the final product. The finer dendritic structure in DS components translates to finer γ^1 structure in the material after heat treatment, although faster cooling rate in DS components after solutionization could have also contributed to the fineness of γ^1 structure.

Conclusions

1) The directionally solidified materials showed much lower volume fraction of un-dissolved γ -γ^1 eutectic and finer γ^1 structure after heat treatment as compared to its counterpart with equiaxed structure.
2) Effective solutionizing, finer γ^1 structure and absence of micro-porosity in DS components is reflected in its superior tensile, stress rupture and fatigue properties.

References

1) C.T.Sims, N.S.Stoloff, W.C.Hagel, "SUPERALLOYS II", Wiley-Inter-Science Publication, (1987), 230-233.
2) V.S.Verma, et al, Workshop on "Recent Trends in Production and Processing, Quality Assurance and Airworthiness Certification of Aeroengine hot end components" Oct 2008.
3) B.Gopala Krishna et al, DMRL Technical Report, TR 94180 (1994)
4) R.F.Decker, "Strengthening Mechanisms in Nickel-Base Superalloys", INCO publication, (1969), 275 – 298.

Ni-Co 2013
Edited by: Thomas Battle, Michael Moats, Violina Cocalia, Harald Oosterhof, Shafiq Alam, Antoine Allanore, Rodney Jones, Nathan Stubina, Corby Anderson, and Shijie Wang
TMS (The Minerals, Metals & Materials Society), 2013

DEVELOPMENT OF NICKEL BORON ALLOYS FOR BRAZING MATERIALS

Kerem Can Tasyürek, Murat Alkan, Onuralp Yücel

Istanbul Technical University, Faculty of Chemical and Metallurgical Engineering, Department of Metallurgical and Materials Engineering, ITU Ayazaga Campus, Maslak, 34469, Istanbul, Turkey.

Keywords: Nickel Boron, Ni-based alloys, Brazing alloys, Carbothermic process

Abstract

Nickel-base filler materials are generally used in applications where resistance to corrosion and heat resistance required. Their description as high-temperature brazing filler materials is amply justified. In this study, nickel boron (NiB) alloys is produced via carbothermic reduction starting from boric acid (H_3BO_3) together with high purity nickel (NiO), charcoal and sawdust by using DC arc furnace which the boron-containing composition of the basic metal is tapped at the bottom. In the arc furnace experiments, different starting mixtures were carried out, and optimum starting composition for producing 15-20 wt. % B containing NiB alloys were investigated. Alloy compositions, metal recoveries, graphite electrode and energy consumptions were performed. The obtained products were characterized by using wet chemical analysis (AAS/ICP), XRD and SEM/EDS techniques.

Introduction

The development of the technologies has showed us for the different materials being worked together. For a long time, lots of methods have been developed to connect the same or different materials. Brazing can be described such a connecting two or more metallic materials by melting another material whose melting temperature is lower than those of connected materials [1]. Brazing with filler metal, which has been widely range of used such a cost effective appliance in high technology industries, has been showed to be able to manufacture high performance links with ideal strength and well corrosion resistance. There is an essential situation after the strong metallurgical process and the base metal in order to prevent the precipitation of brittle intermetallic phases like silicides, borides and phosphides from the relatively high contents of respective metalloids in brazing alloy [2]. Also if the dissolution process was not well controlled, there exits some problem that the over-dissolution of base metal could be have unwanted corrosion pits being brazed [3]. The invention of carbothermal method of producing nickel-boron composition by the reduction of boron-containing raw material, e.g. an oxide of the base materials and the carbon material applying and electrical furnace and boron comprise material of the basic metal is tapped at the bottom [4].

Our work aims to developed a scientific and engineering background in the production of Nickel boron alloys (NiB) which can be used as a brazing material, wear-corrosion-oxidation resistive applications via carbothermic reduction that is the effective and attractive process technique regarding high mass of production for industry such as brazing, automotive, electronics, aircrafts, coatings etc.

Experimental Procedure

In this study, nickel-boron alloy was produced from a mixture of boric acid, nickel oxide, charcoal and wood charcoal. The purity of granular boric acid was 99.5%, and nickel oxide has a 99.0% purity, respectively. Chemical analysis of charcoal and wood charcoal are given in Table I.

Table I. Chemical Analysis of Charcoal and Wood Charcoal (wt. %).

Reductant	C_{fix}	Ash	Volatiles
Charcoal	65.18	9.52	25.30
Wood Charcoal	10.16	0.35	89.49

The laboratory type 270 kVA DC arc furnace used in this work was graphite lined and 250 kg charge capacity. The diameter of the furnace was 36 cm and 90 cm depth. The voltage and current readings were measured by manually. During the taping of the alloy, temperature measurements were done by using of an optical pyrometer.

Charcoal, wood charcoal and selected ratios of H_3BO_3 and NiO were mixed and 60 kg of these mixtures were fed to the open arc heated furnace. H_3BO_3/NiO ratio was selected as 1.11 for the experiments, and the amount of charcoal addition was selected between 12.80 – 17.80 % of charge by weight. After 1 hour of the experimental run, the liquid metal was tapped and the arc stopped. The furnace was left to cool for checkup. After cooling, the samples were taken from sinter sectors in the furnace with the range of 20 cm. These samples were analyzed with XRD (Rigaku, CuKα) technique.

Results and Discussions

Chemical analysis of the alloys obtained by addition of the different ratios of the charcoal to the initial mixtures were given in Table II. As H_3BO_3/NiO ratio of 1.11 fixed, boron contents in the final alloy was increased with increasing in ratio of fixed C to the total initial mixtures. Alloy weights were measured as 5.4 kg in first experiment and as 3.9 kg in second experiment, respectively. Alloy weight was decreased with increasing in ratio of fixed C to the total initial mixtures. The casting temperatures were measured as 1182°C in the first experiment and as 1145°C in the second experiment, respectively.

Table II. Chemical Analysis of Alloys (wt. %).

Exp. No	Charcoal / Total Mixture Ratio, %	Compositions, wt. %						
		Ni	B	Al	S	Si	Fe	C
1	12.80	90.06	9.13	0.007	0.006	0.35	0.26	0.07
2	17.80	85.25	13.76	0.280	0.003	0.23	0.33	0.06

After the cooling of arc furnace, at the bottom of the arc furnace there was a formation of alloy and slag with weight of 12.1 kg. Chemical analysis showed that, alloy had a composition of 75.05 % Ni, 18.40 % B, 0.35 % Al, 0.007 % S, 0.68 % Si, 0.23 % Fe, 1.62 % C by weight, respectively. Alloy weights and energy consumption values were given in Table III. At optimum metal production conditions, total energy consumption was measured 200 kW and electric energy was consumed 11.62 kWh/kg product.

Table III. Weights of Alloys and Energy Consumptions of the Experiments.

Alloy	Weight, g	Total Energy Consumption, kWh/kg	Total Electrode Consumption, gr/kg product	Total Electrode Consumption, cm/hour	Duration of Casting, min.
1	5.4	11.62	52.57	3.77	64
2	3.9				79
Tailed alloy in the furnace	7.9				-

In order to understand possible mechanism of nickel-born formation, samples from different sectors of the charge have been taken out. These samples have been carefully examined by XRD techniques. The comparative XRD graphs of the samples was given in Figure 1, and compounds list formed at different sectors in furnace was given in Table IV. The stoichiometry of the reactions and the corroding of the upper electrode tip have proven that reduction reactions take place at the upper electrode tip.

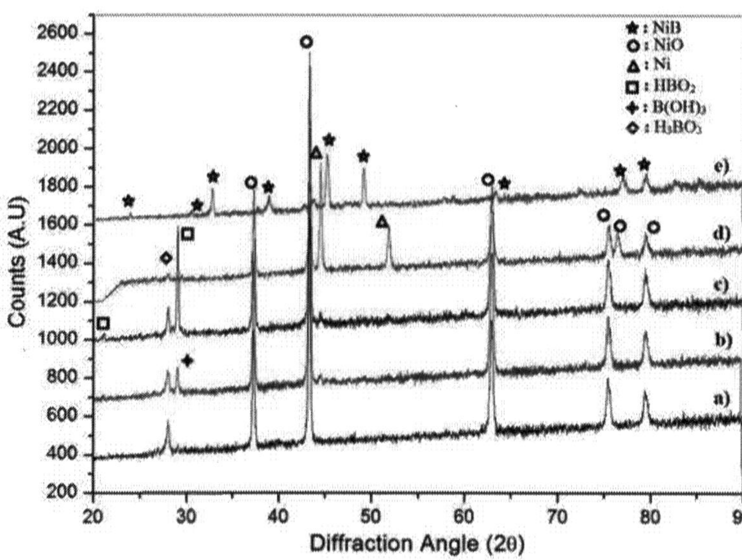

Figure 1. Comparative XRD graphs of the samples taken from a) 20 cm depth, b) 40 cm depth, c) 60 cm depth, d) 80 cm depth, e) at the bottom of the furnace.

Table IV. Weights of Alloys and Energy Consumptions of the Experiments.

Compounds	Sectors				
	20 cm depth	40 cm depth	60 cm depth	80 cm depth	At the bottom of EAF
NiB					X
NiO	X	X	X	X	X
Ni				X	
HBO_2			X		
$B(OH)_3$		X			
H_3BO_3	X	X	X		

Conclusions

In the production of nickel-boron alloy in an electric arc furnace, as the charge containing boric acid, nickel oxide, charcoal and wood charcoal dehydrate during the feeding into the furnace and while boron oxide melts, the charge is sintered but forms the porosity needed, by using carbonized charcoal. This permits the maintenance of the reduction reaction. With the help of high temperature, boron and nickel are reduced together from a glassy slag to produce 15-20 % B containing nickel-boron alloys.

It's an ongoing study, further studies have been carried out with using different initial mixture ratios (NiO/H_3BO_3, etc.), with using different ratios of fixed C to initial mixtures. Also the micro-structural studies will be realized by using SEM/EDS technique.

Acknowledgement

The authors wish to express their sincere appreciation to SentesBIR and ITU-BAP (Istanbul Technical University, Office of Scientific Researches Projects) for financial support of this study.

References

1. M. Uzkut, N.S. Köksal, B.S. Unlü, "The determination of element diffusion in connecting SAE 1040/WC material brazing" *Journal of Materials Processing Technology*, 169 (2005) 409-413.
2. X. W. Wu, R.S. Chandel, H.P. Seow, H. Li, "Wide gap brazing of stainless steel to nickel-based superalloy" *Journal of Materials Processing Technology*, 113 (2001) 215-221.
3. X.P Zhang, Y.W. Shi, "A dissolution model of base metal in liquid brazing filler metal during high temperature brazing", *Scripta Materialia,* 50 (2004) 1003-1006.
4. R. Hahn, H.J. Retelsdorf, R. Fichte, S. Sattelberger, *Carbothermal Method of Producing Cobalt-Boron and/or Nickel-Boron*, Germany, United State Patent No: 4623386, 1986.

Ni-Co 2013

Ni and Co Hydrometallurgy

Session Chairs

Violina Cocalia
Harald Oosterhof

Ni-Co 2013
Edited by: Thomas Battle, Michael Moats, Violina Cocalia, Harald Oosterhof, Shafiq Alam, Antoine Allanore, Rodney Jones, Nathan Stubina, Corby Anderson, and Shijie Wang
TMS (The Minerals, Metals & Materials Society), 2013

CONTINUOUS CO-PRECIPITATION BEHAVIOUR AND STABILITY OF ARSENIC(V) FROM Fe(II,III)-Al(III)-Ni(II) SULPHATE EFFLUENT SOLUTIONS

Christoph Doerfelt[1], George P. Demopoulos [1*]

[1]McGill University, 3610 University, H3A 2B2 Montreal, Québec, Canada
(*Corresponding author: george.demopoulos@mcgill.ca)

Keywords: arsenic, removal, ferrous, ferric, effluent solutions, stability, co-precipitation

Abstract

Several non-ferrous metal ores and concentrates contain significant amounts of arsenic that upon mobilization during hydrometallurgical processing necessitates its effective removal and disposal in environmentally stable tailings. The most common method to accomplish this operation involves lime neutralization in the presence of excess iron. It is known that for its effective co-precipitation, arsenic must be in its pentavalent state and iron as ferric at a molar ratio Fe(III)/As(V)>3. Upon selection of the right pH profile the precipitates produced in this way are stable under oxic conditions. This may be not true however under sub-oxic or anoxic conditions that might develop in tailings ponds. In this context the potential stabilizing role of other co-ions becomes important. As such, this paper investigates the removal and redox stability of arsenic with ferric and various co-ions, Fe(II), Al(III), Ni(II), in a 2-stage continuous co-precipitation process.

Introduction

Processing of non-ferrous ores like nickel and copper ores has to often deal with arsenic contamination and the challenge of disposing arsenic in an environmentally stable manner. Nowadays arsenic is precipitated with excess ferric from sulphate containing effluent solutions in continuous co-precipitation circuits. A molar ratio of Fe(III)/As(V)>3 has been shown to work well [1]. The effluent solutions are neutralized with lime, which results in up to 25 times better arsenic removal and stability of the products compared to neutralization with NaOH [2]. The final pH of the neutralization depends on other contaminants. Arsenic co-precipitation is favored at lower pH but the pH has to be high enough if nickel is present in order to ensure the removal of the latter as hydroxide [3]. *Chen et al.* investigated tailings of the McClean Lake mine with EXAFS/XANES and compared the spectra to model compounds. They found that up to pH 7 and molar ratio of Fe(III)/As(V)=4 poorly crystalline ferric arsenate ($FeAsO_4 \bullet xH_2O$) would form as the principal arsenic phase with arsenate-adsorbed ferrihydrite being the minor phase [4]. At pH 8 or higher arsenical ferrihydrite becomes the dominant arsenic carrier [5]. Using zeta potential measurements De Klerk et al. arrived at similar observations [6]. Co-precipitates of Fe(III)/As(V) depending on the Fe(III)/As(V) ratio [1, 3, 7]. are stable over a relatively wide pH range. The ferrihydrite transformation to the thermodynamically more stable goethite/hematite phases predicted by thermodynamics is severely blocked, when tetrahedral sorbates like arsenate are present [8]. The Vale (formerly Inco) plant in Sudbury has deposited arsenical iron (III) co-precipitates in their tailings management facility for many years. No significant arsenic release has been detected [9]. However, there are concerns that arsenic may be released from ferrihydrite or ferric arsenate in reducing environments [10]. These can develop in the tailings management facilities by microbial/bacterial activity. Sub-oxic environments (oxygen starvation) can develop if sulfides react with soluble O_2 [11]. Laboratory studies have demonstrated the release of arsenic during reductive dissolution of ferrihydrite. *Erbs et. al* used hydrochinone to investigate the reduction of co-precipitated arsenical ferrihydrite and ferrihydrite with post- precipitation adsorbed arsenate. The co-precipitated arsenical ferrihydrite exhibited greater resistance to reduction, which was attributed to the occupation of active sites by adsorbed arsenate[12]. *Rochette et. al* demonstrated the dissolution of scorodite at slightly acidic pH and $E_h < 0$ mV [13]. These concerns and the fact that many of the effluents already contain ferrous and many other ions have led to the investigation of effect of ferrous iron on co-precipitation and stability of As(V) under oxic conditions by Daenzer [14]. This paper complements that research by extending the stability study to reducing environments. Furthermore supporting ions like Al(III) that are redox stable and the influence of nickel are also investigated.

Materials and Methods

In the scope of this paper two types of experiments were conducted. The first type of experiment consists of the production of solids and the second part is comprised of stability testing of the products.

The solids were produced in a continuous two-reactor setup by neutralizing the acidic sulfate feed solution with slaked lime to a final pH=8; the pH of reactor 1 was maintained at either 4 or 6. The feed composition was made of 1400 ppm arsenic(V), Fe(II,III)/As(V) molar ratio equal to 4 and where applicable co-ions like Al(III) and Ni(II), all as sulfate salts. Previous publications by De Klerk et al. provide a very detailed guide on the continuous co-precipitation circuit that was used for these experiments [6, 15].

For stability testing only the steady-state product was used after doubling the solid/liquid ratio by decanting overlaying liquid. Two reference samples, drift (no pH adjustment) and oxic (pH=8), were prepared. The reducing environment was generated and maintained by frequent NaHS (0.125M) addition to a third sample (pH=8, E_h=0mV). NaOH (1M) was used for regular pH adjustment.

The filtrates were analyzed for As, Fe, Al and Ni by ICP-OES using an iCAP 6000 series ICP spectrometer by Thermo Scientific. Fe(II) was determined by dichromate titration.

Results and Discussion

Co-Precipitation

To measure the effect of ferrous on arsenic(V) removal three experiments with increasing Fe(II) partition up to complete ferric substitution were conducted with Fe(tot)/As(V)=4. The average As(V) concentrations during steady-state for reactor 1 and reactor 2 respectively can be found in Table I. The co-precipitation of ferrous and arsenate (see CD2 [Fe(II)/As(V)=4]) yields arsenate concentrations of 13 ppb, which compares to the co-precipitation behavior of ferric and arsenate (CD4 [Fe(III)/As(V)=4]). The pH in the first reactor was set to six for CD2, because a hydrolysis experiment by Daenzer showed pH=6 to be a very effective initial precipitation pH for ferrous arsenate formation-precipitation [14]. In experiment CD3 (Fe(II)/Fe(III)/As(V)=2/2/1) the arsenic concentrations are somewhat higher reacching 100 ppb in reactor 1 and to 1 ppm in reactor 2.

Table I. Average Arsenic Concentrations During Co-Precipitation at Steady-State

Experiment	Fe/As molar ratio	pH set points	R1 avg. [As] mg/L	R2 avg. [As] mg/L
CD2	Fe(II)/As=4	6 & 8	0.017	0.013
CD3	Fe(II)/Fe(III)/As=2/2/1	4 & 8	0.094	0.976
CD4	Fe(III)/As=4	4 & 8	<MDL	0.009

The performance of the 2-stage co-precipitation process can be further evaluated with the data shown in Figures 1 and 2 corresponding to the test with Fe(II)/As(V)=4 (CD2). Figure 1 shows the concentration profiles for Fe(II) and As(V). Figure 2 displays the E_h and pH profiles during co-precipitation. The reactors reached steady-state after 6.5 hours. It is only after that point that slurry was collected for stability testing.

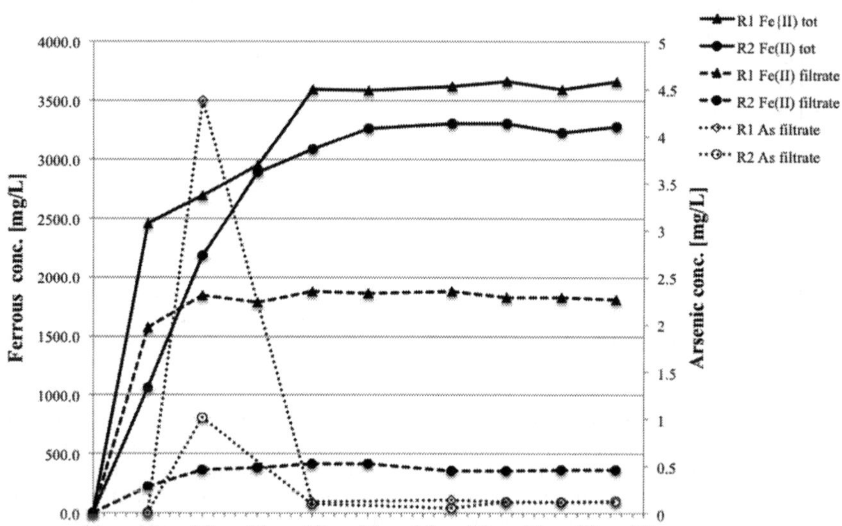

Figure 1: Fe(II)/As(V)=4: Fe(II) and arsenic concentrations during co-precipitation.

The initial ferrous concentration was 4076 ppm. Roughly 10% of ferrous oxidizes during steady-state operation due to apparent air intake in reactors 1 and 2. In the first reactor 1800 ppm of ferrous and 1400 ppm of arsenic precipitate which equals to a molar ratio of Fe(II)/As(V)=1.72 that is very close to the stoichiometry of ferrous arsenate ($Fe_3(AsO_4)_2$) formation. The average E_h during steady-state in reactor 1 (pH=6) and reactor 2 (pH=8) was 84mV and -478mV respectively (see Figure 2 for E_h and pH profiles).

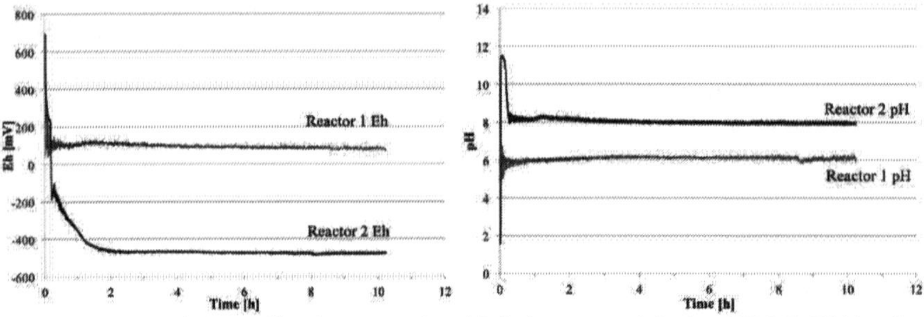

Figure 2. E_h and pH profiles for reactor 1 and 2 during co-precipitation (CD2:Fe(II)/As=4)

Table II. Average E_h and pH Values During Steady-State

Experiment	Fe/As molar ratio	pH set points	Avg. E_h R1 [mV]	Avg. E_h R2 [mV]
CD2	Fe(II)/As=4	6 & 8	84	-478
CD3	Fe(II)/Fe(III)/As=2/2/1	4 & 8	141	-118
CD4	Fe(III)/As=4	4 & 8	757	415

Table II summarizes the mean E_h and pH values for all three experiments. Increasing amounts of ferrous result in decreasing E_h with most notable the very low E_h for the second reactor in CD2.

Stability Testing

In this series of experiments the behavior of the solids in oxic and anoxic environments is studied. This is work in progress providing here only a sample of collected data. There are two data points for each sampling day. This is so because samples were taken right before and after the adjustment of pH and E_h (see Figure 3).

The adjustment of the samples, which is done while the slurry is stirred with a magnetic stirrer, seems to have some cascade effect on arsenic release. Before the adjustment is done, meaning at lower pH and higher E_h, less arsenic is in solution. After increasing the pH to pH=8, arsenic is released into solution. This spike in arsenic concentration in the filtrate is drastically reduced after only two days with further decrease the longer the interval between sampling.

This "cascading" behavior is partially explained by the increase in pH, which results in increased arsenic solubility. The first data points (days 130, 132 and 140) did not yet require sulfide addition as the E_h was still very low (compare production E_h=-478mV) and decreased to 0 mV solely by the pH adjustment. Figure 4 shows the E_h/pH progression of the anoxic sample. The opposite behavior –increase in pH, decrease in E_h – is apparent. Also notable is that the E_h is increasing up to 100 mV between sampling. The sampling interval does seem to have a limited influence on this, because short sampling intervals also experience an increase in E_h by more than 80 mV.

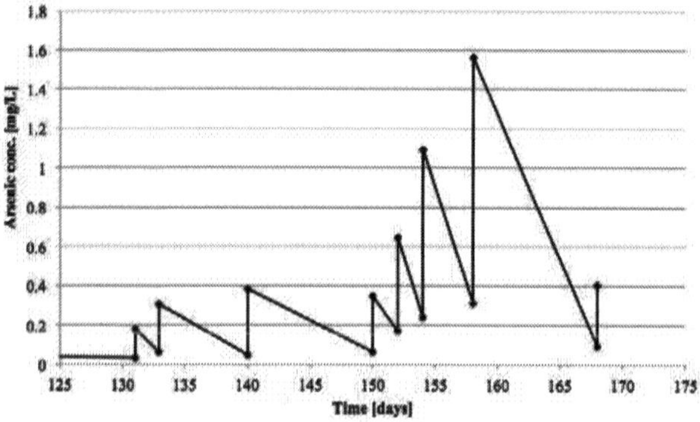

Figure 3. Fe(II)/As(V)=4: As release in anoxic environment after 132 days of pre-aging

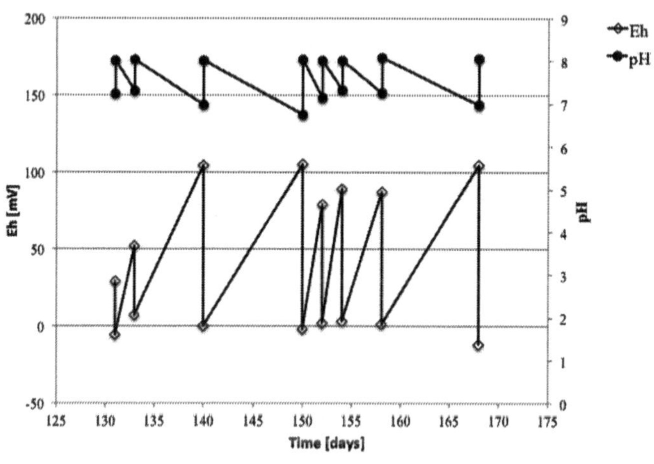

Figure 4. E_h and pH profile during ageing for the anoxic sample (Fe(II)/As(V)=4)

Table II summarizes the stability testing data that has been acquired up to this point. For the ferrous/arsenate production sample the oxic sample shows 0.89 ppm of arsenic, while the sulfide adjusted sample has 0.4 ppm of arsenic in solution. The mixed experiment CD3 (Fe(II)/Fe(III)/As(V)=2/2/1) and the ferric/arsenate experiment CD4 released higher levels of arsenic in the presence of sulfides. Therefore it can be said tentatively that ferrous has a stabilizing effect in anoxic environments. This issue will be investigated further with longer ageing data (work in progress) and appropriate characterization.

Table II. Stability Testing Results of the Latest 'After-Adjustment' Sample

Experiment	Age [days]	Oxic[mg/L]	Anoxic As [mg/L]
Fe(II)/As(V)=4	168*	0.89	0.40
Fe(II)/Fe(III)/As=2/2/1	42	0.44	0.85
Fe(III)/As(V)=4	110	0.50	1.21

*includes 125 of column pre-aging without sampling/adjustment

Effect of Nickel and Aluminum Co-Ions

Previous work in our laboratory [16] has shown the presence of co-ions like aluminum and nickel in the Fe(III)/As(V) co-precipitation system to have a stabilizing effect on As(V) [17]. Thus for Fe(III)/As(V)=4 containing Ni(II)/As(V)=0.5 the As(V) concentration in solution after 210 days of pH=8 adjusted aging was down to 0.4 mg/L from 0.63 mg/L without nickel. The origin of the stabilizing role of Ni(II) is currently under investigation [18]. The partial substitution of aluminum for ferric (Fe(III)/Al(III)/As(V)=3/1/1) also exhibited better stability behavior than Fe(III)/As(V) (=4) alone. Thus after 225 days only 0.44 mg/L As(V) was found in solution in contrast to 0.65 mg/L As(V) without aluminum. Further work is planned to evaluate the stabilizing roles of Ni and Al under anoxic conditions.

Acknowledgements

This work is supported via a NSERC CRD grant. Areva Resources Canada and Cameco Corporation, the industrial sponsors of this research, are also thanked for their support.

Literature

1. E. Krause, V.A. Ettel, "Solubilities and Stabilities of Ferric Arsenate Compounds," *Hydrometallurgy*, 22 (1989), 311–337.

2. Y. Jia, G.P. Demopoulos, "Coprecipitation of arsenate with iron(III) in aqueous sulfate media: effect of time, lime as base and co-ions on arsenic retention," *Water Research* 42 (2008), 661–668.

3. D. Langmuir et al., "Predicting arsenic concentrations in the porewaters of buried uranium mill tailings," *Geochimica et Cosmochimica Acta*, 63 (1999), 3379–3394

4. N. Chen et al., "Structural characterization of poorly-crystalline scorodite, iron(III)–arsenate co-precipitates and uranium mill neutralized raffinate solids using X-ray absorption fine structure spectroscopy," *Geochimica et Cosmochimica Acta*, 73 (2009), 3260–3276.

5. L.G. Twidwell, R.G. Robins, J.W. Hohn, "The Removal of Arsenic From Aqueous Solution by Coprecipitation with iron(II)," *Arsenic Metallurgy: Fundamentals and Applications*, ed. R.G. Reddy and V. Ramachandran, (Warrendale, PA, The Minerals, Metals & Materials Society, 2005), 3-24.

6. R.J. de Klerk et al., "Continuous circuit coprecipitation of arsenic(V) with ferric iron by lime neutralization: Process parameter effects on arsenic removal and precipitate quality," *Hydrometallurgy*, 111-112 (2012), 65–72.

7. D. Langmuir, J. Mahoney, J. Rowson, |Solubility products of amorphous ferric arsenate and crystalline scorodite (FeAsO4·2H2O) and their application to arsenic behavior in buried mine tailings," *Geochimica et Cosmochimica Acta*, 70 (2006), 2942–2956.

8. R.G. Ford, "Rates of hydrous ferric oxide crystallization and the influence on coprecipitated arsenate," *Environmental Science & Technology*, 36 (2002), 2459–63.

9. G.B. Harris and E. Krause, "The Disposal of Arsenic from Metallurgical Processes: Its Status Regarding Ferric Arsenate", *Extractive Metallurgy of Copper, Nickel and Cobalt, vol. I (Fundamental Aspects)*, ed. R.G. Reddy and R.N Weizenbach (Warrendale, PA: The Minerals, Metals and Materials Society, 1993), 1221-1237.

10. G.B. Harris, "The Removal and Stabilization of Arsenic from Aqueous Process Solutions: Past, Present and Future," *Minor Elements 2000,* ed. C.A. Young, (Littletown, CO, 2000), 3-20.

11. A.P.D. Wilmot et al., "Kinetics of Sulfide Oxidation by Dissolved Oxygen," *Water Pollution Control Federation*, 60 (1988), 1264–1270.

12. J.J. Erbs et al. " Reductive dissolution of arsenic-bearing ferrihydrite," *Geochimica et Cosmochimica Acta*, 74 (2010), 3382–3395.

13. E.A. Rochette, G.C. Li, S.E. Fendorf, "Stability of Arsenate Minerals in Soil under Biotically Generated Reducing Conditions," *Soil Science Society of America Journal*, 62 (1998), 1530-1537.

14. R. Daenzer, "Investigating the role of ferrous iron in the arsenic (V) -iron(II, III) coprecipitation process system," (M.Eng. thesis, McGill University, 2011).

15. R.J. de Klerk, G. P. Demopoulos, "Continuous Circuit Production and Accelerated Ageing of Iron(III)-Arsenic(V) Coprecipitates - Probing Process-Stability Relationships", *EPD Congress 2008: Proceedings of Sessions and Symposia Sponsored by the Extraction and Processing Division (EPD)*, ed. S.M. Howard, (Warrendale, PA: The Minerals, Metals and Materials Society, 2008), 3-10

16. R.J. de Klerk, |Investigating the Continuous Circuit Coprecipitation of Arsenic (V) with Ferric Iron in Sulphate Media," (M.Eng. thesis, McGill University, 2008).

17. R. Daenzer, R.J. De Klerk, G.P. Demopoulos, "Arsenic Control in Process Tailings: Continuous Co-Precipitation of As(V) with Iron Sulphate Media", in Proc. Uranium 2010, Eds. E.K. Lam et. al., (CIM, 2010), Vol. 2, 365-375

18. N. Chen et al., "XAFS Study of Arsenical Nickel Hydroxide," (Paper presented at the 15th International Conference on X-ray Absorption Fine Structure, Beijing, China, 22-28 July 2012)

Ni-Co 2013
Edited by: Thomas Battle, Michael Moats, Violina Cocalia, Harald Oosterhof, Shafiq Alam, Antoine Allanore, Rodney Jones, Nathan Stubina, Corby Anderson, and Shijie Wang
TMS (The Minerals, Metals & Materials Society), 2013

HYDROCHLORIC ACID REGENERATION VIA CALCIUM SULFATE CRYSTALLIZATION FOR NON-FERROUS CHLORIDE LEACHING PROCESSES

Thomas Feldmann[1], George P. Demopoulos[1]

[1]McGill University
Department of Mining and Materials Engineering
3610 University Street, Montreal, QC, Canada, H3A 0C5

Keywords: Lixiviant regeneration, hydrochloric acid, calcium sulfate, reactive crystallization, crystal growth, chloride hydrometallurgy

Abstract

Regeneration of hydrochloric acid constitutes a key operation in many new chloride hydrometallurgical processes, including polymetallic Ni-Cu sulfides, laterite ores or extraction of rare earths from apatite ores. For leaching processes that generate a spent calcium chloride solution, one novel option for regenerating HCl involves reaction of $CaCl_2$ liquor with concentrated H_2SO_4 at low temperatures (T<100°C). Due to the favorable price and availability of sulfuric acid this constitutes a viable option. Therefore, key considerations concern the process kinetics of the crystallization carried out by stage-wise mixing of solutions to control the precipitation reaction, which leads to formation of different calcium sulfate phases depending on concentrations and temperature. Crystal growth and phase transformation kinetics determine the efficiency of required solid-liquid separation steps and optimum residence time in individual process stages. In this paper highlights from our work defining appropriate operating conditions and results from crystallization kinetic studies will be presented.

Introduction

The use and development of chloride metallurgical processes continues to be of interest to a number of hydrometallurgical applications. This can be explained by the high solubilities of chloride salts and the option to carry out fast and effective leaching at ambient pressures and relatively low (below the boiling point) temperatures [1]. A number of processes have been developed in the past. Examples include the leaching of low grade silver ore tailings, titanium dioxide bearing slags, nickel laterite ores or the treatment of complex sulfidic base metal ores [1-5]. A key to the successful operation of such operations is the ability to regenerate the hydrochloric acid used in a cost effective way. A chloride based flowsheet for nickel laterite leaching has been developed by Intec Ltd. As can be seen in Figure 1 a calcium chloride rich solution is reacted with sulfuric acid to regenerate HCl and form calcium sulfate anhydrite (AH).

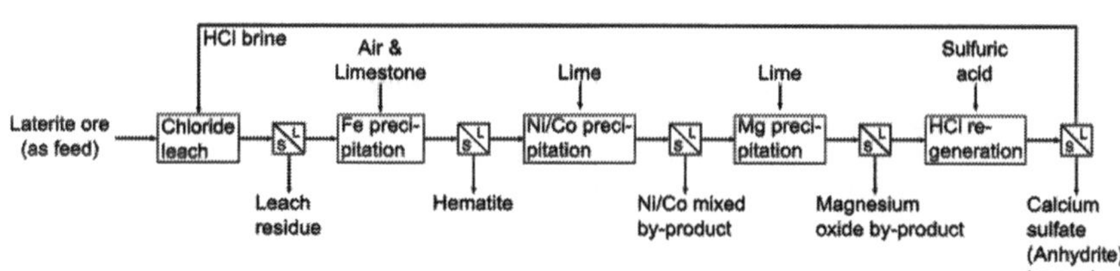

Figure 1. Intec nickel laterite process flowsheet with HCl regeneration (adapted from [6])

Unlike the common treatment of $FeCl_2/FeCl_3$ steel pickling solutions by pyrohydrolysis, this technique has prohibitive energy requirements if applied to $CaCl_2$ solutions. This is caused by the low vapor pressures of the salt even at temperatures above 1000°C [7]. Previous research at McGill University conceptualized a low temperature (T<100°C) aqueous regeneration process for hydrochloric acid. The process involves mixing of concentrated $CaCl_2$ solution with concentrated H_2SO_4 solution in a multistage continuous reactor set up, schematically shown in Figure 2. The principal chemical reaction is given in Equation 1.

$$CaCl_2 + H_2SO_4 + xH_2O \rightarrow 2HCl + CaSO_4 \cdot xH_2O \quad (1)$$

The viability of the process is based on the low solubility of calcium sulfate phases in aqueous solutions. Upon controlled mixing of the solutions calcium sulfate crystals are produced and hydrochloric acid is regenerated [8]. Research on the determination of process operating windows such as temperature and concentration levels for the production of high quality calcium sulfate dihydrate (DH) or the valuable calcium sulfate α-hemihydrate (HH) phase instead of formation of low value anhydrite (AH) [9,10] has been undertaken.

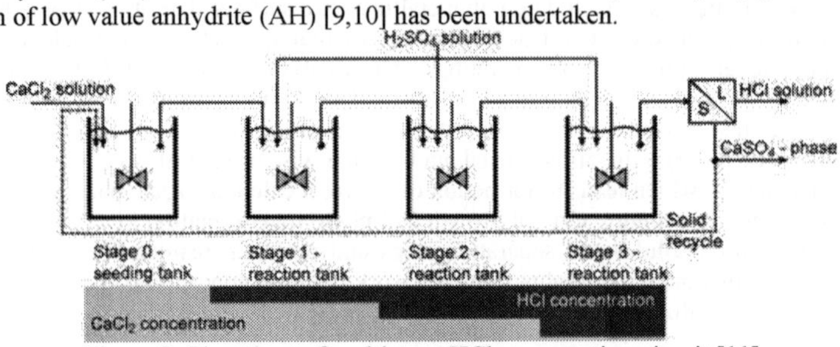

Figure 2. Flowchart of multistage HCl regeneration circuit [11]

The practical implementation of this process depends strongly on the crystallization kinetics of the system. Firstly, in order to reuse the regenerated HCl it is necessary to have efficient solid-liquid separation. Key influencing factors are the crystal size and its morphology [12]. Large and compact crystals usually filter better than small, fine and complex shaped ones such as AH. Secondly, the presence of multiple calcium sulfate phases requires knowledge of their relative stabilities over time in order to design a process that produces a pure solid product. Calcium sulfate forms three principal phases in aqueous solution depending on the amount of crystal water bound in the lattice these are DH/gypsum ($CaSO_4 \cdot 2H_2O$), AH/anhydrite ($CaSO_4$) as thermodynamically stable phases or HH/hemihydrate ($CaSO_4 \cdot 0.5H_2O$), which is metastable. Thermodynamic modeling and experiments showed that especially in the presence of high strength HCl, AH is the stable phase [13]. Its formation is undesirable due to its fibrous morphology, which results in extremely poor filtration rates compared to high filtration rates obtained for DH or HH [9,14]. As a consequence of this questions arise about the Intec laterite process (Figure 1) that features production of anhydrite. In order to design an optimized process crystallization kinetic studies were undertaken, namely investigation of the crystal growth rate and time dependent phase (meta)stability, highlights of which are presented in this report.

Previous research

In the past there have been several investigations of the growth kinetics of calcium sulfate dihydrate [15-17] observing a second order growth rate law but very few on calcium sulfate hemihydrate [18-20]. These studies mostly focused on dilute solutions or a different matrix than the $CaCl_2$-HCl system investigated here. So far only the growth rate of HH in concentrated $CaCl_2$ solutions was studied [20], however the simultaneous presence of HCl was never studied before although HCl is known to influence the stability/solubility of HH [10,21]. Hence the need to include it in the context of the industrial acid regeneration application.

A similar lack of knowledge exists on the influence of strong $CaCl_2$/HCl solutions on phase transformation. For other hydrometallurgical systems it was found that the DH to HH, DH to AH and HH to AH transformations typically proceeded via a dissolution-precipitation mechanism [22-26]. Increases in temperature [24], the presence of seed of the more stable phase [23-25,27] and low water activity [25] accelerate the transformation process. The time signifying the onset of the conversion process (otherwise known as induction time) has been reported to vary from hours to several days depending on solution composition and temperature.

Experimental

Experimental Setup

The experiments were conducted in a glass semi-batch reactor allowing for temperature control, reagent addition via peristaltic feed pumps and mechanical agitation. Furthermore, the 2L reaction vessel had baffles to improve mixing conditions.

The growth rate experiments involved addition of $CaCl_2$ and Na_2SO_4 solutions to ensure an equimolar Ca^{2+} and SO_4^{2-} ion inflow to a seeded background solution mimicking steady state conditions of the acid regeneration process. Sodium sulfate instead of sulfuric acid as a source of sulfate ions for the crystal growth process was chosen in order to not significantly change the acid concentration in the reactor. This method resulted in nearly constant supersaturation conditions during the crystal growth process. Details are reported elsewhere [28].

The phase transformation experiments were conducted under conditions mimicking steady state stages of the acid regeneration process. A slurry of DH or HH was prepared by adding 75g solid to the solution (1L), which had been heated previously to the desired temperature. For some experiments a small amount of AH (5mol% of the total solid) was added to the slurry as seed. For further details refer to reference [11].

Table I gives a summary of the experimental conditions chosen for both types of experiments. In the growth rate experiments the reagent inflow rate (0-0.6 mol/h), temperature (70-95°C) and specific power input of stirring (0.02-1.29 W/kg) were varied. For the phase transformation experiments temperature (40-90°C) and solution composition (see Table II) was varied. All solutions and the solid material used in these experiments were prepared from ACS reagent grade chemicals.

Table I. Standard conditions of crystal growth experiments

Standard conditions		
Parameter	*High $CaCl_2$ condition*	*High HCl condition*
Concentration of $CaCl_2$ [mol/L]	2.8	0.7
Concentration of HCl [mol/L]	1.4	5.6
Temperature [°C]	80	80
Specific power input stirring [W/kg]	0.16	0.16
Duration of experiment [min]	60	60
Initial volume of solution [mL]	1000	1000

Table II. Solution composition corresponding to a 3-stage steady-state HCl regeneration circuit

Conditions	*HCl conc. [mol/L]*	*$CaCl_2$ conc. [mol/L]*
Stage 0	0.0	4.97
Stage 1	3.14	3.39
Stage 2	6.29	1.82
Stage 3	9.43	0.25

Characterization Methods

In order to determine the crystal growth rate, volume-based particle size distributions were measured with a Horiba Laser Scattering Particle Size Distribution Analyzer LA-920. Additionally, BET surface area measurements of the seed crystals were undertaken with a Micromeritics Tristar Surface Area Analyzer. The crystal morphology was analyzed with a Philips XL30 FEG SEM. XRD analysis was carried out on a Bruker AXS powder diffractometer. Finally, chemical analysis was conducted with a Dionex ICS 5000 ion chromatograph and a Thermo Scientific iCAP 6000 ICP-MS apparatus. Thermodynamic calculations were conducted with the OLI Stream Analyzer [29]. To ensure reproducibility, the growth kinetics experiments were repeated three times and arithmetic averages were employed in the analysis of the data.

Results and Discussion

Crystal Growth Kinetics

Determination of Crystal Growth Rates The crystal growth rate was determined via measurements of the change of particle size over time. For the calculation of the growth rate the measured particle size distributions were fitted to a log-normal probability distribution. The parameters of the fitted distribution were used to calculate the mode of the distribution as a measure of the characteristic length of the particles. The calculation of the mode was done for each of the five samples taken from a single experiment and regressed against time with a linear function. The slope of this function represents the change in particle size with time and hence the crystal growth rate. This approach is schematically described in Figure 3.

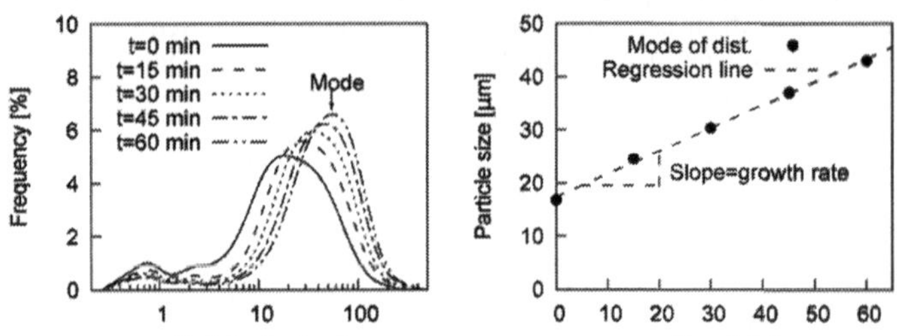

Figure 3. Schematic representation of the derivation of the crystal growth rate from measured particle size distribution data. Example: high $CaCl_2$ condition, T=80°C, specific power input=0.16 W/kg, molar inflow rate=0.3 mol/h [28]

Finally recognizing that crystal growth predominantly takes place at active surface sites, the measured growth rates were made independent of the specific surface area available for growth by taking into account the initial surface area of the seed crystals (average 0.548±0.072 m^2/g). In the following the results for each of the varied experimental parameters are discussed.

Reagent Inflow Rate Figure 4 (a) shows the crystal growth rate of HH as a function of the reagent inflow rate. This opens in principle the opportunity for crystal size control via inflow rate regulation. The increase in growth rate with inflow rate is not monotonic. The increase of the reagent inflow rate from 0 to 0.3 mol/h resulted in a nearly linear increase of the crystal growth rate, thereafter reaching a maximum at 0.3 mol/h. At molar inflow rates above 0.3 mol/h the growth rate plateaus. This dependency of the growth rate with increasing molar inflow rate (supersaturation) can be explained on the basis of von Weimarn's second law. It states that for a given crystallization time the average particle size passes through a maximum as a function of initial supersaturation [30-32]. As a consequence the inflow rate corresponds to different initial supersaturation levels; hence explaining the observed behavior of the growth rate as a function of molar inflow rate. No significant effect of solution composition was found.

Stirring Rate In the case of diffusion-controlled processes changes in the stirring rate (expressed as specific power input) should influence the measured crystal growth rate [33]. Experiments done at a molar inflow rate of 0.3 mol/h and a temperature of 80°C showed that a variation of the specific power input did not have any noticeable effect on the crystal growth rate within the range of investigated conditions (see Figure 4 (b)). This behavior is attributed to the turbulent flow regime that characterized the experimental conditions under which the growth of HH crystals was studied (Reynolds numbers of ~15000 to ~60000). These turbulent flow conditions most likely resulted in a reduction of the diffusion boundary layer down to its limiting thickness hence the lack of a noticeable influence of stirring rate.

Temperature In case of a reaction controlled process, changes of reaction temperature would influence the rate of the process [34]. However, experiments with a reagent inflow rate of 0.3 mol/h and a specific energy input of 0.16 W/kg showed no distinct influence of temperature for either of the two solution types tested (see Figure 4 (c)). Based on the evidence presented above the crystal growth of HH is concluded to be a diffusion controlled linear process under the investigated conditions. A more detailed discussion can be found in a previous publication [28].

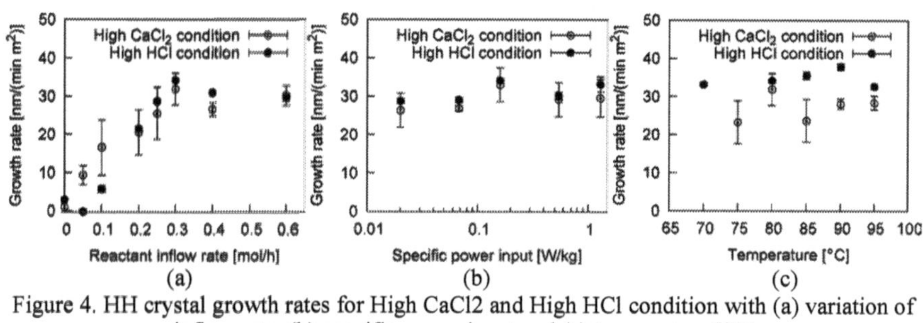

(a) (b) (c)

Figure 4. HH crystal growth rates for High CaCl2 and High HCl condition with (a) variation of inflow rate, (b) specific power input and (c) temperature [28]

Phase Transformation Kinetics

Quantification of Calcium Sulfate Phase Transformations In order to quantitatively describe the progress of phase conversion ratios of absolute intensities of characteristic peaks of each phase's XRD pattern were compared. This method is generally considered suitable for quantitative phase analysis of mixtures [35]. To aid phase quantification, calibration curves based on prepared mixtures with known composition were used and the respective intensity ratios calculated. The intensity ratios were calculated from the intensity of peaks at 2θ 29.10°/25.43°, 29.71°/25.43° and

29.10°/29.71° respectively. The accuracy of this method was within ±5% of the actual composition of a sample. The phase compositions reported in the following section are based on the average composition obtained by using two different intensity ratios for each binary system.

Phase Transformation of Calcium Sulfates and Influencing Factors A typical set of phase transformation curves obtained from the various experiments is given in Figure 5. The three graphs show the evolution of the fraction of DH, HH and AH in the transforming solid over time at different temperatures under stage 1 condition. It can be seen that the share of HH increases at the expense of DH and at temperatures of 70°C and 80°C the formation of AH is seen to take place at the expense of HH. Similar observations were made with the other experimental conditions tested.

Upon careful examination of the transformation curves key differences become evident. It is generally seen that the phase conversion from DH to HH started to increase at first slowly but soon after continued at an accelerated rate and thereafter slowing down before all DH is consumed. The transformed fraction of DH to HH over time followed under all investigated conditions a sigmoidal curve type, which can be described with a general logistic function. In contrast the transformation of HH to AH is seen to proceed in a linear fashion signaling a different mechanism

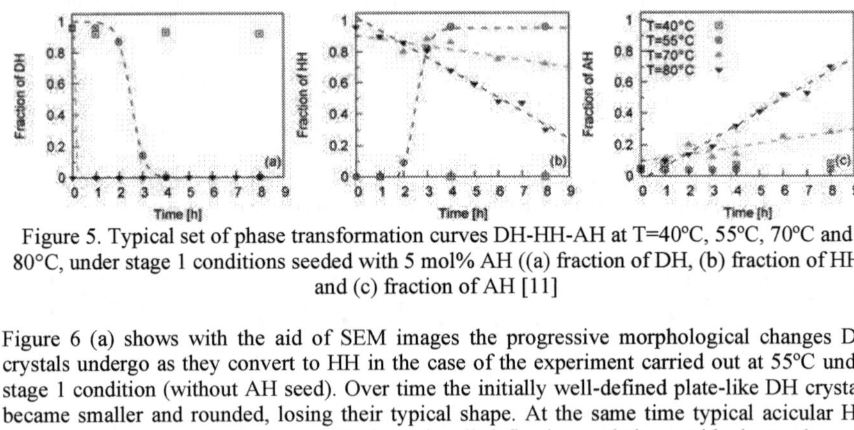

Figure 5. Typical set of phase transformation curves DH-HH-AH at T=40°C, 55°C, 70°C and 80°C, under stage 1 conditions seeded with 5 mol% AH ((a) fraction of DH, (b) fraction of HH and (c) fraction of AH [11]

Figure 6 (a) shows with the aid of SEM images the progressive morphological changes DH crystals undergo as they convert to HH in the case of the experiment carried out at 55°C under stage 1 condition (without AH seed). Over time the initially well-defined plate-like DH crystals became smaller and rounded, losing their typical shape. At the same time typical acicular HH crystals appeared next to DH. The HH showed well defined crystal shape with sharp edges. At the end of the experiment all DH crystals had disappeared. Figure 6 (b) shows the progressive conversion of HH to AH at 70°C under stage 3 conditions without the presence of AH seed crystals. As can be seen, over time the fibrous morphology of AH developed on the surface of HH crystals. An interesting feature is that even the final AH product found after 8 h retained the overall acicular shape of the initially present HH crystals to a certain degree. In contrast to the DH to HH conversion no rounding of the original HH crystals is seen. From the evaluation of the results obtained it was concluded that the overall conversion process of DH to AH proceeds via a two-step process (DH to HH and HH to AH) with each step following a different mechanism.

(a) DH to HH conversion (b) HH to AH conversion

Figure 6. SEM images showing the typical evolution of the phase transformations over time [11]

A dissolution-precipitation based phase conversion mechanism of DH to HH was found in calcium chloride salt solution in the absence of HCl [20]. This conversion mechanism is characterized by a sigmoidal phase fraction against time relationship [36]. A similar sigmoidal behavior was found in the current work. This justifies the assumption that also in solutions containing 0.25 to 4.97 mol/L $CaCl_2$ and 0 to 9.43 mol/L HCl this mechanism characterizes the DH to HH phase transformation process.

The sigmoidal behavior, which is a characteristic feature in auto-catalytic processes [37], can be explained by the role of HH crystals, which act apparently as seed ("catalyst") promoting the transformation. After the initial formation of HH crystals these act as seed. Over time the surface of these crystals increases hence providing an increasing number of active surface sites for incorporation of ions into the crystal lattice structure. This results in acceleration of the conversion process until all DH has been converted to HH.

Understanding the conversion of HH to AH is of special interest. Owing to its fibrous morphology the formation of AH can lead to a number of operational inefficiencies, e.g., reduced filtration rates or blockages of pipes and pumps. The evidence presented above (for more detail see [11]) indicates a topotactic transformation mechanism. This means the new phase (AH) nucleates and grows on the surface of the parent phase (HH). A condition for such a process are elements of structural similarity between the two phases, which is fulfilled by the one-dimensional similarity of their Ca-SO_4 chains due to their similar lattice structure [38,39]. Crystal structural investigations have shown that during the conversion from the HH to the AH lattice, the structure of HH needs to change only slightly to yield the AH lattice [40]. An investigation of the effect of temperature showed that increasing temperatures generally speed up the phase transformation process, while for example the transformation from HH to AH under stage 3 conditions took 5.6 hours at 65°C it was completed within 2 hours at 80°C. Furthermore, especially the onset of the HH to AH conversion was dependent on the presence of AH. The presence of small amounts of AH lead to a significantly early onset of this phase transformation, however in all cases where DH converted to AH, HH was formed intermediately.

Conclusions

The research conducted with reference to crystallization kinetic aspects influencing the performance of a novel HCl regeneration process involving reaction of $CaCl_2$ with H_2SO_4 led to the following conclusions:

- The crystal growth rate of HH depends on the addition rate of reagents to the process. This allows to effectively control the final particle size distribution of the solid product.

- If DH is targeted as the preferred crystal variety, then the operating temperature of the crystallization process should be set to a value of 40°C or lower.
- For the stages with higher HCl concentration, e.g., stage 2 (6 M) and stage 3 (9 M) the retention time should be limited to 2 hours or less to avoid the undesirable conversion of HH (target solid) to AH if the process is operated at temperatures above 65°C.
- The reactor design needs to avoid dead zones in the fluid flow, as small amounts of AH (as low as 5mol% of the total solid amount) were found to induce the phase transformation of the slurry to AH.

A general concept of the multistage regeneration process is given in the flowsheet of Figure 7. It includes a section for incoming solution concentration by evaporation, which helps to achieve the desired final HCl concentration. Also the strength of the sulfuric acid can be adjusted. This is followed by the stagewise sulfuric acid addition into continuous stirred tank reactors as presented in Figure 2. This is followed by solid liquid separation equipment and washing of the calcium sulfate product.

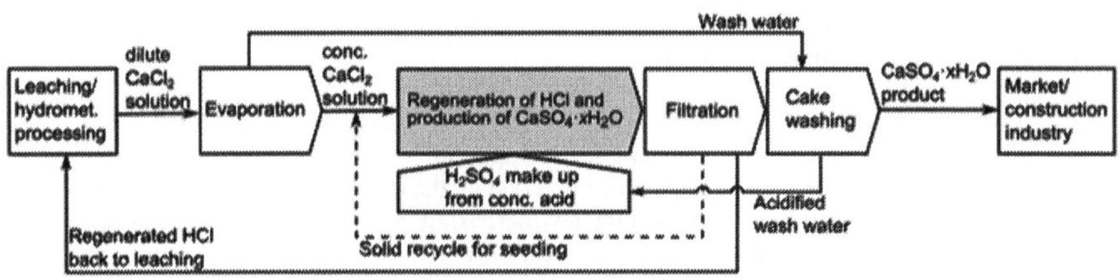

Figure 7. Flowchart of hydrochloric acid regeneration circuit and calcium sulfate production

Acknowledgement

The financial support of Arafura Resources, Ltd., Perth, WA, Australia is gratefully acknowledged.

References

[1] D. Flett., "Chloride Hydrometallurgy for Complex Sulphides: A review" In: E. Peek, G. Van Weert eds., *International Conference on the Practice and Theory of Chloride/Metal Interaction, 32nd Annual Hydrometallurgy Meeting, Montreal.* (Canadian Institute of Mining, Metallurgy and Petroleum, Montreal, 2002), 255-276.

[2] M. Dry et al., "Development of a Novel High-Chloride Circuit for the Starfield Resources' Ferguson Lake Project" In: S.N. Monteiro, D.E. Verhulst, P.N. Anyalebechi, J.E. Pomykala eds., *TMS 2011 140th Annual Meeting & Exhibition - Symposium Chloride 2011: Practice and Theory of Chloride-Based Metallurgy, San Diego,* (TMS The Minerals, Metals & Materials Society, Warrendale, 2011), 415-424.

[3] G.B. Harris et al., "Recovery of Copper from a Massive Polymetallic Sulfide by High Concentration Chloride Leaching" In: P.A. Riveros, D.B. Dreisinger, M.J. Collins eds., *The John E. Dutrizac Symposium on Copper Hydrometallurgy,* (Canadian Institute of Mining, Metallurgy and Petroleum, Montreal, 2007), 21-42.

[4] D.E. Verhulst, V.I. Lakshmanan, „Hydrometallurgy of Chlorides: A Review of Recent

Developments", In: S.N. Monteiro, D.E. Verhulst, P.N. Anyalebechi, J.E. Pomykala eds., *TMS 2011 140th Annual Meeting & Exhibition - Symposium Chloride 2011: Practice and Theory of Chloride-Based Metallurgy, San Diego,* (TMS The Minerals, Metals & Materials Society, Warrendale, 2011), 405-413.

[5] R. Winand, "Chloride Hydrometallurgy," *Hydrometallurgy*, 27 (1991), 285-316.

[6] P. Wood and A. Tong "The Intec Nickel Laterite Process", URL: http://www.intec.com.au/uploaded_files/document_uploads/INP_presentation.pdf, Accessed: Sep 26, 2012

[7] W. Steinbach, and F. Baerhold, "Comparison of Spray Roasting and Fluidised Bed Granulation for the Recovery of Hydrochloric Acid from Metallurgical Processes through Pyrohydrolysis" In: E. Peek, G. Van Weert eds., *International Conference on the Practice and Theory of Chloride/Metal Interaction, 32nd Annual Hydrometallurgy Meeting, Montreal.* (Canadian Institute of Mining, Metallurgy and Petroleum, Montreal, 2002), 643-655.

[8] G. Demopoulos et al., "New Technologies for HCl Regeneration in Chloride Hydrometallurgy," *World of Metallurgy – Erzmetall*, 61 (2008), 89-98.

[9] A. Al-Othman and G. Demopoulos, "Gypsum Crystallization and Hydrochloric Acid Regeneration by Reaction of Calcium Chloride Solution with Sulfuric Acid," *Hydrometallurgy*, 96 (2009), 95-102.

[10] S. Girgin, and G. Demopoulos, "Production of the High Value Material alpha-$CaSO_4$ Hemihydrate out of Spent $CaCl_2$ Solutions by Reaction with H_2SO_4" In: M.E. Schlesinger ed., *EPD Congress 2004*, (TMS (The Minerals, Metals & Materials Society), Warrendale 2004), 627-639.

[11] T. Feldmann and G.P. Demopoulos, "Phase Transformation Kinetics of Calcium Sulfate Phases in Strong $CaCl_2$-HCl solutions," *Hydrometallurgy*, 2012 (Article In Press), DOI: 10.1016/j.hydromet.2012.08.015

[12] R. Wakeman, "The Influence of Particle Properties on Filtration," *Separation and Purification Technology*, 58 (2007), 234-241.

[13] Z. Li, and G. Demopoulos, "Model-based Construction of Calcium Sulfate Phase Transition Diagrams in the HCl-$CaCl_2$-H_2O System between 0 and 100°C," *Industrial Engineering and Chemistry Research*, 45 (2006), 4517–4524.

[14] S. Girgin, "Crystallization of Alpha-Calcium Sulphate Hemihydrate by Aqueous Reaction of Calcium Chloride with Sulphuric Acid" (PhD thesis, McGill University, 2006).

[15] S.-T. Liu, G.H. Nancollas, "The Kinetics of Crystal Growth of Calcium Sulfate Dihydrate," *Journal of Crystal Growth*, 6 (1970), 281-289.

[16] G.H. Nancollas, M.M. Reddy, F. Tsai, "Calcium Sulfate Dihydrate Crystal Growth in

Aqueous Solution at Elevated Temperatures," *Journal of Crystal Growth*, 20 (1973), 125-134.

[17] G.J. Witkamp, J.P. van Der Eerden, G.M. van Rosmalen, "Growth of Gypsum I. Kinetics," *Journal of Crystal Growth,* 102 (1990), 281-289.

[18] M. El Moussaouiti, et al., "Agglomeration Kinetics of Calcium Sulphate Hemihydrate Crystals in Sulpho-Phosphoric Solutions," *Journal of Crystal Growth,* 169 (1996), 118-123.

[19] M. El Moussaouiti, et al., "Crystallization of Calcium Sulphate Hemihydrate in Concentrated Phosphoric Acid Solutions," *Chemical Engineering Journal*, 68 (1997), 123-130.

[20] L. Yang et al., "Growth Rate of α-calcium Sulfate Hemihydrate in K-Ca-Mg-Cl-H_2O Systems at Elevated Temperature," *Journal of Crystal Growth,* 311 (2009), 4518-4524.

[21] Z. Li and G.P. Demopoulos, "Solubility of $CaSO_4$ Phases in Aqueous HCl + $CaCl_2$ Solutions from 283 K to 353 K," *Journal of Chemical Engineering Data*, 50 (2005), 1971-1982.

[22] A. Austin, et al. „Precipitation of Calcium Sulfate from Sea Water at High Temperatures," *Desalination*, 16 (1975), 331-344.

[23] G. Azimi, and V. Papangelakis, „Mechanism and Kinetics of Gypsum - Anhydrite Transformation in Aqueous Electrolyte Solutions," *Hydrometallurgy*, 108 (2011), 122-129.

[24] H. Farrah, G. Lawrance, and E. Wanless, "Gypsum-Anhydrite Transformation in Hot Acidic Manganese Sulfate Solution. A Comparative Kinetic Study Employing Several Analytical Methods," *Hydrometallurgy,* 75 (2004), 91-98.

[25] L. Hardie, "The Gypsum-Anhydrite Equilibrium at One Atmosphere Pressure," *The American Mineralogist*, 52 (1967), 171-200.

[26] J. Kontrec, D. Kralj, and L. Brecevic, "Transformation of Anhydrous Calcium Sulphate into Calcium Sulphate Dihydrate in Aqueous Solutions," *Journal of Crystal Growth,* 240 (2002), 203-211.

[27] J. Dutrizac and A. Kuiper, "The Solubility of Calcium Sulphate in Simulated Nickel Sulphate-Chloride Processing Solutions," *Hydrometallurgy*, 82 (2006), 13-31.

[28] T. Feldmann, T., G.P. Demopoulos, "The Crystal Growth Kinetics of Alpha Calcium Sulfate Hemihydrate in Concentrated $CaCl_2$-HCl Solutions," *Journal of Crystal Growth*, 351 (2012), 9-18.

[29] OLI Stream Analyzer, *http://www.olisystems.com/new-streamanalyzer.shtml*, Accessed: September 30, 2012.

[30] P.P. von Weimarn, "The Precipitation Laws," *Chemical Reviews,* 2 (1925), 217-242.

[31] J.W. Mullin and H.-M. Ang, "Supersaturation and crystal size distribution changes during the precipitation of nickel ammonium sulphate hexahydrate," *Kristall und Technik,* 12 (1977), 105-115.

[32] D.A. Barlow, J.K. Baird and C.-H. Su, "Theory of the von Weimarn Rules Governing the Average Size of Crystals Precipitated from a Supersaturated Solution," *Journal of Crystal Growth,* 264 (2004), 417-423.

[33] J.A. Dirksen and T.A. Ring, "Fundamentals of Crystallization: Kinetic Effects on Particle Size Distributions and Morphology," *Chemical Engineering Science*, 46 (1991), 2389-2427.

[34] T. Havlik, *Hydrometallurgy - Principles and Applications*, (Woodhead Publishing Ltd., 2006).

[35] R. Jenkins and R. Snyder, *Introduction to X-ray Powder Diffractometry,* (New York, John Wiley & Sons, Inc., 1996).

[36] B. Guan, et al., "Alpha-Calcium Sulfate Hemihydrate Preparation from FGD Gypsum in Recycling Mixed Salt Solutions," *Chemical Engineering Journal*, 174 (2011), 296-303.

[37] J. Carberry, *Chemical Engineering Series: Chemical and Catalytic Reaction*, (New York, McGraw-Hill Book Company, 1976).

[38] J. Günter, and H.-R. Oswald, "Attempt to a Systematic Classification of Topotactic Reactions," *Bulletin of the Institute of Chemical Research Kyoto University*, 53 (1975), 249-255.

[39] D. Freyer, and W. Voigt, "Crystallization and Phase Stability of $CaSO_4$ and $CaSO_4$-based Salts," *Monatshefte Chemie*, 134 (2003), 693-719.

[40] M. Carbone, P. Ballirano, and R. Caminiti, "Kinetics of Gypsum Dehydration at Reduced Rressure: An Energy Dispersive X-ray Diffraction Study," *European Journal of Mineralogy*, 20 (2008), 621-627.

Ni-Co 2013
Edited by: Thomas Battle, Michael Moats, Violina Cocalia, Harald Oosterhof, Shafiq Alam, Antoine Allanore, Rodney Jones, Nathan Stubina, Corby Anderson, and Shijie Wang
TMS (The Minerals, Metals & Materials Society), 2013

HYDROMETALLURGICAL NICKEL LATERITE PROCESSING: A REVIEW OF CURRENT SX FLOWSHEETS AND INDUSTRY TRENDS

Adam Fischmann, Shane Wiggett, Troy Bednarski, Violina Cocalia and Cyril Bourget

Cytec Industries Inc., 1937 West Main Street, Stamford, CT 06902 USA

Keywords: nickel, laterite, solvent extraction

Abstract

The state of the art of nickel and cobalt solvent extraction in the nickel laterite industry is assessed, covering the reagents in commercial use (organophosphorus reagents, carboxylic acids, chelating oximes and amines), research developments and chemical fundamentals that result in the different approach taken with each of the reagent types.

Introduction

Nickel ore deposits are found either as sulfides or laterites. Although 40% of the known nickel reserves are associated with sulfidic ores, such ores represent approximately 60% of nickel processing [1], due mainly to their much lower processing costs. However, depletion of these resources and increased concern over environmental issues regarding the processing of sulfur and the handling of sulfur dioxide has made the processing of lateritic ores much more attractive, as reflected by the increasing share of Ni produced from laterites from the 1950s onwards.

Exploitation of nickel laterite resources has been characterized by the diversity of the processing routes utilized. In terms of hydrometallurgical processing, HPAL (high pressure acid leaching) is the main leaching technology adopted to date but there has been increased interest in heap leaching. The PLS resulting from heap leaching has a different composition to that derived from HPAL (higher Fe:Ni and Mg:Ni) [2, 3], which would affect the downstream processes. The current methods for treating the PLS from HPAL result in either an intermediate precipitation product or a high value-added metal or oxide product derived from solvent extraction (SX).

The intermediate precipitation products, known in the industry as MHP (mixed hydroxide product) and MSP (mixed sulfide product) contain Ni, Co and other impurity metals. For MHP, there is no selectivity for Ni and Co over Fe, Cr, Al, Cu or Zn, limited selectivity over Mn and good selectivity only against Mg, Ca and Na, while achieving about 90% Ni recovery [2]. The product also contains 35-45% moisture. For MSP, selectivity is good, but the process requires hazardous reagents under pressure. The on-site production of H_2S requires first converting a hydrocarbon (oil-price dependent) to H_2 and then reacting this with elemental sulfur. Furthermore, the sulfide precipitation tank requires exotic materials of construction. The main drawback of the precipitation routes is that Ni and Co are not separated, resulting in a lower value product. Low recovery and high recirculating loads of Ni and Co are also a concern.

Separation of Ni and Co and rejection of the impurity metals can be accomplished by solvent extraction. The increasing application and acceptance of SX for Ni and Co extraction and separation in the Ni laterite industry has been driven by the success of SX for other metals, e.g. Cu and U [4]. This review will summarize the SX reagents and flowsheets in commercial use for Ni and Co extraction and separation and describe the underlying chemistry of their function, with examples from industrial practice.

Chemistry of Nickel and Cobalt Separation

Ni and Co are adjacent in the periodic table and exhibit very similar chemical behavior. There are, however, several differences that can be exploited in Ni and Co SX; a) Co^{2+} has a tendency to form tetrahedral complexes whereas Ni^{2+} prefers octahedral or square planar geometry; b) Co^{2+} readily forms the anionic tetrachloro complex $CoCl_4^{2-}$ in chloride media, whereas Ni^{2+} only weakly associates with chloride (except under special circumstances that are not encountered in hydrometallurgy, e.g. saturated LiCl solution [5]); c) Co can exist in two oxidation states, +2 and +3, depending on the ligand environment, whereas only divalent Ni is present under hydrometallurgical conditions; and d) the intrinsic rate of ligand exchange is slower for Ni^{2+} than for Co^{2+}, and for Co^{3+} it is so slow as to be considered a kinetically inert ion on the time scale of SX (Figure 1).

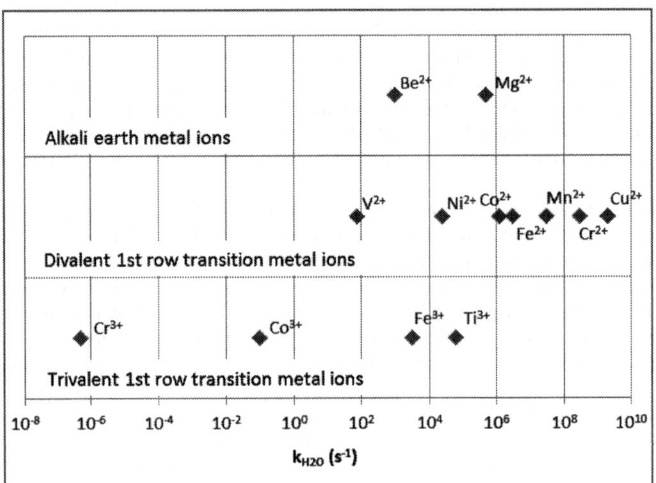

Figure 1. Water exchange rate constants for selected metal ions (adapted from Helm and Merbach [6])

Commercially Used SX Reagents for Nickel and Cobalt Extraction

There are three classes of SX reagents, viz. acidic, basic and solvating. Acidic extractants exchange H^+ for metal ions, basic extractants are positively charged and extract anionic metal complexes and solvating extractants solubilize neutral metal complexes or ion pairs in the organic phase. For Ni/Co extraction and separation in nickel laterite processing, only acidic and basic extractants are used. Table I summarizes these in terms of structure, type of medium, stoichiometry of the extracted complex(es) and lists commercially available products. Cytec supplies the CYANEX™ 272, CYANEX™ 301 and ACORGA™ K2000 reagents for Ni and Co solvent extraction.

Table I: Extractants for Ni and Co SX in Ni Laterite Processing

Extractant class	Extractant type	Extractant structure		Commercial examples	Extraction medium	Extracted complex(es)	Reference
Acidic	Phosphinic acid	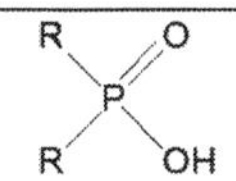	(LH)	CYANEX™ 272 Ionquest™ 290	Sulfate	$[Co(LHL)_2]$	[7]
Acidic	Dithiophosphinic acid	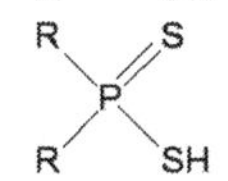	(LH)	CYANEX™ 301	Sulfate	$[NiL_2(H_2O)_2]$ $[CoL_2]$	[7]
Acidic	Carboxylic acid	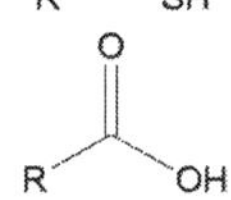	(LH)	Versatic™ 10	Sulfate	$[CoL_2].4LH$ $[CoL_2.2LH]_2$ $[NiL_2].4LH$ $[NiL_2.2LH]_2$	[8]
Acidic	Phenolic oxime	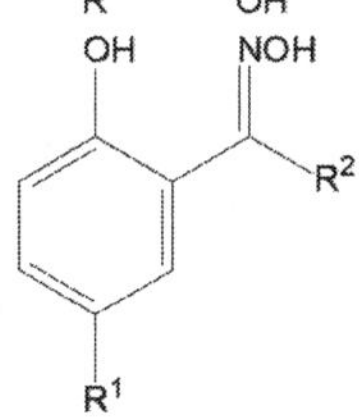	(LH)	ACORGA™ K2000 Lix™ 84-I	Ammonia	$[NiL_2]$	[9]
Basic	Tertiary amine	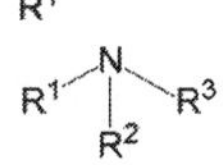	(L)	Alamine™ 308	Chloride	$(LH)_2[CoCl_4]$	[10]

Acidic extractants

Organophosphorus Reagents

Currently, phosphinic acids, such as CYANEX 272, represent the most attractive option for Ni/Co separation [11, 12]. The Ni content of a laterite ore body is typically 10 to 100 times greater than Co. The higher affinity of CYANEX 272 for Co over Ni permits the extraction of the lower concentration metal, which lowers the operating and capital costs of the SX plant (less NaOH or NH_4OH required for neutralization and a smaller initial organic fill). Eqn. 1 shows the extraction equilibrium for Co^{2+}, in which two hydrogen-bonded phosphinic acid dimers each lose one proton to form the extracted complex. In the Ni laterite industry, however, the concentrations of Mn and Mg are higher than encountered in other Ni/Co separation applications. It is therefore important to review the processing options available to reject the impurity metals. Cytec has the capability to simulate Ni/Co separation CYANEX 272 SX circuits, enabling higher reagent utilization while maximizing metal separation [13, 14].

$$Co^{2+}_{aq} + 2(LH)_{2\,org} = [Co(LHL)_2]_{org} + 2H^{+}_{aq} \quad (1)$$

Mg, Ca and Mn rejection can be achieved by using H_2S to form MSP, containing Ni, Co and Zn. The MSP can then be leached under pressure with O_2 and H_2SO_4 to recover Ni and Co into solution, and residual Fe^{3+} can be precipitated by NaOH addition. Zn can then be removed in an initial CYANEX 272 circuit, with a second CYANEX 272 circuit used to separate Co from Ni by extracting Co, using base for pH control. Using sulfuric acid to strip generates a Co-rich solution (pH 2.0-2.5) suitable for electrowinning or hydrogen reduction. The Ni-containing raffinate (pH 5.0-5.5) is at a suitable pH for metal recovery by hydrogen reduction [4], while the pH must be adjusted to ~3.5 for Ni electrowinning [15]. The MSP-releach-CYANEX 272-H_2 reduction route is used at Murrin Murrin [4]. This and other processing options are available to remove Mg and Mn from Co solutions, and are described in the context of African Cu-Co processing [16]. Mg removal prior to Co extraction is critical, since otherwise a large portion of the CYANEX 272 capacity would be taken up by Mg.

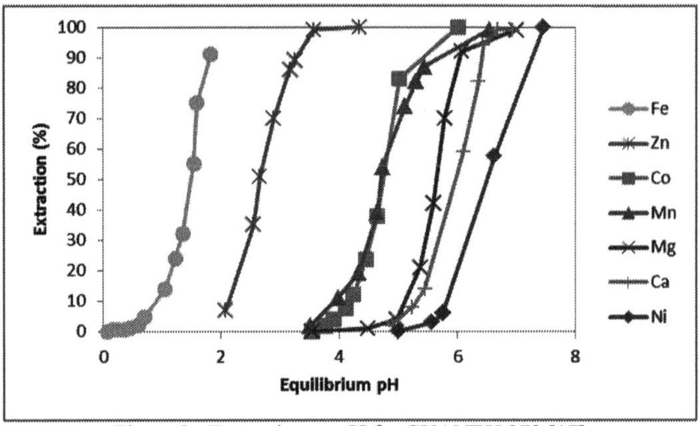

Figure 2: Extraction vs pH for CYANEX 272 [17]

Unlike phosphinic acids, dithiophosphinic acids, such as CYANEX 301, can be used directly on a laterite PLS after Fe removal, due to the excellent rejection of Mn, Mg and Ca (Figure 3). The low pH of Ni and Co extraction eliminates the need for pH adjustment during extraction [18]. The strong co-extraction of Ni and Co does however necessitate a secondary Ni/Co separation step. At the Goro plant in New Caledonia, this is achieved by changing the medium from sulfate to chloride and separating Ni and Co using a tertiary amine reagent (see below). The change to chloride medium is accomplished by stripping Ni and Co from the loaded CYANEX 301 with 6 mol/L HCl at 60°C using a five minute mixing time [19]. These harsh stripping conditions are necessary due to the slow stripping kinetics of Ni.

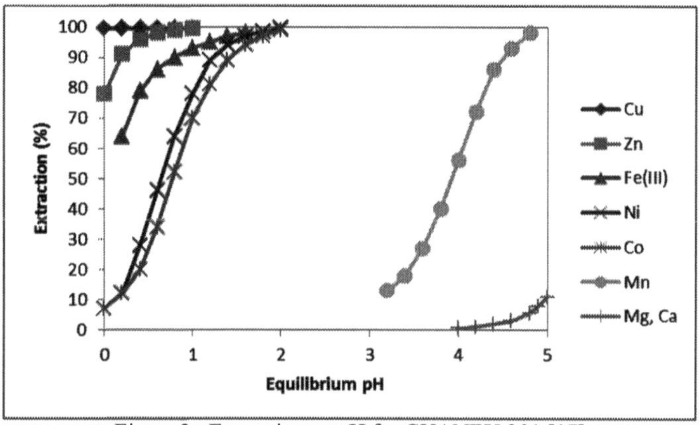

Figure 3: Extraction vs pH for CYANEX 301 [17]

Molecular modeling calculations have been carried out on a simple analog of CYANEX 301 ($H_2P(S)SH$), and confirmed that the structures of the complexes consist of tetrahedral $Co(H_2P(S)S)_2$ and octahedral $Ni(H_2P(S)S)_2(H_2O)_2$ [20], as shown in Figure 4. Vapor pressure osmometry measurements showed that the CYANEX 301 complexes are present in the organic phase as $[CoL_2]$ and $[NiL_2(H_2O)_2]$ [7]. In contrast to CYANEX 272, CYANEX 301 does not exist as dimers in the organic phase [7], therefore, the extraction equilbria are different (Eqn.2 and Eqn. 3):

$$Ni^{2+}_{aq} + 2LH_{org} + 2H_2O = [NiL_2(H_2O)_2]_{org} + 2H^+_{aq} \quad (2)$$
$$Co^{2+}_{aq} + 2LH_{org} = [CoL_2]_{org} + 2H^+_{aq} \quad (3)$$

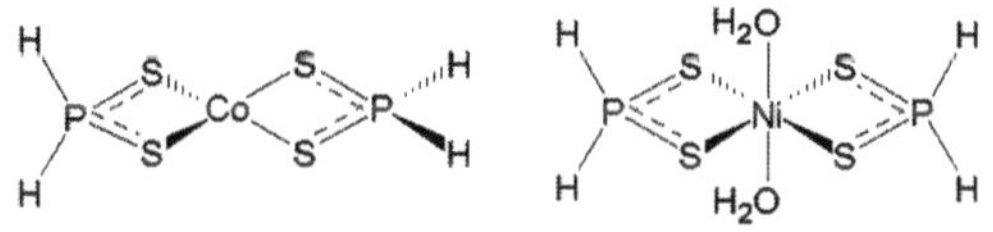

Figure 4: Calculated structures of Co and Ni complexes with $H_2P(S)SH$ [20]

A significant challenge to the commercial use of CYANEX 301 is its reduced stability in the presence of oxidizing agents, including O_2 and oxidizing metals such as Fe^{3+} [18]. Oxidation proceeds from dithiophosphinic acid to monothiophosphinic through to phosphinic acid (Eqn. 4), which results in a loss of selectivity and strength. The Goro plant manages this issue by using very large pulsed columns to maintain an inert atmosphere [21] rather than open-air mixer-settlers to carry out the extraction and stripping.

$$R_2P(S)SH \rightarrow R_2P(S)OH \rightarrow R_2P(O)OH \quad (4)$$

An additional consideration includes the susceptibility of Co to oxidation when extracted by dithiophosphinic acids, as evidenced by a change to the UV-visible spectrum and the difficulty of stripping Co after exposure to air [22]. Cu^{2+} is also very difficult to strip if loaded, and consequently must be removed from the PLS down to very low levels by ion exchange prior to contact with the organic to avoid poisoning of the reagent [23].

Carboxylic Acids

Carboxylic acids extract Ni by a proton exchange reaction, forming a complex containing undissociated acid molecules (Eqn. 5) which can then dimerize in the organic phase (Eqn. 6); Co behaves similarly to Ni [8].

$$Ni^{2+}_{aq} + 3(LH)_{2\ org} = [NiL_2 4HL]_{org} + 2H^+_{aq} \quad (5)$$

$$[NiL_2 4HL]_{org} = [NiL_2 2HL]_{2\ org} + 2(HL)_{2\ org} \quad (6)$$

Carboxylic acids have good selectivity for Ni and Co over Mg, but suffer from poor selectivity against Ca. Co-extraction of Ca can cause gypsum precipitation in H_2SO_4 strip solutions. Significant aqueous solubility of the carboxylate at the extraction pH can lead to high losses of extractant unless a reagent recovery stage is included, which adds to the operating cost.

Chelating Oxime Reagents

In ammoniacal media, Ni and Co exist as mixtures of cationic ammine complexes (Eqn. 7) [24]. Depending on the concentration of ammonia, $[Co(NH_3)_n]^{2+}$ is not stable in ammoniacal solutions due to the presence of the strong-field ammine ligands, which results in a filled t_{2g} subshell when low-spin d^7 Co^{2+} is oxidized to low-spin d^6 Co^{3+}, removing the orbital degeneracy of the low-spin d^7 state (Figure 5). This in fact forms the basis of Ni/Co separation in ammoniacal media, which relies on the kinetic differences between Ni^{2+} and Co^{3+}. Co^{3+} is kinetically inert (see Figure 1) and is not extracted during the residence time in the mix-box (despite a high distribution ratio at equilibrium). In practice, e.g. at the Yabulu refinery in Townsville, Queensland [25], $[Co(NH_3)_n]^{2+}$ is deliberately oxidized to $[Co(NH_3)_n]^{3+}$ with H_2O_2 to avoid extraction of Co.

$$M^{2+} + nNH_3 = [M(NH_3)_n]^{2+} \quad (7)$$

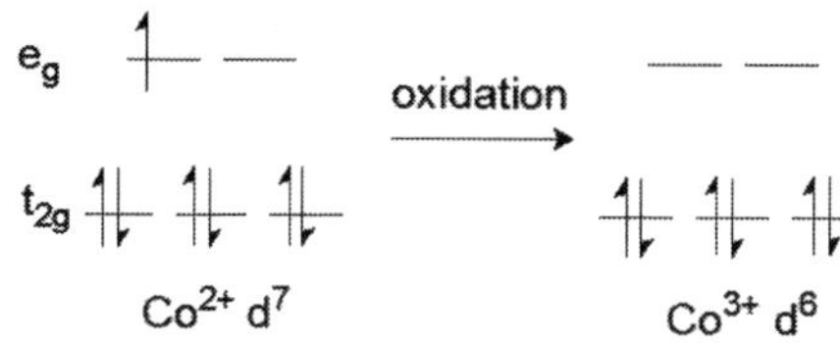

Figure 5: Electronic configurations of Co^{2+} and Co^{3+} in ammonia (adapted from Cotton and Wilkinson [26])

Ni extraction proceeds according to Eqn. 8, via deprotonation of the phenol and chelation of the phenolate oxygen and the oxime nitrogen. Since the extraction is in ammoniacal medium, ammonium ions are formed rather than protons, in a 2:1 L:M stoichiometry [10]. Ni stripping can be effected by reversing the equilibrium with strong NH_3 solution, or with dilute H_2SO_4.

$$[Ni(NH_3)_n]^{2+}_{aq} + 2LH_{org} = [NiL_2]_{org} + (n\text{-}2)NH_{3\,aq} + 2NH_4{}^{+}_{aq} \qquad (8)$$

In practice, a small amount of Co^{2+} is extracted. The phenolic oxime acts like ammonia and exerts a strong ligand field favoring oxidation to Co^{3+}. Co^{3+} can only be removed from the organic by reductive stripping in acid with a reducing metal such as Zn or Fe, which converts inert Co^{3+} to labile Co^{2+} (Eqn. 9). An unavoidable consequence of this treatment is that a significant amount of the oxime is degraded to ketone by the aggressive conditions of the reductive strip, so re-oximation is necessary before the organic can be returned to the circuit.

$$[CoL_3]_{org} + M_s + 3H^+ \rightarrow Co^{2+}_{aq} + M^{2+}_{aq} + 3LH_{org} \qquad (9)$$

The Yabulu refinery in Townsville, Queensland, operates an ammoniacal SX circuit with ketoxime, and the Co level in the organic is maintained at < 3 g/L by treating a bleed stream to remove Co [25]. At this level of Co loading, 0.15 mol/L of ketoxime is coordinated to Co^{3+} assuming a 3:1 stoichiometry, poisoning 30% of the ketoxime reagent (typical concentration is 15% v/v).

Other limitations of phenolic oximes in ammoniacal media include slow extraction and stripping kinetics for Ni, slow phase separation, and Cu build up. The strong ammonia strip that recovers Ni does not fully recover Cu, so a bleed stream of the organic must be treated with H_2SO_4 to prevent accumulation of Cu on the organic [23].

Basic extractants

Amine Reagents

There are four types of amine reagents, primary (RNH_2), secondary (R_2NH), tertiary (R_3N) and quaternary (R_4N^+), although only tertiary amines are used in Ni laterite processing. Separation of Ni and Co from chloride media using tertiary amines operates by an entirely different mechanism to the acidic extractants. In chloride solution, Co^{2+} readily forms a variety of chloride complexes, and at sufficiently high chloride concentrations, the tetrahedral anionic complex $[CoCl_4]^{2-}$ is dominant. Conversely, Ni^{2+} does not easily form anionic chloride complexes in HCl solution (although $[NiCl_4]^{2-}$ can form in aqueous Me_4NCl [27] and saturated LiCl solution [5]).

Ni extraction from chloride solutions using a tertiary amine has been reported [28], but high extraction (>80%) was only possible from solutions containing >7 g/L Ni, > 3 mol/L HCl using >50% v/v amine in xylene. In 5 mol/kg HCl solution, with 1 mol/kg Ni (~60 g/L Ni), the dominant species is Ni^{2+}_{aq}, with about 20% present as $NiCl^{+}_{aq}$, and at 1 mol/kg HCl, the fraction of $NiCl^{+}_{aq}$ does not exceed 10% in the range 0.01-1.0 mol/kg Ni [29]. Therefore, at low pH, the tertiary amine is protonated (R_3NH^+) and two such ammonium cations form an ion pair with $[CoCl_4]^{2-}$, selectively extracting Co into the organic phase (Eqn. 10), while Ni remains in the raffinate [10].

$$2R_3N_{org} + 2H^{+}_{aq} + [CoCl_4]^{2-}_{aq} = (R_3NH)_2[CoCl_4]_{org} \qquad (10)$$

As mentioned earlier, tertiary amines are utilized in the Goro flowsheet to separate Co and Ni from the CYANEX 301 strip solution. The Ni-HCl solution is subjected to pyrohydrolysis to recover NiO and HCl [19]. The use of chloride media and amine extractants creates a corrosive environment requiring special materials of construction. Slow phase separation is also of concern, requiring the use of an all-aromatic (and consequently low flash point) diluent to maintain low viscosity and minimize aqueous in organic entrainment [23].

RESEARCH DEVELOPMENTS AND IMPLICATIONS

<u>Synergistic SX</u>

Flett [30] was the first to demonstrate a synergistic effect on metal extraction between an aliphatic α-hydroxyoxime and a carboxylic acid. In the following decades, extensive research on the application of synergistic mixtures applied to Ni/Co separation in the context of Ni laterite exploitation has been carried out, most recently by CSIRO researchers [31]. The most promising combination consists of an aliphatic α-hydroxy oxime, a carboxylic acid and tributyl phosphate (TBP). This synergistic mixture possesses good selectivity against Mn, Mg and Ca and extracts Ni and Co in the pH 5-6 range. The extracted Ni complex consists of two oximes and two carboxylates, and TBP is essential to achieve sufficiently rapid Ni extraction and stripping kinetics. Another approach has been to exploit this kinetic difference to separate Co from Ni in extraction and stripping [32, 33], although in general kinetic separations are more difficult to control in real processes than thermodynamic ones.

The major drawback of this approach is the instability of the oxime, which is susceptible to hydrolysis and oxidation (Figure 6), both under acidic stripping conditions (forming acyloin and diketone) [34] and metal-loaded extraction conditions (forming diketone and keto-oxime) [35]. It is particularly vulnerable to upset conditions resulting in high Mn loading which leads to more rapid degradation [36]. Keto-oxime formation is also promoted by Co extraction. Acyloin formation is reversible via re-oximation, and while reduction of the keto-oxime back to the hydroxyoxime is technically possible with sodium borohydride (but not with metallic Zn or Fe) [37], the cost of this treatment would be quite high.

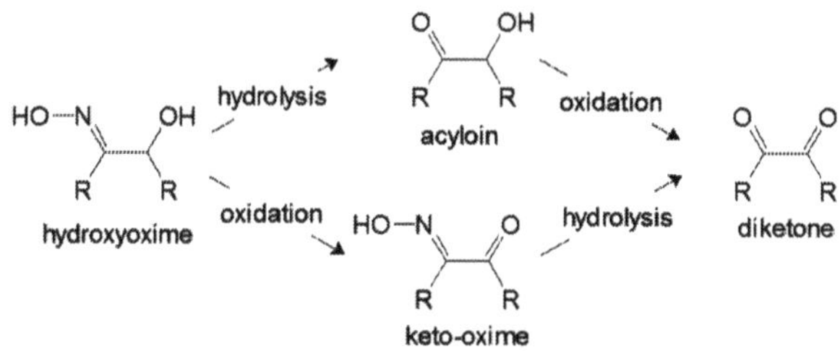

Figure 6: Degradation pathways for aliphatic α-hydroxyoximes

Pre-loaded Carboxylate SX

Ni and Co extraction from laterite PLS using a Mg-loaded carboxylic acid was recently proposed [38]. According to this patent application, the Mg-loaded carboxylic acid extracts Ni, Co, Zn, Cu and Mn from the laterite PLS after first removing Fe. The loaded organic is then contacted with a concentrated $NiCl_2$ solution, which scrubs Co, Mn, Zn, Cu, Mg and Ca. Ni is stripped with sulfuric acid and recovered by electrowinning or another technique. Co and Mn are subsequently extracted away from Zn, Cu, Mg and Ca using another SX circuit (containing an amine extractant), and Co is selectively precipitated as a sulfide from the strip solution. This process has been considered for the Agata project in the Philippines [39].

While the reagent is low cost, the flowsheet is complicated, containing numerous scrubbing and washing stages as well as Mg-loading stages. Furthermore, the use of different extractants (carboxylic acid and amine) runs the risk of reagent cross-contamination.

CONCLUSIONS

Solvent extraction continues to have an important role in the Ni laterite industry, with Murrin Murrin in operation since 1996, the SX portion of the Yabulu refinery operating since 1989 and Ambatovy and Goro due to become fully operational in 2012. Both Murrin Murrin and Ambatovy separate Ni from Co using phosphinic acid SX after first rejecting Mg, Ca and Mn via sulfide precipitation. Metallic Ni and Co are then recovered from the purified solutions by H_2 reduction. The Yabulu refinery treats a variety of feedstocks by leaching with ammonia, oxidizing Co^{2+} to Co^{3+}, extracting Ni and producing high purity Ni metal, carbonate and oxide, and mixed Co oxide/carbonate. The Goro plant will reject the impurities Mn, Mg and Ca by extracting Ni and Co directly from the PLS with CYANEX 301, and then separate Co from Ni by extraction with a tertiary amine from the HCl strip solution, producing Co carbonate and Ni oxide.

Despite the range of available reagents and flowsheets, none possess the ideal combination of high selectivity for Ni and Co against Mn, Mg and Ca and high resistance to chemical degradation. The development of an SX route that avoids precipitation and does not require special measures to minimize degradation would simplify flowsheets, improve recovery and produce higher quality, higher value products. Ideally, such a reagent would recover and separate Ni and Co simultaneously from the major impurities that are present in large amounts in high concentration in laterite PLS (Mn, Mg and Ca). Development continues by the major players to balance the compromises and provide a solution to the clear need for a more selective and robust extractant for the Ni laterite industry.

REFERENCES

1. Mudd, G.M.; (2010) "Global trends and environmental issues in nickel mining: Sulfides versus laterites" *Ore Geology Reviews* **38:** 9-26

2. Willis, Boyd. (2007) "Downstream Processing Options for Nickel Laterite Heap Leach Liquors" *ALTA 2007 Nickel/Cobalt 12*, ALTA Metallurgical Services, Perth, Australia, 25 pp.

3. Steemson, M.L.; and Smith, M.E. (2009) "The development of nickel laterite heap leach projects" ALTA 2009 Nickel/Cobalt 14, ALTA Metallurgical Services, Perth, 21 pp.

4. Crundwell, F.K.; Moats, M.S.; Ramachandran, V.; Robinson, T.G.; and Davenport, W.G. (2011) Extractive Metallurgy of Nickel, Cobalt and Platinum-Group Metals, Elsevier, Amsterdam, 610 pp.

5. Florence, T. M.; and Farrar, Yvonne J. (1968) "Liquid-Liquid Extraction of Nickel with Long-Chain Amines from Aqueous and Nonaqueous Halide Media" Analytical Chemistry 40: 1200-1206

6. Helm, L.; and Merbach, A.E. (1999) "Water exchange on metal ions: experiments and simulations" Coordination Chemistry Reviews 187: 151-181

7. Tait, B.K. (1993) "Cobalt-nickel separation: the extraction of cobalt(II) and nickel(II) by Cyanex 301, Cyanex 302 and Cyanex 272" Hydrometallurgy 32: 365-372

8. Tanaka, M.; Nakasuka, N.; and Sasane, S. (1969) "Extraction of nickel with capric acid" Journal of Inorganic and Nuclear Chemistry 31: 2591-2597

9. Szymanowski, J.; and Borowiak-Resterna, A. (1991) "Chemistry and Analytical Characterization of the Effect of Hydroxyoxime Structure upon Metal-Complexing and Extraction Properties" Critical Reviews in Analytical Chemistry 22: 519-566

10. Cox, M. (2004) "Solvent extraction in hydrometallurgy" in Solvent Extraction Principles and Practice, 2nd Ed., Marcel Dekker, New York, p. 455-505

11. Flett, Douglas S. (2004) "Cobalt-Nickel Separation in Hydrometallurgy: a Review" Chemistry for Sustainable Development 12: 81-91

12. Flett, D.S. (2005) "Solvent extraction in hydrometallurgy: the role of organophosphorous extractants" Journal of Organometallic Chemistry 690: 2426-2438

13. Soderstrom, M.; Bourget, C.; Jakovljevic, B.; and Bednarski, T. (2010) "Development of process modeling for CYANEX® 272" ALTA 2010 Ni/Co/Cu Proceedings, ALTA Metallurgical Services, Perth, Australia, 17 pp.

14. Soderstrom, M.; Bourget, C.; and Jakovljevic, B. (2011) "CYANEX™ 272 extractant process modelling update" ALTA 2011 Ni/Co/Cu Proceedings, ALTA Metallurgical Services, Perth, Australia, 8 pp.

15. Pavlides, A.G. (2008) "Developments in cobalt and nickel electrowinning technology" Anglo American Mining Conference, Johannesburg, 11 pp.

16. Tinkler, O.; Flett, D.; and Bourget, C. (2007) "Flowsheet considerations for copper-cobalt projects in the DRC" ALTA 2007 Nickel/Cobalt 12, ALTA Metallurgical Services, Perth, Australia, 21 pp.

17. Cytec Industries, Inc., (2010) Mining Chemicals Handbook

18. Mihaylov, I. (2003) "Solvent Extractants for Nickel and Cobalt: New Opportunities in Aqueous Processing" JOM 55(7): 38-42

19. Mihaylov, I.; Krause, E.; Colton, D.F.; Okita, Y.; Duterque, J.-P.; and Perraud, J.-J. (2000) "The development of a novel hydrometallurgical process for nickel and cobalt recovery from Goro laterite ore", CIM Bulletin 93: 124-130

20. Otero-Calvi, Alexis; Montero, Luís A.; and Stohrer, Wolf-Dieter (2008) "DFT modelling of cobalt and nickel complexes with dithiophosphinic acid", Journal of Molecular Structure: THEOCHEM 859: 93-97

21. Fisher, K.G. (2011) "Cobalt processing developments", Proceedings of the 6th Southern African Base Metals Conference, 18-20 July, 2011, Phalaborwa, South Africa, p. 237-258

22. Fleitlikh, I. Yu.; Pashkov, G. L.; Grigorieva, N. A.; Nikiforova, L. K.; Pleshkov, M. A.; and Shneerson, Y. M. (2011) "Cobalt and Nickel Recovery from Sulfate Media Containing Calcium, Manganese, and Magnesium with a Mixture of CYANEX 301 and a Trialkylamine", Solvent Extraction and Ion Exchange 29: 782-799

23. Bacon, G.; and Mihaylov, I. (2002) "Solvent extraction as an enabling technology in the nickel industry" Journal of the South African Institute of Mining and Metallurgy 102: 435-443

24. Smith, R.M.; and Martell, A.E. (1974) Critical Stability Constants, Vol. 4., Plenum, New York

25. Fittock, John E. (2007) "Nickel and cobalt refining by QNi Pty Ltd, Yabulu, QLD" in Monograph 19 — The Sir Maurice Mawby Memorial Vol 3rd Ed, Australian Institute of Mining and Metallurgy (Melbourne), p. 1-47

26. Cotton, F.A.; and Wilkinson, G. (1988) Advanced Inorganic Chemistry, 5th Ed., New York, John Wiley and Sons, 1455 pp.

27. Scarrow, R. K.; and Griffiths, T. R. (1967) "The Formation of Tetrahedral Tetrachloronickelate(II) in Aqueous Solution" Chemical Communications 1967: 425

28. Sayar, N.A.; Sayar, A.A.; and Filiz, M. (2009) "Simulation and optimisation-oriented modelling considerations for the extraction of Ni(II) from its acidic aqueous chloride solutions into Alamine 336-m-xylene systems" Hydrometallurgy 95: 280-284

29. Lee, Man Seung; and Nam, Sang-Ho (2009) "Chemical equilibria of Nickel Chloride in HCl Solution at 25°C" Bulletin of the Korean Chemical Society 30: 2203-2207

30. Flett, D.S.; and Titmuss, S. (1969) "Synergistic effect of LIX 63 on the extraction of copper and cobalt by naphthenic acid" Journal of Inorganic and Nuclear Chemistry 31: 2612-2613

31. Cheng, Chu Yong; Barnard, Keith R.; Zhang, Wensheng; and Robinson, David J. (2011) "Synergistic solvent extraction of nickel and cobalt: a review of recent developments" Solvent Extraction and Ion Exchange 29: 719-754

32. McCoy, T.; Mayhew, K.; Jones, D.; Barnard, K.R.; Zhang, W.; Cheng, C.Y.; and Robinson, D. (2011) "Separation of Co from Ni in an impure sulphate solution, part 1–extraction" Proceedings of the 19th International Solvent Extraction Conference, 3-7 October, 2011, Santiago, Chile, 8 pp.

33. McCoy, T.; Mayhew, K.; Jones, D.; Barnard, K.R.; Zhang, W.; Cheng, C.Y.; and Robinson, D. (2011) "Separation of Co from Ni in an impure sulphate solution, part 2–extraction"

Proceedings of the 19th International Solvent Extraction Conference, 3-7 October, 2011, Santiago, Chile, 7 pp.

34. Barnard, K.R.; and Turner, N.L. (2011) "The effect of temperature on hydroxyoxime stability in the LIX 63 – Versatic 10 – tributyl phosphate synergistic solvent extraction system under synthetic nickel laterite conditions" Hydrometallurgy 109: 245-251

35. Turner, N.L.; and Barnard, K.R. (2012) "The effect of individual complexed metals on hydroxyoxime stability in the LIX 63–Versatic 10–tributyl phosphate synergistic solvent extraction system under synthetic nickel laterite extract conditions" Hydrometallurgy http://dx.doi.org/10.1016/j.hydromet.2012.02.006

36. Barnard, K. R.; Turner, N.L.; and Shiers, D.W. (2010) "LIX® 63 stability in the presence of Versatic 10 under proposed commercial extract and strip conditions, part III: Effect of manganese and cobalt loading on oxime stability at 30°C" Hydrometallurgy 104: 268-277

37. Barnard, K.R. (2008) "Isolation and characterization of the LIX®63 degradation products 5,8-diethyl-6,7-dodecanedione monooxime (keto-oxime) and 5,8-diethyl-6,7-dodecanediol (diol) and regeneration of oxime from keto-oxime" Hydrometallurgy 90: 147-153

38. Nakon, D. (2006) "Process for preparing nickel loaded organic extractant solution" PCT Application, WO 2006/029439 A1, 32 pp.

39. Willis, B.; and Gifford, M.G. (2010) "Technical report on the Agata Nickel Project scoping study" Boyd Willis Hydromet Consulting, November 19, 2010, 276 pp.

Ni-Co 2013
Edited by: Thomas Battle, Michael Moats, Violina Cocalia, Harald Oosterhof, Shafiq Alam, Antoine Allanore, Rodney Jones, Nathan Stubina, Corby Anderson, and Shijie Wang
TMS (The Minerals, Metals & Materials Society), 2013

Studies on Refining Cobalt Salt Solution by Extraction Chromatography to Prepare High Purity Cobalt

[1]Chen Song, [1]Zhang Li , [1]Lang Shuling , [1]Cai Zhenping, [1]Wang Lijun

[1]Division of Rare Metals Metallurgy& Materials;
General Research Institute for Non-ferrous Metals;
Beijing 100088, China

Keywords: Extraction Chromatography, Levextrel Resins, Ni-Co Separation, High Purity Cobalt

Abstract

The key to preparing high-purity cobalt is purifying of cobalt solution. Cobalt sulfate solution was refined by PC88A levextrel Resins which produced from HPD-100 type of ethylene benzene -divinylbenzene porous adsorption resin and PC88A extractant. The results indicate that under the condition which the solvent impregnated resins containing PC88A is 55% in Φ65 mm ×700mm chromatography column, extraction pH value is 3.7, buffer elution pH value is 2.9, velocity of flow is1~1.5 chromatography column volume/h, when solution volume is three times of column volume, stripping solution is collected and stripping by pH 1.0 hydrochloric acid. The high purity cobalt chloride solution is obtained. Removing organic impurity form cobalt chloride solution and concentrating, the high purity cobalt which obtained by electrolyze the cobalt chloride solution is tested by GDMS. The results indicate that the total impurity of twenty elements is less than 10 ppm.

Introduction

High purity cobalt with a purity greater than 99.99% is mainly used in large and very large scale integration, microelectronic devices, magnetic recording media, dynamic memory, semiconductors coating target and other fields, meanwhile, it has the potential to become an alternative material for conductor material and aluminum for wire, thus it has a very bright application prospects [1, 2, 3].

The key to get high purity cobalt is the preparation of highly refined cobalt salt solution, in such a process, the separation of nickel from cobalt solution is the core-technology. The commonly used hydrometallurgical separation method, such as chemical precipitation method [4], ion exchange method [5, 6, 7, 8] and extraction method [9, 10, 11, 12] all can not meet the requirements to prepare high-purity salt solution.

The extracting chromatography [13] is a separation technology which combines the high selectivity of the extractant with the multi-stages of ion exchange separation technique. Chromatography separation means to fill the porous resin with extractant as fillers into the chromatographic column, when the feed solution passes the chromatographic column, different metal ions in the feed solution will be distributed between the aqueous phase and the extractant, and thus the valuable metal may be separated from impurities. Extract various metal ions on the resin, when the chromatographic column is stripped with some solvent, such ions will move with the eluant in different speeds, so that to achieve the purpose of further separation.

In the study about purifying cobalt solution with the extraction chromatography, Yo-shizuka et. al [14] used levextrel resin with Cyanex272 to separate Co(II) from Ni(II) in nitrate and sulfate medium, meanwhile, the behaviors of various metal ions in the resin were studied. Zhou

Chunshan, Wangyan et. al [15, 16] used P204,the transformed levextrel resin to separate Co^{2+} from Cu^{2+}, Sn^{2+},Mn^{2+}, Zn^{2+},Fe^{3+} and other metal ions.

In this paper, the process which purifies the cobalt salt solution with the PC88A levextrel resin and prepares high-purity cobalt through electrolysis is studied.

Experimental

Reagents and Equipments

Reagents used in the test, including sodium hydroxide, hydrochloric acid, sulfuric acid and cobalt sulfate were analytical grade without farther purification. PC88A extractant was industrial-grade products with a purity of 93%. Acetic acid-sodium acetate buffer solution was prepared on our laboratory. The resin used in the test was the HPD-100 Styrene-divinyl benzene porous adsorbing resins. The cobalt raw materials used in the test are listed in Table 1.

Table I: Component of $CoSO_4$ for experiment raw material

Impurity	Na	Ca	Mn	Fe	Ni	Cu	Zn
Content wt%	0.001	0.005	0.005	0.001	0.01	0.001	0.005

Extraction Chromatography used Φ 65mm × 700mm glass tubes as the chromatographic columns .Cobalt electrolytic experiments were implemented inside the 2L polypropylene electrolyzer. Titanium material coated with ruthenium oxide was used as the anode and titanium plate was used as the cathode. Electrolysis was achieved through constant current generated from MKF-30A24V rectifier power supply.

Experiment Process and Characterization

The porous adsorbing resins was cleaned by ethanol and 6 $mol\cdot L^{-1}$ hydrochloric acid and then repeatedly washed with distilled water, dried in low temperature and sieved. Mixed the resin with a particle size of 180~250μm with a certain amount of PC88A extractant, as well as sodium hydroxide solution in a certain concentration. The concentration of the extractant in the resin was controlled in a gravimetric method. Stirred mixture for 2 hours before filled into the columns in wet method. After that, stripped with 6 $mol\cdot L^{-1}$ HCl solution umtil there was no Fe^{3+} found, then stripped with distilled water until the pH value arrived at 5.0~5.5.

A certain amount of cobalt sulfate solution in which a certain pH value was prepared and the concentration of Co^{2+} was 1.0 $mol\cdot L^{-1}$. The chromatographic column was stripped by this solution to start extraction. When it got balanced, the chromatographic column was stripped by HAc-NaAc buffer solution with a certain pH value. When the stripped liquid reached a certain eluate volume, HCl solution in which a certain concentration was used as eluant and stripped cobalt. Finally the metal cobalt was obtained by concentrate and electrolyzes eluate.

Chromatographic column were regenerated with 6 mol $\cdot L^{-1}$ HCl solutions, until no Fe^{3+} found in eluate. Then chromatographic column were stripped with distilled water to the pH value of eluate was 5.0~5.5.

The concentration of the metal ions in the solution was determined by atomic absorption spectrometry and ICP-MS method.

Results and Discussion

Cobalt Distribution Ratio, Extraction Rate and Optimum Extraction pH Value

The mass of resin in the test was 2g, cobalt solution with different volumes was mixed with resin and the extraction lasted for 15 minutes, checked the concentration of cobalt ions in the remained liquid. Based on the cobalt ion distribution ratio (D) and the extraction rate (E), a D~pH curve was obtained.

It is showed in Figure 1 and Figure 2 that: in a certain pH value condition, the extraction rate of cobalt ions in PC88A impregnated resin and the distributed rate of cobalt ions in water phase increased with the increase of pH value, however when the pH value exceeds 4.5, this increase trend begin to slow down.

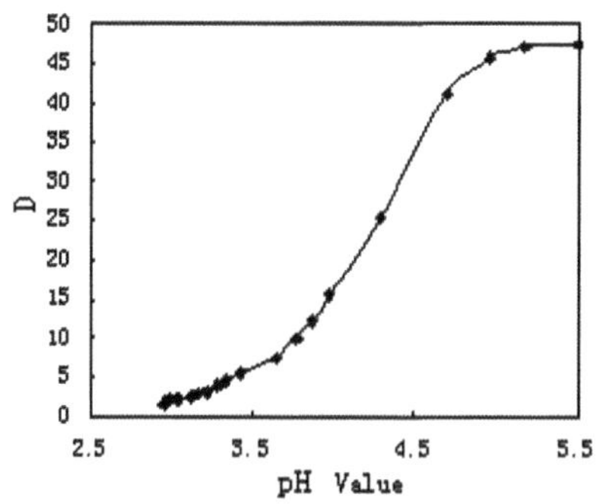

Fig1. D-pH curve of cobalt

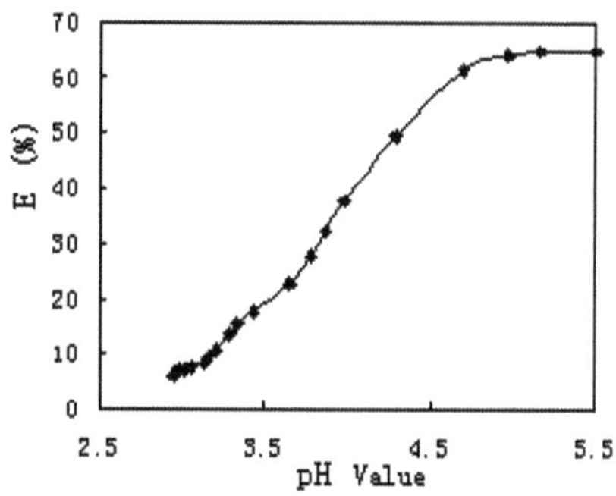

Fig2. E-pH curve of cobalt

It is showed in Figure 3 that when the mass of the resin is 2g and the pH value is smaller than 4.5, the cobalt extractive content increased with the increase of pH value. When the pH value of the solution is larger than 4.5, the cobalt extractive content is trends to become stable. At this moment, both the cobalt distribution ratio and the extraction rate get the best results. However, when the pH value is too high, the extracted residual liquid is easy to emerge precipitation. Therefore, the pH value of the initial solution should be controlled at 4.6~4.8.

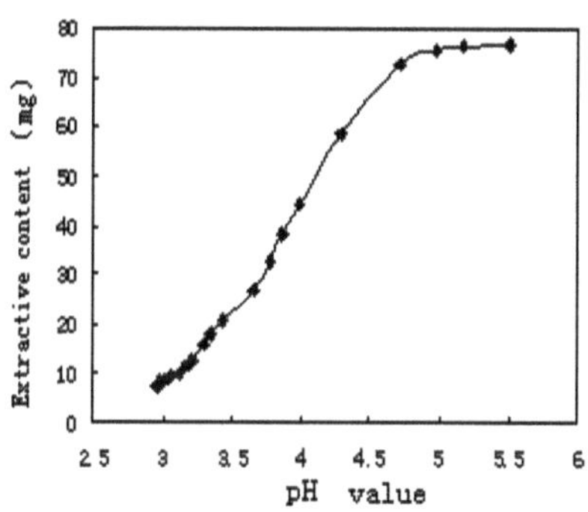

Fig3. Relation between PC88A resin extractive content and pH value

Eluting Curve of Cobalt and Nickel

The chromatographic column of the loaded cobalt is stripped with HCl solution with pH 1.5. The eluting liquid is collected once began to elute the column, collected 10ml per time, measured the concentration of cobalt and nickel ions in solution. The eluting curve is shown in Figure 4 and 5.

It is showed in Figure 4 and 5 that: when the eluting volume is smaller than 50ml, the elution amount is the empty volume of chromatographic column, when the eluting volume reaches 50ml, the concentration of nickel ions could decrease to $13\times10^{-6}g\cdot L^{-1}$. This shows that the separation between cobalt and nickel may realize by used eluting liquid with a certain pH value. To improve eluting efficiency and separation effect, the pH value in the column must be maintain at a constant value during elution process. Since the buffer solution is able to stabilize the pH value within a certain range, the acetic acid-sodium acetate buffer solution with a certain pH value is studied to use eluting test.

Acetic acid-sodium acetate buffer solution of $0.10mol\cdot L^{-1}$ whose pH value was respectively 2.02 and 2.98 was used as eluting liquid. When the eluting volume was 140ml, the extraction band in the chromatographic column which effective length is 700mm respectively was decreased to 10.5cm and 7.5cm, as shown in Figure 6. The [Ni]/[Co] in the collected cobalt solution was measured to be 4.5×10^{-3} and 2.9×10^{-3}. The nickel-cobalt separation effect of acetic acid-sodium acetate with a pH value of 2.98 is better than that with a pH value of 2.02.

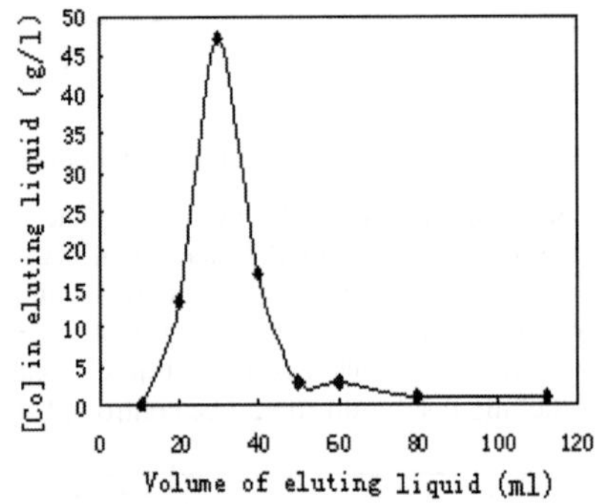

Fig4. Cobalt concentration in eluting liquid with pH1.5

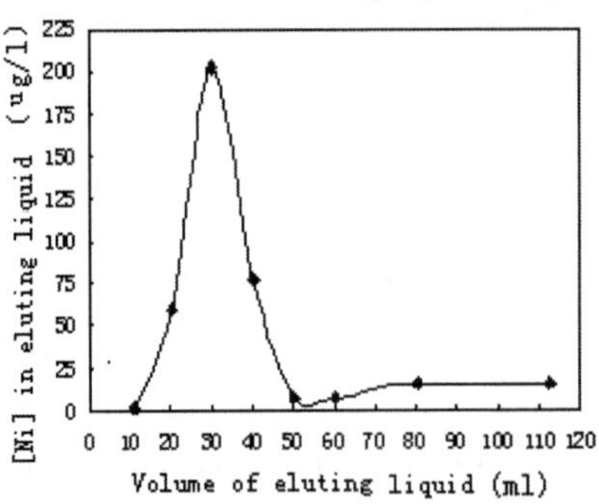

Fig5. Nickel concentration in eluting liquid with pH1.5

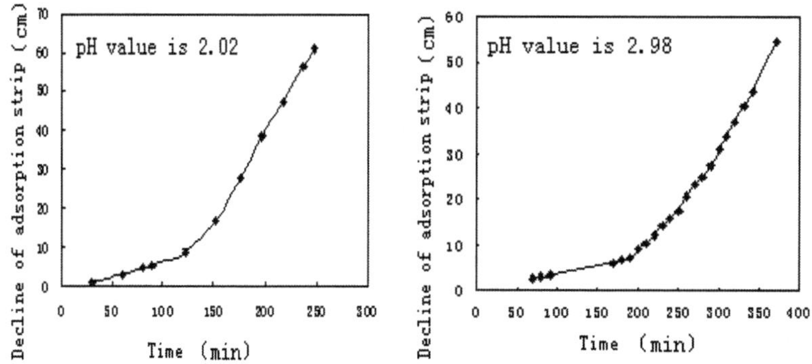

Fig. 6: Relation between decline of adsorption strip and time in NaAc-Hac buffer system

<u>Determination of eluting volume and eluting velocity</u>

The cobalt-nickel eluting curve shows that the eluting volume determines the nickel content in the cobalt solution obtained. Chromatographic columns with different flow rates were used in the test. The eluting liquid was collected when the eluting volume was respectively 3, 4 and 5 times of the empty volume of the chromatographic column. The concentration of cobalt and nickel ions in the eluting liquid was measured and listed in Table II.

Table II: Result of buffer eultion in acetic acid-sodium acetate solution with pH 2.98

No.	Volume of eluting liquid /mL	Decline of adsorption strip /cm	[Co] $g \cdot L^{-1}$	[Ni] $g \cdot L^{-1}$	[Ni]/[Co] 10^{-6}
4-1	95	6.4	6.22	$<1.0\times10^{-4}$	<16
4-2	140	7.5	6.02	$<2.0\times10^{-4}$	<33
4-3	168	9.5	8.75	$<2.5\times10^{-4}$	<28.6
5-4	97	6.4	6.13	$<2.0\times10^{-4}$	<33
5-5	140	8.0	9.15	$<3.0\times10^{-4}$	<33
5-6	177	9.5	10.21	$<3.0\times10^{-4}$	<29.4

*The flow rate in test No. 4-1, No.4-2 and No. 4-3 was 0.35~0.39ml•min^{-1}, the flow rate in test No.5-4, No.5-5 and No. 5-6 was 0.60~0.74ml•min^{-1}

It is showed that when the eluting liquid volume reaches 3 times of the empty volume of the chromatographic column, the content of nickel in cobalt solution reaches the lowest value.

The flow rate of eluting liquid is an important parameter for extraction chromatography, for it determines the contact duration between the fixed phase and the movable phase, reflects the fluid form of the chromatographic separation process, affects the value of the theoretical plate height [13]. When the flow rate is low, the theoretical plate height mainly depends on the flow pattern and the mobile phase lateral diffusion, the flow rate is easily affected by irregular fill, size and shape of the support. When the flow rate is relatively high, the theoretical plate height mainly depends on the transformation particles effect, extraction kinetics and the lateral diffusion of the organic phase [13]. The test shows that: when the flow rate is too low, the components are gathered on the top of the chromatographic column and can not be fully expended, thus the separation time may be greatly extended. When the flow rate is too high, the elements to be separated in the feed solution can not be timely distributed in balance, thus the separation can not be completely implemented.

It is showed in the test results that when the flow rate is within the scope of 1～1.5 times of the empty volume of the chromatographic column, the flow rate does not have significant influence on the separation effects. It was determined in the test that the solution should be collected once the extracted eluting liquid volume reaches 3 times of the empty volume of the chromatographic column; the eluting speed was determined to be the one empty volume of chromatographic column per hour.

Continuous experiment and cobalt electrolysis

Continuous experiment on chromatography separation and cobalt electrolysis was implemented in chromatographic separation column made of Φ65mm×700mm glass tubes.YZ1515 pump was used for feeding. The chromatographic column was eluted with 6 $mol \cdot L^{-1}$ of HCl solution, until no iron was detected in the eluting liquid. Following, the chromatographic column was rinsed with distilled water until the pH value of the eluting liquid became 5.0～5.5.

Eluted the chromatographic column with cobalt sulfate in which the [Co] was $40g.L^{-1}$ until the pH value of the eluting liquid was 3.73, at this moment, the chromatographic column from which the cobalt ions were extracted became blue from pink.

Eluted the chromatographic column with NAc-NaAc buffer solution whose pH value was 2.98 as the eluting liquid, begun to collect the eluting liquid when the eluting liquid was 3 times of the empty volume of the chromatographic column, rapidly washed the cobalt ions in the decolorized chromatographic column with HCl solution with pH value of 1.0 as the elution liquid.

The experiment results on the eluting liquid are listed in Table III. The results show that: the cobalt solution is deeply purified; the contents of the main metal impurities are significantly reduced.

The collected cobalt chloride solution was removed organic impurities and concentrated to an extent at which the [Co] was larger than $100g \cdot L^{-1}$for electrolysis.

TableIII: Analysis results of eluting liquid in serise experimentation

No.	Co ($g \cdot L^{-1}$)	Ni ($m\ g \cdot L^{-1}$)	Fe ($m\ g \cdot L^{-1}$)	Cu ($m\ g \cdot L^{-1}$)	Zn ($m\ g \cdot L^{-1}$)	Mn ($m\ g \cdot L^{-1}$)	Ca ($m\ g \cdot L^{-1}$)
3-1#	34.17	0.41	<0.25	2.70	1.01	1.42	15.22
3-2#	30.54	<0.5	<0.25	<0.25	<0.25	<0.25	2.82
3-3#	37.2	<0.5	<0.25	<0.25	<0.25	<0.25	0.60

The electrolytic process conditions are as follows: the anode was titanium material coated with ruthenium oxide and the cathode was pure titanium, the temperature was 60℃; the cathode current density was $300A \cdot m^{-2}$, the cobalt content in spent electrolyte was $20g \cdot L^{-1}$, the pH value was greater than 4.

The GDMS detection results on meal cobalt obtained through electrolytic process are listed in Table Ⅳ. Meanwhile, the RRR (residue resistance ratio) test was implemented and the obtained RRR value was 109.

Table IV: Analysis results of electrolyze cobalt by GDMS

Name of element	Ni	Fe	Cu	Al	Ca
Content 10^{-6} wt%,	7.36×10^{-3}	1.66×10^{-2}	2.27×10^{-2}	3.09×10^{-3}	5.80×10^{-3}
Name of element	K	Na	Sn	Pb	Bi
Content 10^{-6} wt%,	6.85×10^{-3}	0.691	2.89×10^{-2}	3.87×10^{-3}	1.30×10^{-3}
Name of element	Ga	Ge	Si	Ba	B
Content 10^{-6} wt%,	3.05×10^{-3}	6.85×10^{-3}	1.50×10^{-2}	3.85×10^{-4}	3.54×10^{-4}
Name of element	C	N	O	P	S
Content 10^{-6} wt%,	0.364	2.45×10^{-2}	4.265	1.04×10^{-3}	2.17×10^{-3}

Conclusion

Deep purification of cobalt solution may be realized by using PC88A extraction resin with an extractant content of 55% through extraction chromatography.

The distribution ratio, extraction rate and extraction volume, optimum extraction pH were determined, cobalt and nickel elution curves were obtained and the process conditions for chromatographic separation were defined.

Series experiment was conducted with chromatographic columns of Φ65mm×700mm. The process conditions were as follows: cobalt sulfate in which the cobalt content was $1.0mol\cdot L^{-1}$ with an adjusted pH value of 4.6～4.8 rinsed into chromatographic column until the pH value of the eluate became 3.73, then used HAc-NaAc buffer solution with a pH value of 2.98 as chromatographic column stripping liquid, begun to collect the eluting liquid when eluate volume became 3 times of the empty volume of the chromatographic column, and then stripping the cobalt ions in the chromatographic column with HCl solution with pH 1.0.

The purified cobalt solution was removed the organics impurity and concentrated following by electrolytic deposition. The electrolysis temperature was 60℃.The cathode current density was $300A.m^{-2}$. The high purity electrolytic cobalt which total content of the impurity elements was proved to be smaller than $10mg\cdot kg^{-1}$ by GDMS was obtained.

References

1. Chen Rong et al., “High Purity Cobalt Manufacturing”, *China Journal of Rare Metals*, 29(5) (2005), 797-802

2. Guangxin Wang, Daniel M.Hydock, and John Lehman, *Method for Processing Metals* [p]. US: 6818119 B2. 2004.11.

3. Isshiki M, Fukuda Y, and Igaki K., “Preparation of high purity cobalt”, *Journal of the Less-Common Metals*, 105(2) (1985), 211-220

4. He Huanhua, Cai Qiaofang, *Nickel &Cobalt Metallurgy in China* (Beijing, Metallurgical Industry Press, 2000), 359.

5. John E. Fittock et al. *Solvent extraction of cobalt from cobalt sulphate solution*[p]. US: 5605668. 1997.

6. K. Minura et al., “Preparation of high-purity cobalt”, *Materials Science and Engineering*, A334(2002), 127-133

7. Jiang Chuanfu, Sun Shuqing and Yu Jinyan, "Ion Exchange and Measurement by Srectrophotometric Method of Cobalt and Nickel", *Journal of Jinzhou Teachers College,* 21(1) (2000), 34-37

8. Wang Ronggeng, Li Xueping and Zhai Xueliang, "The Technical Progress in Recovery of Ni and Co by Ion-exchange Resin", *Inorganic Chemical Industry,* 37 (11) (2005), 11-13

9. N.BDevi, K.C.Nathsarma and V.Chakravortty, "Separation and Recovery of Cobalt（II）and Nickel (Ⅱ) from Sulphate Solutions Using Sodium Salts of D2EHPA, PC88A and Cyanex272" , *Hydrometallurgy,* 49(1998) ,47-61.

10. N.BDevi, K.C.Nathsarma and V.Chakravortty, "Separation of Divalent Manganese and Cobalt Ions from Sulphate Solutions Using Sodium Salts of D_2EHPA, PC88A and Cyanex272, *Hydrometallurgy,* 54(2000), 117-131.

11. Rane M V et al.,"Recovery of High Purity Cobalt from Spent Ammonia Cracker Catalyst", *Hydrometallurgy,* 77(2005),247-251

12. N.V. Thakur, S.L. Mishra, "Separation of Co, Ni and Cu by Solvent Extraction Using Di-(2-ethylhexy) Phosphonic Acid, PC88A",*Hydrometallurgy,* 48(1998),277-289

13. Sun Suyuan, Li Baoan, *Extraction Chromatography and Its Application* (Beijing, Atomic Energy Press, 1982),630.

14. Kazuharu Yoshizuka et al., "Distribution Equilibria in The Adsorption of Cobalt(Ⅱ) and Nickel(Ⅱ)on LevextrelResin Containing Cyanex 272", *Hydrometallurgy,* 23(2-3) (1990),309-318

15. Zou Chunshan et al., "Study on Cobalt Separation with Anion-exchange and CL-P204 Extraction Chromatography", *Mining &Metallurgy,* 8(3) (1999),97-100.

16. Wang Yan et al., "Study on Separation of Co^{2+} from Cu^{2+}, Cd^{2+}, Mn^{2+}, Zn^{2+} and Fe^{3+} by Extraction Chromatography", *Chinese Journal of Chromatography,* 17(1) (1999), 55-57

Ni-Co 2013
Edited by: Thomas Battle, Michael Moats, Violina Cocalia, Harald Oosterhof, Shafiq Alam, Antoine Allanore, Rodney Jones, Nathan Stubina, Corby Anderson, and Shijie Wang
TMS (The Minerals, Metals & Materials Society), 2013

THE STARVED ACID LEACHING TECHNOLOGY (SALT) FOR RECOVERY OF NICKEL AND COBALT FROM SAPROLITES AND CARON PLANT RESIDUES

David Dreisinger and Jim Clucas

Search Minerals Inc.,1320 – 885 West Georgia Street, Vancouver, Canada V6C 3E8

Keywords: Nickel, Cobalt, Atmospheric Leaching, Acid, Saprolite, Caron Plant, SALT

Abstract

The conventional processing of high grade nickel saprolite deposits is typically by ferro-nickel or matte smelting. Low grade nickel saprolites are not treated. The Caron process for nickel laterite treatment produces residues that still contain significant quantities of un-leached nickel and cobalt. The Starved Acid Leach Technology (SALT) was developed to recover nickel and cobalt from these lean resources. Relatively small amounts of acid are applied to ground laterite or residue in order to selectively leach nickel and cobalt. The leaching is performed atmospherically under low free acid conditions. Nickel and cobalt may be recovered via resin-in-pulp processing or by solid-liquid separation and recovery from solution.

Introduction

Commercial processes for laterite ore treatment vary according to laterite type and local conditions. The two general categories of laterite ore are saprolites and limonites. Saprolites are generally treated by pyrometallurgical processes for reduction of nickel to matte or ferro nickel alloys. The high cost of drying, chemical reduction and electric furnace smelting requires that the saprolite grade for economic processing is of the order of 1.8% nickel or higher. There are of course large amounts of saprolite that fall below this "cut-off" grade that are not currently economic. Limonites can be treated by the Caron Process which involves drying of the ore, a reduction roast of the whole ore to reduce nickel and cobalt (and some iron) to the metallic state followed by ammonium – ammonium carbonate solution to leach nickel and cobalt. Nickel and cobalt recovery are low (70-85% for Ni and 20-50% for Co). The Caron plant tailings therefore contain high levels of unrecovered nickel and cobalt. Relatively few plants have used the Caron Process and new installations are not anticipated due to the high energy cost of drying and reduction roasting. Nevertheless, plants in Brazil (Niquelandia), Australia (Yabulu) and Cuba continue to operate.

In contrast to the success of the smelting and Caron plant processes, hydrometallurgical extraction of nickel and cobalt has generally been marked by difficult plant startups and financial failure. The high pressure acid leach (HPAL) process was pioneered at Moa Bay in Cuba in the 1960's in the treatment of limonites. In the 1990's the HPAL process was touted to revolutionize the nickel industry with low capital and operating costs for new installations. Murrin Murrin, Cawse and Bulong in Australia were started and in each case the technical difficulties of operating at extreme temperature and pressure conditions and the challenges in "ramping up" a complex hydrometallurgical facility conspired to cause these plants to fail economically. At present, Cawse and Bulong are closed and Murrin Murrin continues to operate after undergoing a financial restructuring.

Since the beginning of this century, a number of additional HPAL plants have been built and are either operating (Coral Bay Nickel), or restarting (Ravensthorpe), or "ramping up" (Goro) or commissioning (Ambatovy and others). The success of these projects is varied. The Coral Bay facility in the Philippines is a technical and financial success. The plant was constructed at a brownfields site with careful engineering and scaleup of largely known technology. The plant produces a mixed sulfide product of nickel and cobalt for refining by the Sumitomo refinery in Japan. The Ravensthorpe nickel laterite plant was constructed by BHP Billiton Stainless Steel Materials as a showcase for the Enhanced Pressure Acid Leach (EPAL) process incorporating upgrading of laterite, HPAL for limonite and atmospheric leaching of saprolites. The plant encountered technical difficulties during startup during a period of low nickel prices and BHP Billiton elected to close the facility. The plant has been sold to First Quantum Minerals who report (www.first-quantum.com);

> *First Quantum Minerals acquired Ravensthorpe as a decommissioned mine in February 2010 for US$ 340 million and proceeded to make significant modifications and improvements to the process and facilities in the following 18 months. Nickel was first produced in October 2011 during commissioning and first exports were shipped on 25 November 2011.*

This plant is expected to be technically and financially successful under First Quantum but following the Australian laterite plant pattern, lost substantially all of the investment for the original owner (BHP Billiton).

On April 25, 2001, INCO announced "INCO to Proceed with a US$1.4 Billion nickel-cobalt project at Goro, New Caledonia". This announcement was based on many years of metallurgical testwork, a $50 Million USD pilot plant in New Caledonia and detailed engineering and cost studies. As the plant construction started, costs began to increase excessively and the project stopped. The project has been re-engineered and restarted and is currently in active commissioning (2012). The total capital cost of the facility has now risen to at least $6 Billion USD for a facility to produce 60,000 tonnes of nickel per annum (when it reaches operating capacity). The specific investment is of the order of $100,000 USD per annual tonne of nickel capacity.

In a similar vein, the Ambatovy nickel project in Madagascar will use HPAL to process a limonite ore. The capital cost (www.sherritt.com) is now reported to be of the order of $5.5 Billion for an annual production of 60,000 tonnes of nickel per annum. This is of the same order of magnitude as the Goro facility.

Based on this review, there are a number of critical technical and economic issues that need to be summarized.

1. Caron plant tailings represent a potential source of value. At a grade of 0.4-0.5% Ni and ~0.1% Co, there is significant in-situ value in previously mined, ground, surface tailings deposits at Caron facilities.
2. Economic processing of "below cut-off" grade saprolites represent an economic target for consideration. At say 1.5% Ni grade, the in-situ value of below cut-off grade saprolites is $330 USD/t using a nickel price of $20,000/t.
3. The economic model for HPAL treatment of greenfield nickel laterite deposits seems fatally flawed. At $100,000 investment per annual tonne of nickel production, a combination of high nickel price and low operating cost are required for economic attractiveness. Long term nickel prices are not expected to be "high" and the operating costs for HPAL plants, while often projected to be low in engineering feasibility studies, seldom achieve this result.

Against this techno-economic backdrop Search Minerals has attempted to find a process that could be technically and economically successful. I.e. the process must be simple to engineer, construct, commission and "ramp up" and must do so at a profit.

There are a number of reference points for such a concept. The gold industry economically processes materials that are often even less than 1 g/t of Au content. This equates to an in-situ ore value of around $50/t at current gold prices of $1650-1700/ounce of gold. Perhaps closer to home is the case of the Sepon Copper Project in Laos, which finished construction and started in March 2005. Oxiana Limited were the developer of the project (successor companies include Oz Minerals and now MMG Resources). The project treats a weathered copper ore by atmospheric acid leaching [1,2] followed by SX-EW for cathode production. The project cost $227 Million US for a production rate of 60,000 tpa of Cu. This equates to $3,780/annual tonne of copper production, fully 1/26th of the current going rate for a nickel HPAL plant development. The plant was designed to treat 170 dry tonnes per hour of ore. This is equivalent to $152/annual tonne of ore treated. The ramp up for Sepon was 2/3 of nameplate production for period April – December 2005 and then at or above nameplate from 2006 to present. The analysis presented here is simplified and building a nickel recovery plant is not the same as either a copper or a gold plant. However, the lessons learned from these examples ought to provide some guide to the future processing of nickel ores.

Search Minerals has developed what is called the Starved Acid Leach Technology (SALT) in order to (1) meet the opportunities related to Caron Plant tailings and (2) develop a method to treat "below cut off" grade saprolite ores [3, 4]. The technology has progressed from the bench to the pilot plant. Search Minerals announced that it signed an agreement in August 2011 with Votorantim Metais Níquel S.A. and Votorantim Novos Negócios S.A., operating subsidiaries of Votorantim Participações S.A. ("Votorantim"), a privately owned Brazilian conglomerate headquartered in the city of São Paulo, Brazil, by which Votorantim will undertake testwork on the Company's Starved Acid Leaching Technology process. The testwork will be carried out at a pilot plant to be designed and assembled by Votorantim at Brazil's Centro de Tecnologia Mineral.

Search Minerals and PT ANTAM (Persero) Tbk ("ANTAM") have signed a Heads of Agreement ("HOA") to cooperate in developing a Starved Acid Leach Technology ("SALT") plant to apply the Company's SALT technology in processing low grade saprolite nickel ore from ANTAM's mining permits in one or more of Halmahera, Pomalaa or other mining projects located within Indonesia.

The results of the initial bench testing of the SALT leaching process applied to Caron Plant tailings and below cut-off grade saprolites are described in this paper. The results of the pilot plant testing will be reported elsewhere when complete.

Starved Acid Leaching Technology Development

Caron Plant Tailing Leaching

A sample of Caron plant tailings was obtained. The analysis is shown below. The sample was used as received without milling.

Table I. Caron Plant Tailings Sample Analysis

SiO_2	Cu	Zn	Co	Ni	Cr	Mn	Al	Ca	Fe	Mg
%	ppm	ppm	ppm	ppm	ppm	ppm	%	%	%	%
29.76	1144	436	538	5072	14364	7452	3.39	0.44	30.86	3.40

Leaching Procedure. A 100 g sample was slurried with 1 L of solution containing 10, 25, 50 or 100 g/L of sulphuric acid. This corresponds to acid additions of 100, 250, 500 or 1000 kg/t. The slurry was heated to the target temperature and leached for 48 h under stirring conditions in a 2 L leach reactor. The temperature was maintained constant at 50 °C for the first 4 experiments (at 4 acid levels) and then increased from 50 to 80 °C while maintaining 100 kg/t acid addition and addition of oxygen (to oxidize ferrous to ferric). Samples of slurry were taken at 1, 2, 4, 8, 24, 32 and 48 h. Each of the intermediate samples was filtered to produce a solution for assay. At the end of the 48 h the remaining slurry was filtered and washed. The filtrate and wash solution were collected separately. The washed solid residue was dried and weighed. Note that for the first 4 tests with variable acid addition, the solids removed from the intermediate samples were NOT returned to the leach. This procedure was modified for the final 4 tests to ensure that all solids were leached for the full 48 h.

All solution and solid samples were sent to International Plasma Laboratory (IPL) in Vancouver for analysis. The solids were analyzed using a multi-acid digestion/whole rock assay suite and the solutions were analyzed by the ICP 30 water package analysis.

Caron Plant Tailing Leach Results. An example of the type of data obtained from leaching of Caron plant tailings is summarized in Table 2 below. The nickel and cobalt leaching results of the 8 acid leach experiments are summarized in Table 3 below.

Experiments 1-4 show increasing nickel and cobalt extraction with increasing acid addition over the range of 100-1000 kg/t of tailings. Clearly addition of acid to ever higher levels produces only an incremental benefit in overall metal extraction. At 10 X the acid addition (1000 kg/t versus 100 kg/t) the nickel extraction doubles and the cobalt extraction is only about 50% higher. The iron extraction rises from 5% to 89% over the same range of acid addition. The terminal pH and calculated acid consumptions (for dissolved species) indicates that most of the acid was consumed in each case. As the acid addition was increased to 1000 kg/t, the main consequence was the dissolution of iron.

Table II. Results of the Caron Tails Leaching Test with 100 kg/t Acid Addition. 80 °C and 48 h leach time.

		Analysis (% or mg/L)								
		Al	**Ca**	**Co**	**Cu**	**Fe**	**Mg**	**Mn**	**Ni**	**Zn**
Input	**Mass (g)/ Vol (mL)**									
Feed Sample	100	3.39	0.44	538	1144	30.86	3.4	7452	5072	436
Units (g)	**Total**	3.39	0.44	0.054	0.114	30.86	3.4	0.745	0.507	0.044
Output										
Residue	88	3.42	0.43	213	365	35.28	3.12	3453	3425	338
Solutions										
1	10	332	25	26	54.5	1398	453	334	127.8	2.8
2	10	442	31	35	71.8	1724	589	423	170.3	3.7
3	10	467	29	35	73.8	1774	642	450	190.4	4.2
4	10	481	29	37	75	1723	696	444	200.8	5.1
5	10	472	31	37	75.1	1442	838	466	227.9	4.7
6	10	481	32	38	77.1	1379	983	516	264.8	4.9
7	800	446	32	37	72.9	823	1100	560	290	4.9
8	1150	14	6	1	3	34	37	17	8.9	0.3
Units (g)										
Residue		3.01	0.378	0.019	0.032	31.046	2.746	0.304	0.301	0.03
Solutions										
1		0.003	0	0	0.001	0.014	0.005	0.003	0.001	0
2		0.004	0	0	0.001	0.017	0.006	0.004	0.002	0
3		0.005	0	0	0.001	0.018	0.006	0.004	0.002	0
4		0.005	0	0	0.001	0.017	0.007	0.004	0.002	0
5		0.005	0	0	0.001	0.014	0.008	0.005	0.002	0
6		0.005	0	0	0.001	0.014	0.01	0.005	0.003	0
7		0.357	0.026	0.03	0.058	0.659	0.88	0.448	0.232	0.004
8		0.017	0.006	0.002	0.003	0.039	0.042	0.019	0.01	0
	Total	3.41	0.412	0.052	0.098	31.838	3.71	0.797	0.555	0.034
	Balance	-0.60%	6.30%	2.90%	14.20%	-3.20%	-9.10%	-7.00%	-9.50%	21.50%
Acid Cons. (g)		2.18	0.08	0.06	0.1	2.09	3.89	0.88	0.42	0.01
		22.45%	0.82%	0.62%	1.03%	21.52%	40.06%	9.06%	4.33%	0.10%
Total Acid Cons.		9.7	g							
		97.1	kg/t							

For the temperature series with oxygen addition the calculated acid consumption is around 100 kg/t (same as addition) but does decrease slightly with increasing temperature. This is consistent with greater re-precipitation of iron at higher temperature, likely as hydronium jarosite, which would account for additional acid consumption (but in the solids rather than in solution). The major soluble-species acid consumption is due to iron, aluminum, magnesium and manganese. The magnesium component of acid consumption is relatively minor in the case of leaching of Caron plant tailings due to the relatively low levels of magnesium in the solid feed.

Table III. Starved Acid Leach Experiments Treating Caron Plant Tailings.

Test	T (°C)	Acid (kg/t)	O_2 Addn	Ni Ext (%)	Co Ext (%)	Fe Ext (%)	Final pH
1	50	100	No	36.51	65.34	5.00	2.93
2	50	250	No	52.93	76.47	23.75	2.28
3	50	500	No	61.94	82.10	52.72	1.68
4	50	1000	No	72.19	91.32	88.92	0.94
5	50	100	Yes	41.90	64.56	4.49	2.58
6	60	100	Yes	45.74	64.11	2.49	2.41
7	70	100	Yes	51.13	69.73	1.67	2.32
8	80	100	Yes	53.29	70.66	0.93	2.41

The Ni, Co, Zn, Cu, Mn levels increased during each leach experiment. The Fe levels in solution peak and stay high for most of the first series of four experiments with variable acid addition. For the second series of four higher temperature experiments (with oxygen sparging) there was a marked decrease in the iron in solution with time due to oxidation and precipitation as hydronium jarosite. The jarosite precipitation is a favourable reaction as the acid released upon Fe precipitation continues to leach Ni and other metals. The nickel extraction rose with increased temperature of leaching.

$$3Fe_2(SO_4)_3 + H_2O = 2(H_3O)Fe_3(SO_4)_2(OH)_6 + 5H_2SO_4$$

Saprolite Leaching

A series of 9 saprolite samples were obtained and tested. Table IV shows the composition of the major elements. The nickel content ranged from 0.97 to 2.05%. The magnesium content ranges from 12 to 21%, consistent with saprolite mineralogy. The iron levels are relatively low compared to the Caron plant tailings.

Table IV. Chemical Analysis (%) of Saprolite Samples Tested.

Sample	Ni	Co	CO_3	SiO_2	Al	Ca	Cr	Fe	Mg	Mn
1	2.05	0.025	0.03	38.31	0.85	0.68	0.43	10.08	12.75	0.15
2	1.68	0.011	0.06	43.17	0.44	0.33	0.19	5.32	20.60	0.07
3	1.25	0.015	<0.01	43.18	0.70	0.68	0.33	7.62	17.27	0.13
4	1.04	0.015	0.10	42.12	0.47	0.57	0.24	7.60	21.01	0.13
5	1.12	0.017	<0.01	47.81	0.63	0.81	0.31	8.37	12.35	0.14
6	1.02	0.014	<0.01	42.73	0.48	0.67	0.21	7.05	19.51	0.12
7	1.35	0.023	<0.01	44.52	0.79	0.64	0.37	10.35	13.94	0.19
8	1.45	0.018	<0.01	43.27	0.51	0.44	0.27	8.81	17.59	0.15
9	0.97	0.016	0.23	42.13	0.46	0.67	0.20	7.50	19.68	0.13

The saprolite samples were leached in the same way as the Caron plant tailings above. A volume of 1 L of solution containing sulfuric acid and 250 g of dry saprolite sample were mixed in an aerated reactor for 8 hours at 85 °C. Samples were taken periodically and analyzed for dissolved metals. The acid addition values ranged from 150-450 kg H_2SO_4/t. The initial particle size was approximately 50 μm (P80).

Saprolite Leach Results. The results of the testing are shown in Figures below. Each of the nine samples has a particular "fingerprint" with respect to nickel and acid addition. In general, the nickel extraction is equal to or greater than the magnesium extraction. This is consistent with some portion of the ore consisting of "available" nickel and the balance in the magnesium silicate ore matrix. Sample 1 has the biggest difference between the nickel and magnesium curves (nickel extraction is approximately 20% higher than magnesium extraction. The cobalt extraction is variable but generally trends with nickel extraction. The iron extraction is low and in many cases negligible, indicating that the host ore matrix is highly basic, leading to iron hydrolysis and precipitation.

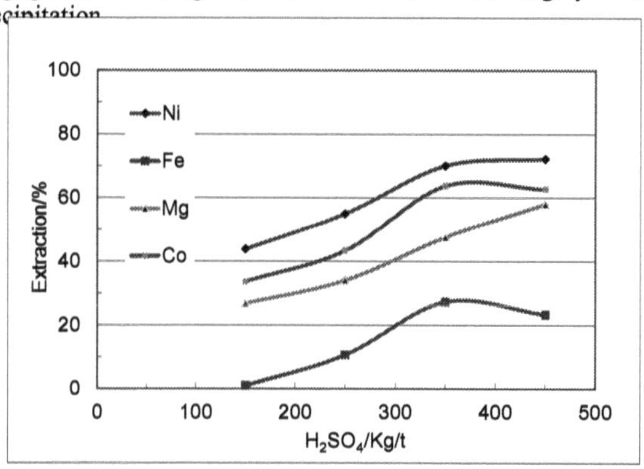

Figure 1. Metal Extraction Versus Acid Addition for Sample 1.

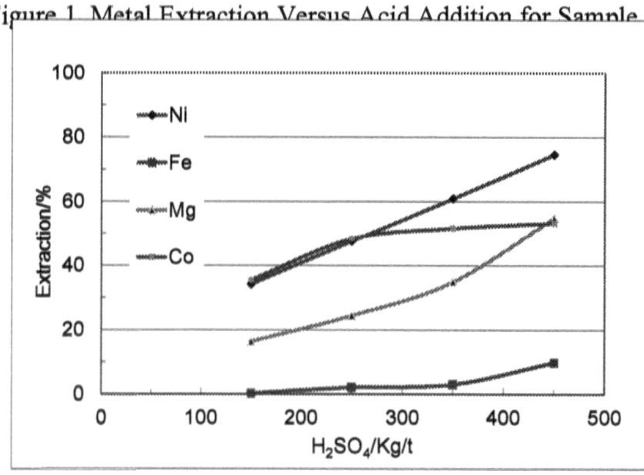

Figure 2. Metal Extraction Versus Acid Addition for Sample 2.

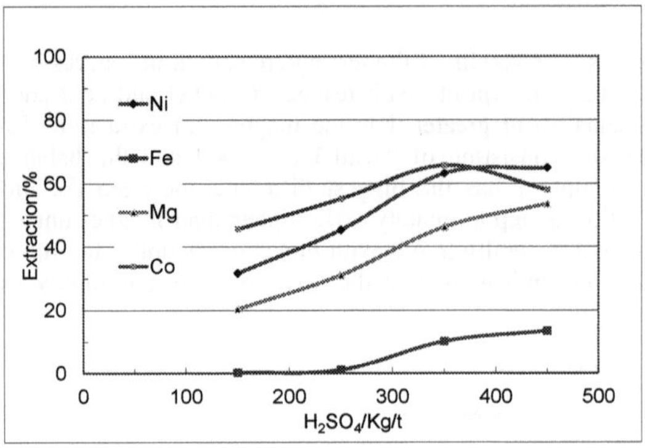

Figure 3. Metal Extraction Versus Acid Addition for Sample 3.

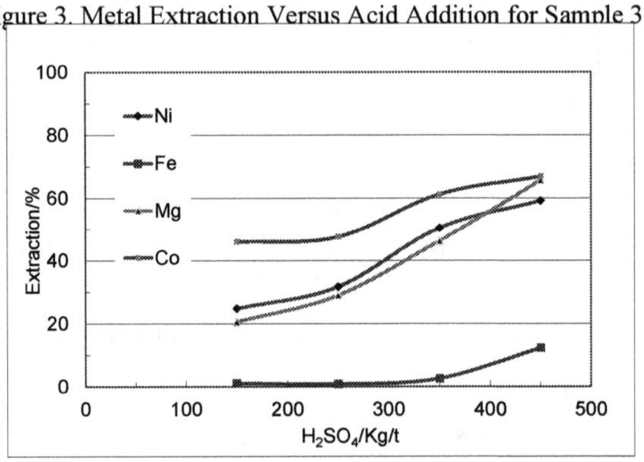

Figure 4. Metal Extraction Versus Acid Addition for Sample 4.

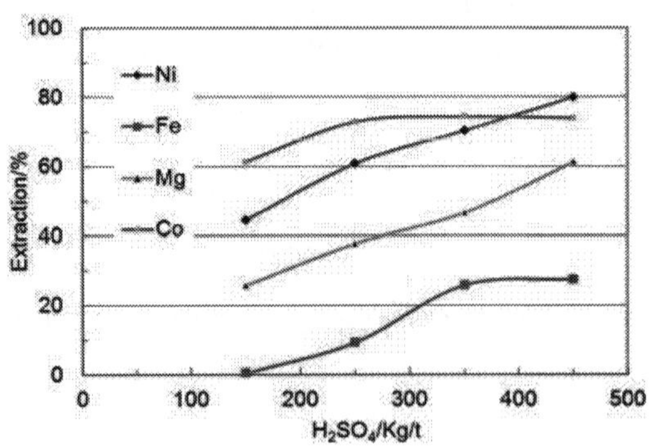

Figure 5. Metal Extraction Versus Acid Addition for Sample 5.

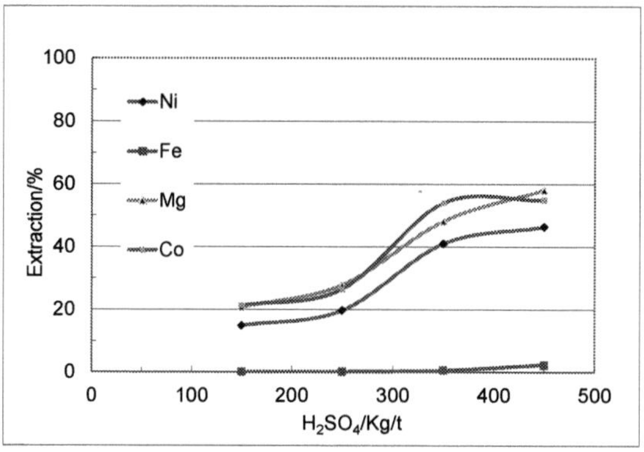

Figure 6. Metal Extraction Versus Acid Addition for Sample 6.

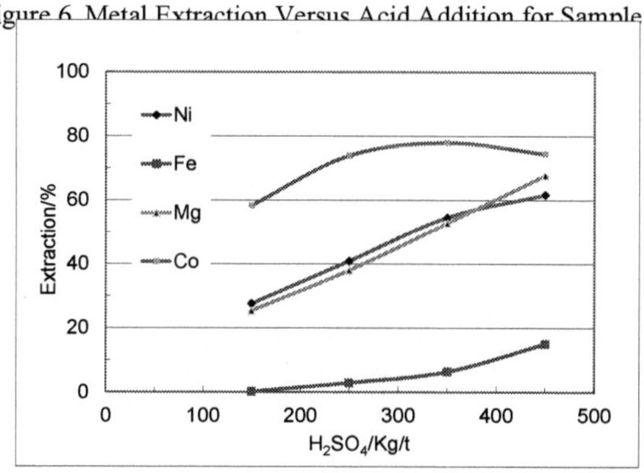

Figure 7. Metal Extraction Versus Acid Addition for Sample 7.

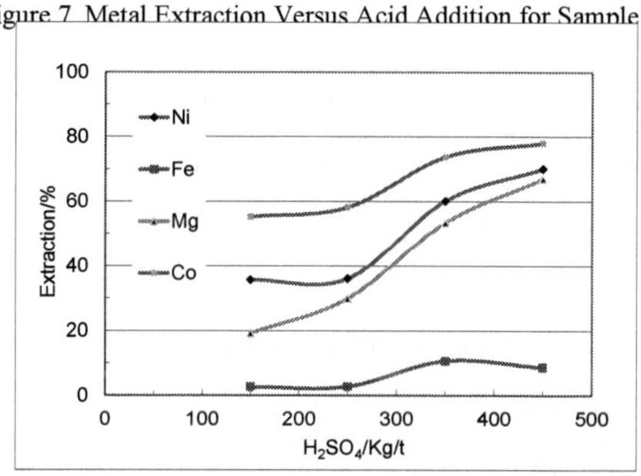

Figure 8. Metal Extraction Versus Acid Addition for Sample 8.

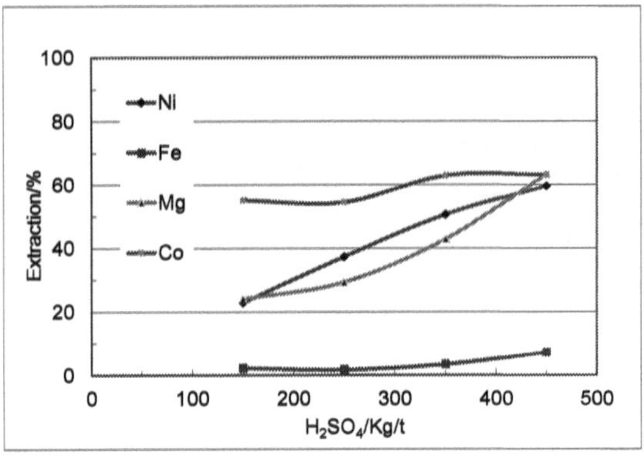

Figure 9. Metal Extraction Versus Acid Addition for Sample 9.

Nickel and Cobalt Recovery

The recovery of nickel and cobalt from the leach slurry arising from the saprolite or Caron tails leach can be accomplished by a number of different routes.

1. Conventional S/L Separation followed by Solution Purification and Ni/Co Recovery.
   The conventional recovery route would involve using a CCD circuit to wash the leached solids and recover a leach solution. The leach solution would then be treated by (for example), pH adjustment and oxidation to precipitate Fe/Al from solution, mixed hydroxide precipitation of Ni/Co (along with Cu, Zn) and finally environmental treatment to precipitate Mn from solution. Mixed sulfide precipitation could also be used. The use of a CCD circuit involves capital and necessitates adding a lot of water to the circuit to wash the solids.

2. Resin-in-Pulp Recovery of Ni/Co
   A second alternative that avoids the capital of the CCD circuit would involve (a) initial pH adjustment and oxidation to precipitate Fe/Al in the slurry, RIP extraction of Ni/Co from solution and then environmental treatment to precipitate Mn from solution. The Ni/Co loaded resin would be eluted with a weak sulphuric acid solution. The eluted Ni/Co would then be precipitated as a mixed hydroxide precipitate with MgO or alternately as a mixed sulfide with a source of soluble sulfide.

   A provisional flowsheet for SALT is shown in Figure 10 below.

Economics Of The SALT Process

The main operating cost in acid leaching of nickel laterites is the cost of acid for leaching. The proof of concept economics therefore starts with consideration of the relative cost of acid and the value of the nickel and cobalt produced in a mixed hydroxide. This is intended to determine "indicative" economics.

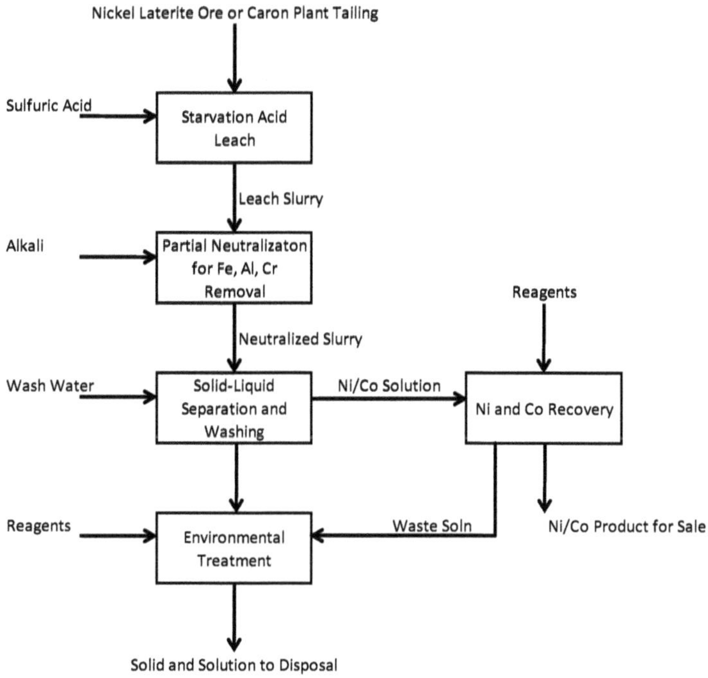

Figure 10. SALT Process Flowsheet

Caron Tailings Treatment

The nickel and cobalt grade and extraction values for the leach at 80°C were 0.51% Ni/53.29% extraction and 0.0538% Co/70.66% extraction. These values were achieved at 100 kg/t acid addition. Assuming 2 million tonnes per year as the rate of tailings treatment, the acid demand would total 200,000 tonnes per annum. Other assumptions are;

- The nickel recovery would be 5435 t/a and the cobalt recovery 760 t/a.
- \$20,000 investment per annual tonne of Ni production
- \$80/t acid cost
- \$8/lb Ni and \$15/lb Co and 85% payable in mixed hydroxide (MHP)
- \$5/t operating cost (\$10,000,000 per annum) for costs other than acid

The revenue from the plant would be \$103 Million. The capital cost would be \$108.7 Million. The acid cost would be \$16 Million and the total operating cost would be \$26 Million. Using these values a simple payback can be calculated at \$108.7/(103-26) = 1.4 years.

If the capital cost were doubled to $217.4 Million, the simple payback would be 2.8 years. This would still be an attractive proposition. Of course these numbers will vary according to the local cost of acid, the prevailing value of nickel and cobalt in mixed hydroxide, etc.

Saprolite Treatment

The average nickel extraction for the saprolites was 57.88% at 1.19% Ni in the head and 350 kg/t of acid addition. Using the same basis for calculation as for the Caron plant tailings, the annual nickel production would be 13,776 t/a. The acid consumption would be 700,000 tonnes per annum with an annual cost of $56 Million. The revenue from the plant (ignore Co) would be $243 Million. The capital cost would be $275.5 Million. The acid cost would be $56 Million and the total operating cost would be $66 Million. Using these values a simple payback can be calculated at $275.5/(243-66) = 1.56 years. Once again, even with a doubling of the capital cost to $551 Million, it would be possible to see a payback in 3.12 years with all other values staying constant.

Conclusions and recommendations

A total of eight leaching experiments were carried out on a Caron plant tailings sample containing 0.51% Ni and 0.054% Co. The extraction of nickel and cobalt were affected by acid addition and temperature (with oxidation). The most attractive economics were obtained with low acid addition (as per the starved acid leach technology concept).

A selection of nine saprolite samples with an average grade of 1.19% Ni were leached at 150-450 kg/t of acid addition. The nickel extraction ranged from 40 to 70% and averaged 57.88% at 350 kg/t of acid addition. Further acid addition beyond this value did not give sufficient additional nickel extraction to warrant the increased acid usage.

The SALT process needs to be subject to a proper engineering study for a specific application to quantify the economics of the process. Nevertheless, the early indications are that the capital cost of a SALT facility should be relatively low and as long as the acid addition is kept to a reasonable level, the payback period could be quite low.

References

1. Baxter, K., Pratt, G. and Dreisinger, D.B., 2003. "The Sepon Copper Project: Development of a Flowsheet", Proceedings of Hydrometallurgy 2003 (the Ritchie Symposium), Eds. Young, C., Alfantazi, A., Anderson, C., James, A., Dreisinger D., Harris, B., 1487-1502, TMS, Warrendale Pennsylvania (2003).
2. Dreisinger, D., Pratt, G., Baxter, K., 2010. "Ore Leach Method with Aqueous Sulfuric Acid Containing Ferric Ions", US Patent 7,799,295, September 21, 2010.
3. Dreisinger, D.B., Clucas, J.D., 2011. "Low Acid Leaching of Nickel and Cobalt from Lean Iron Containing Nickel Ores", PCT Patent Filing, International Publication Number WO 2011/120127 A1, January 20, 2011.
4. Dreisinger, D., Clucas, J., 2012. "The Starved Acid Leach Technology (SALT) for Processing of Nickel Laterites and Caron Plant Tailings", Proc. of WALSIM II, Eds. C.Q. Jia, M. Sudbury, J. Zinck, V. Ramachandran, E.A. Morris, CIM METSOC, Montreal, 39-51.

Ni-Co 2013
Edited by: Thomas Battle, Michael Moats, Violina Cocalia, Harald Oosterhof, Shafiq Alam, Antoine Allanore, Rodney Jones, Nathan Stubina, Corby Anderson, and Shijie Wang
TMS (The Minerals, Metals & Materials Society), 2013

SELECTIVE PRODUCTION OF Co AND Ni POWDERS THROUGH HYDROTHERMAL REDUCTION OF LEACH SOLUTIONS OF A SYNTHETIC MATTE CONTAINING Cu-Ni-Co-Fe-S

D. Mishra[1]*, K-H. Park[2], K. K. Sahu[1], A. Agrawal[1] & C.W. Nam[2]

[1] National Metallurgical Laboratory (CSIR-NML), Jamshedpur, INDIA-831007,
[2] Korea Institute of Geoscience & Mineral Resources (KIGAM), Daejeon, You-Seong Gu, SOUTH KOREA- 305 350, *: Corresponding Author (email: dmishra@nmlindia.org).

Key Words: Hydrothermal Reduction, Nickel Powder, Cobalt Powder.

Introduction

Synthesis of metal powders with tailored characteristics, such as those of Cu, Ni & Co has been the subject of much attention during the past decades. This is due to their potential technological applications in several critical & emerging areas, e.g. in the manufacturing of energy storing devices, optoelectronic and magnetic recording media, high end catalysts, etc. etc. In recent years, nano particles of nickel have found applications as electrodes in multi layer ceramic capacitors (MLCC) due to their excellent electrical conductivity, high melting temperature & low cost [1-3]. Similarly, fine cobalt powders are one of the main constituent for manufacture of rare-earth based permanent magnets[4].

Variety of methods have been reported in literature for producing nickel and cobalt powders from different starting materials; these include: spray pyrolysis [5], ultrasonic spray pyrolysis under reduced atmosphere [6-10], wet chemical (reduction) methods [3,11], electrolysis [12,13] and controlled hydrothermal reduction of aqueous salt solutions of nickel and cobalt [14,15. Among these reported laboratory scale methods, aqueous processing routes such as hydrothermal, wet chemical and electrolysis processes seem economically and technically attractive because of their low temperature operation and control over particle characteristics through easy manipulation of process parameters including use of additives and/or surfactants [16,17].

Majority of the reported processes for production of nickel and cobalt powders employ pure salt solutions as the starting materials, and very few deals with preparation of these metal powders from naturally available raw materials or from various secondary/low grade/waste sources [13, 18-20]. In this type of processes, solution purification to achieve high level of purity/selectivity is being a challenge; e.g. industrial production of nickel powders through Sherrit-Gorden process has low precipitation efficiency due to presence of cobalt, besides difficulty in economical re-generation of ammonia in the process stream [20]. Solvent extraction (SX) methods have been proved to be very efficient particularly for separation of chemically similar elements like Ni & Co. Though literature concerning SX separation of Ni & Co is very rich (detail discussion is beyond the scope of this article), but very few explores mixed solvent systems for better process economy without compromising on the separation efficiency [21]. In this context, the main objective of this study was to explore a mixed solvent system such as PC88A + Cyanex 272 for effective separation of Ni & Co from the Cu depleted acid (sulfuric) leach solution of a synthetic matte containing Cu-Ni-Co-Fe-S for production of high pure nickel and cobalt powders. This study is part of a broader investigation for selective recovery of Cu, Ni & Co from polymetallic sea nodules through reductive smelting – sulfidizatio - acid pressure leaching – hydrothermal reduction route [22,23].

Materials & Methods

The starting material of this investigation was a synthetic sulfide matte containing about 26% Cu, 35% Ni, 4.1% Co, 11.5% Fe & 23% S. The matte was obtained by sulfidization of alloys obtained through reduction smelting of sea nodules. Sea nodules are low grade sources of Cu, Ni & Co (together their content ~ 2.0- 2.5%), therefore, reduction smelting – sulfidization route was adopted to produce low volume mattes for their efficient processing. Details about the reduction smelting – sulfidization – acid pressure leaching can be found elsewhere [24]. In brief, the leach solutions generated has a composition: Cu ~ 25 g/L, Ni ~ 35 g/L, Co ~ 4.0 g/L & Fe ~ 100 ppm which were taken for metal powder precipitation.

Hydrothermal reduction experiments (with H_2) were carried out inside a 1L capacity PARR autoclave (Model No. 4550, Figure 1), having a titanium lined container, which has provisions for temperature control, agitation & sampling. About 600 ml of leach solutions were charged into the autoclave, after adjusting its pH to 3.5 (by addition of either sulfuric acid or 10 % NaOH solution) for precipitation of copper as Cu^0 under the following optimum condition: $PH_2 = 30$ kg/cm^2, pH = 3.5, Temp = 180 °C & duration 2h [25]. Under these conditions more than 99% copper was precipitated as copper powder with purity >99.9%, leaving a solution having Ni & Co concentrations similar to that of the original leach solution.

The generated de-copperrised solutions having Ni & Co concentrations about 30.8 and 3.6g/L respectively were taken for separation studies using the solvent mixture 5% PC-88A + 2% Cyanex-272. For this purpose commercial grade PC88A and Cyanex 272 in distilled kerosene were used as solvent without further purification. 5% iso-decanol was added as phase modifier in all experiments. Individual as well as solvent mixtures were saponified by adding calculated amount of sodium hydroxide. Suitable volumes of aqueous and organic phases were contacted manually in separating funnels (500 ml capacity) for 5 min. After phase separation, the concentration of metals were analyzed in raffinate by directed dilution, while the that of metals in organic phase by complete stripping of known volume of loaded organic using 5% sulphuric acid. All the experiments were carried out at ambient temperature (30±1 ∘C). Metal analysis was done by atomic absorption spectrophotometer (Thermo SOLAAR S-2).

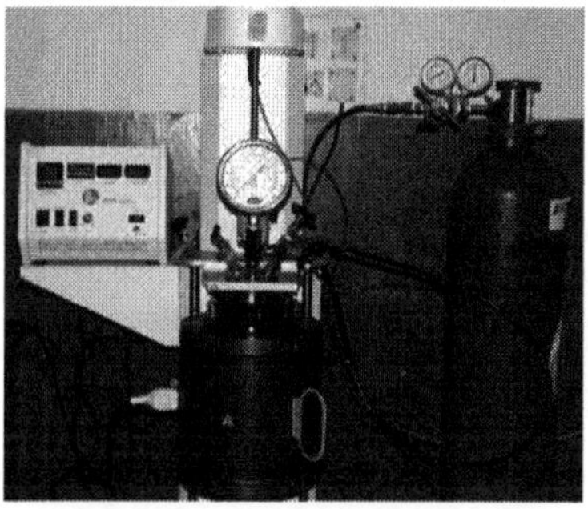

Figure 1. PARR 1L capacity autoclave used for hydrothermal reduction studies.

After each hydrothermal reduction experiments, the autoclave was allowed to cool to room temperature and the solution and metal powders were taken out. The powders were filtered, properly washed with distilled water followed by washing with dilute sodium carbonate (Na_2CO_3) solution to neutralize any free acid and finally with distilled water. Percentage precipitation (reduction) of metal was calculated from the analysis of the filtrate by an AAS mentioned above. The metal powders were treated with sodium potassium tartrate (to protect from surface oxidation) and vacuum dired at 100°C for 6h. These dried powders were stored in desiccators and analyzed for their purity and other properties.

Results & Discussion

The main objective of this study was to synthesize high pure Ni & Co powders from a solution containing Cu, Ni & Co in sulfuric acid medium. It was observed that, copper can be selectively precipitated at lower initial pH (3.5) with good efficiency and purity. However, this was not the case for Ni & Co. Preliminary investigations indicated that, under optimum conditions for Ni precipitation (initial pH >11), variable amount of cobalt was co-precipitated as β-$Co(OH)_2$, and it was difficult to separate Ni^0 from the cobalt hydroxides [26]. Therefore, prior separation of Ni & Co was essential for getting high pure metal powders. A separation scheme for separation of nickel and cobalt was tried with the solvent mixture of PC88A and Cyanex-272 (70% saponified). Cobalt was extracted by the organic solvent selectively, which was subsequently stripped with sulfuric acid after scrubbing of minor amount of co-extracted Ni. Nickel left over in the raffinate was taken for hydrothermal precipitation after adjusting its pH.

Separation Of Cobalt And Nickel From Sulphate Solutions Using A Binary Solvent Mixture

Separation and recovery of divalent cobalt and nickel ions from sulphate solutions containing 30.8g/l Ni, 3.6 g/l Co have been carried out using a mixture of PC 88A and Cyanex 272 in kerosene. Effect of various parameters such as Equilibrium pH, solvent composition, organic to aqueous phase ratio and time etc were studied, details of which are given below

Influence of Solvent Concentration and Equilibrium pH on Co & Ni Separation. Based on earlier experience, solvent mixtures of 5% PC88A + 2% Cyanex 272 and 5% PC88A + 4% Cyanex 272 were used as extractant for separation of cobalt from nickel from the solution of above composition. Equilibrium pH during extraction was varied in the range, 2.5 – 5.5 by controlling the extent of saponification. The extraction data presented in Figure 2 and 3 show that with increase in equilibrium pH, both cobalt and nickel extraction is increasing. With solvent mixture of composition: 5% PC88A + 2% Cyanex 272, a gradual increase in extraction of cobalt with very negligible co-extraction of nickel was observed even up to equilibrium pH of 5.35. At equilibrium pH of 4.4 about 83% of cobalt could be extracted with only 0.7% nickel co-extraction. There was a sharp increase in cobalt extraction with increase in equilibrium pH with the solvent mixture of composition: 5% PC88A + 4% Cyanex 272 and at pH 3.8, about 94% of Co was extracted with about 4.2% nickel co-extraction. Therefore, to avoid higher nickel co-extraction, further experiments were carried out with solvent mixer of composition: 5% PC88A + 2% Cyanex 272.

Effect of O/A Phase Ratio on the Extraction of Metal Ions. Extraction of Co from Ni was carried out at different O/A ratio using 5% PC88A + 2% Cyanex 272 (70% Saponified) at initial aqueous pH of 4 to obtain optimum condition. The result shows (Fig-4) that on increasing O:A ratio, the percentage extraction of Co increases. The McCabe Thiele plot represented in Fig. 5 shows the requirement of 2 stages for the complete extraction of cobalt at O:A = 1:1.

Under equilibrium extraction condition the solvent mixer was loaded in a two stage continuous counter current mode as determined by McCabe Thiele plot using O/A phase ratio of 1:1 (Fig. 5). The raffinate contains only 11 ppm of cobalt along with 30.3 g/L Ni. The loaded organic containing 3.51 g/L Co and 0.45 g/L Ni was scrubbed with 4 g/L cobalt solution at O/A phase ratio of 5. More than 98% nickel scrubbing efficiency was achieved in single stage under the above condition. Only 6 ppm of nickel was present in the loaded organic after scrubbing. The

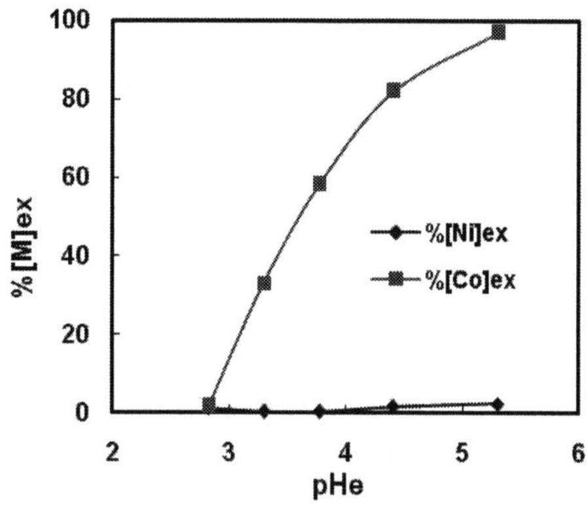

Figure 2. Effect of Eq pH on Ni and Co extraction. Org phase: 5%PC88A + 2% Cyanex 272 in Kerosene oil, Aq feed: 30.8g/l Ni, 3.6 g/l Co,O/A 1:1.

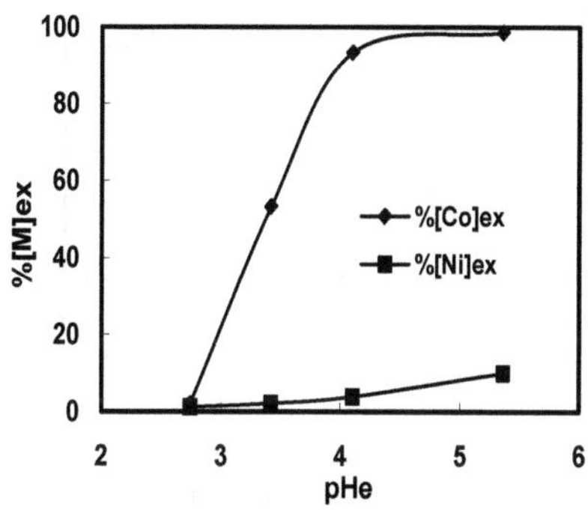

Figure 3. Effect of Eq pH on Ni and Co extraction. Org phase: 5%PC88A + 4% Cyanex 272 in Kerosene oil, Aq feed: 30.8g/l Ni, 3.6 g/l Co,O/A 1:1.

cobalt from the LO was stripped with different concentration of sulphuric acid solution at O/A ratio of 3:1. It was found that 20 g/L sulphuric acid is sufficient to strip more than 99.9% cobalt.

After complete stripping, the strip solution containing 10.51 g/L cobalt was used for making cobalt powder by hydrothermal hydrogen reduction. The raffinate with very negligible concentration of cobalt was also directly used for producing nickel powder by hydrothermal reduction after adjusting the pH.

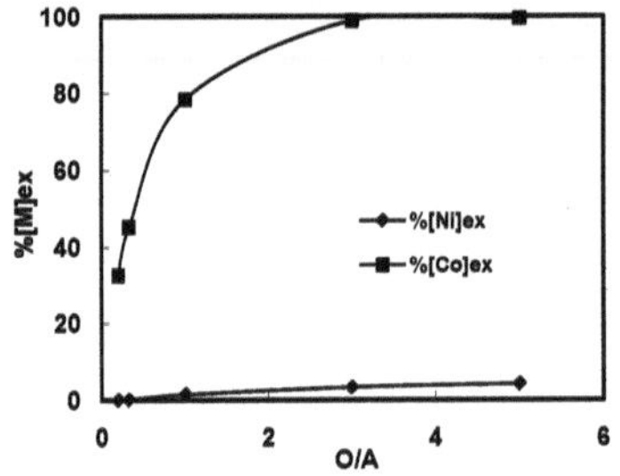

Figure 4. Effect of organic to aqueous variation of Co extraction with solvent: 5% PC 88A+2%Cyanex 272, Aqueous feed: 30.86g/L Ni, 3.6g/L Co.

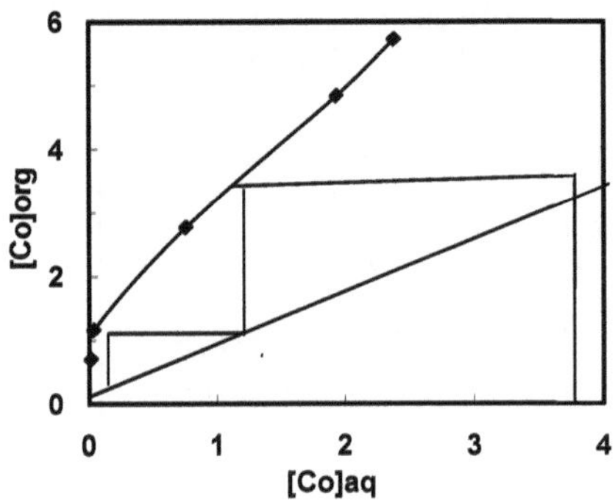

Figure 5. McCabe–Thiele plot for the stage determination for Co extraction with solvent: 5% PC 88A+2%Cyanex 272, Aqueous feed: 30.86g/L Ni, 3.6g/L Co.

Precipitation of Ni^0 from the Raffinate Solution

As mentioned in the previous section, the raffinate containing nickel were taken for hydrothermal reduction, after adjusting its pH to ~12.0 in presence of H_2 gas for precipitation of Ni^0 powder. It may be mentioned that, precipitation parameters were initially optimized by using pure nickel sulfate solutions (results not presented here), and nickel precipitation from the raffinate were carried out under the established conditions: Temperature = 180 °C, pHi = 12.0, pH_2 = 25 kg/cm^2, Duration = 2h & Seed Conc.(Ni^0) = 30 g/L. Figure 7 shows the typical morphology of Ni powders precipitated from the raffinate solutions under above optimum conditions, while Figure8 show its XRD pattern which suggest absence of any impurity phases. Further, the chemical purity of the sample was checked after its complete digestion in nitric acid & analysis by ICP-MS. The sample was found to contain 99.78% Ni with about 0.005% Cu and about 0.187% S (analyzed by CHNS analyser). Further characterization is underway to evaluate its other powder metallurgical characteristics.

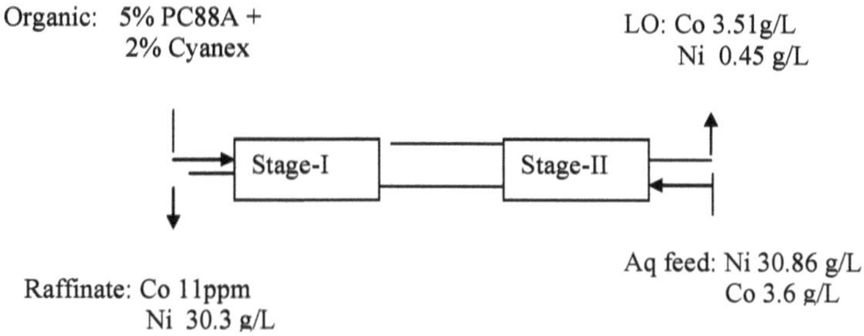

Figure 6. Continuous counter current simulation study of Co extraction with solvent: 5% PC 88A+2%Cyanex 272, Aq feed: 30.86g/L Ni, 3.6g/L Co.

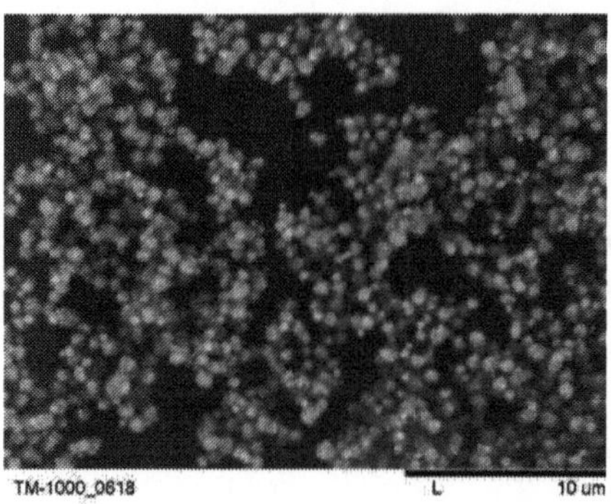

Figure 7. Morphology of Ni powders precipitated from the raffinate liquor; *c.f.* text fro precipitation conditions.

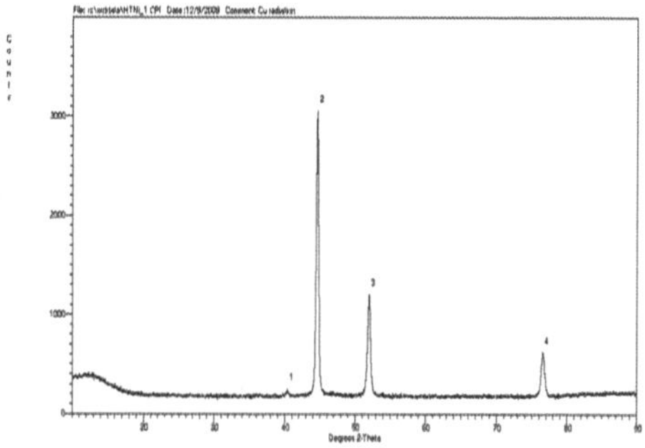

Figure 8. XRD pattern of Ni powders precipitated from the raffinate liquor; *c.f.* text fro precipitation conditions.

Precipitation of Co^0 From Stripped Liquor

The cobalt loaded organic phase was stripped with 20 g/L H_2SO_4 with A/O ratio 3:1, the obtained stripped liquor has a composition Co= 10.51g/L and Ni = 11 ppm. The pH of this solution was adjusted to ~12.0 by addition of 10 % NaOH solution then subsequently taken for hydrothermal reduction. Preliminary cobalt powder precipitation was tired under following conditions: Temperature = 220 °C, pH_i = ~12.0, PH_2 = 35 kg/cm^2, duration = 3h, and Hydrazine (99%) = 0 – 10 ml. Hydrazine was added to initiate the reduction reaction at lower temperature, which may provide seed for rapid precipitation of cobalt. Figure 9 presents the percentage reduction of cobalt under above conditions in presence and absence of hydrazine. It is observed that, even in presence of hydrazine only 50-55% cobalt was reduced under the present conditions. The percentage reduction of cobalt was calculated by analyzing the solid samples after the precipitation reactions. The precipitates were mostly composed of cobalt hydroxides and metallic cobalt. Few experiments were also conducted at higher temperature (250 ^{0}C) and at higher hydrogen partial pressure (45 kg/cm^2). Preliminary results indicated about 80 -85% reduction efficiency. Figure 10 presents the micrograph and XRD analyses of the powder precipitated in presence of 10 g/L hydrazine for 3h. Two different phases [β- $Co(OH)_2$ & Co] are clearly indicated from both the XRD and the micrograph.

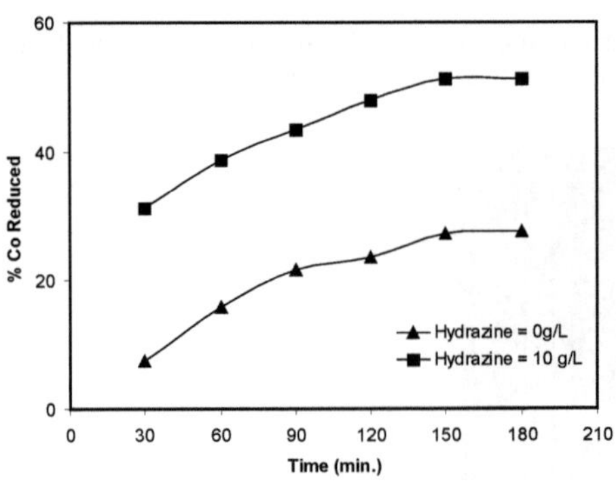

Figure 9. Hydrothermal precipitation of cobalt powder (*c.f.* text for detail precipitation conditions).

Conclusions

The following conclusions can be drawn from the precipitation studies of nickel and cobalt from a copper depleted Cu-Ni-Co sulphate solution generated from acid pressure leaching of a synthetic matte containing Cu-Ni-Co-Fe-S:

- For high purity Ni & Co powders, metal separation was tried by using a solvent mixture PC-88A and Cyanex-272 having good separation factor. Two stage counter-current extraction with 5% PC88A + 2% Cyanex 272 in Kerosene yields more than 99% Co extraction efficiency with very minor Ni co-extraction. Quantitative stripping of Co from loaded organic is possible in single stage with 20 g/L sulphuric acid.

- High pure nickel powders were precipitated (with efficiency >99%) from the raffinate liquor after adjusting its pH to 12.0 under the following conditions: Temp. = 180 °C, pHi = 12.0, pH_2 = 25 Kg/cm^2, Duration = 2h & Seed Conc.(Ni^0) = 30 g/L. The stripped cobalt in sulphuric acid was subjected for hydrogen reduction and under the present conditions about 80 - 85% reduction efficiency was achieved.

- The powder characteristics of precipitated Ni & Co are under evaluation & further fine tuning of process parameters need to be done to get good quality products.

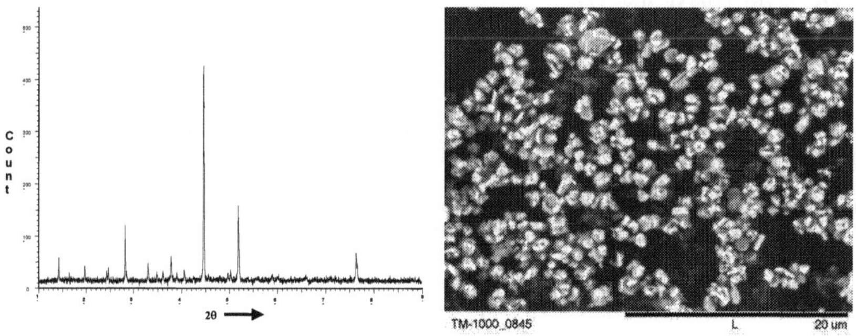

Figure 10. The XRD and electron micrograph of powders obtained after hydrogen reduction of stripped liquor containing 10.51 g/L Co.

Acknowledgement

The authors are thankful to KIGAM, S. Korea for financial assistance in carrying out this work. Authors are also thankful to Director, CSIR-NML for giving permission to publish this work.

References

1. E.L. Abdel-Aal, *et al.*, "Effect of synthesis conditions on preprartion of nickel metal nano powders via hydrothermal reduction technique", Powder Technology, 171 (2007), 63-68.

2. S. Morup, Studies of Superparamagnetism in Samples of Ultrafine Particles, Nanomagnetism, (Kluwer Academic Publishers, Boston, 1993), 93–99.

3. K.H. Kim, et al., "Preparation of fine nickel powders in aqueous solution under wet chemical process", Materials Science & Engineering A, 381, (2004), 337–342.

4. Jin Zhang Gao *et al.*, "Preparation of ultrafine cobalt powder by chemical reductionin aqueous solution", Chinese Chemical Letters, 21 (6) (2001), 555-558.

5.T. Kodas *et al.*, "Nickel Powders, Methods for Producing Powders & Devices Fabricated from the Same", US Patent No.6, 316, 100, 2001.

6. S. Gurmen, S. Stopic, and B. Friedrich, " Synthesis of nanosized spherical cobalt powder by ultrasonic spray pyrolysis", Materials Research Bulletin, 41, (2006), 1882–1890.

7. S. Stopic, I. Llic, D. Uskokovic, "Structural and morphological transformations during NiO and Ni particle generation from chloride precursor by ultrasonic spray pyrolysis", Materials Letters, 24 (6), (1995), 369–377.

8. S. Stopic, I. Llic, D. Uskokovic, "Preparation of nickel sub-micron powder by ultrasonic spray pyrolysis", Int. J. Powder Metall., 32(1), (1996), 59–65.

9. J.H. Kim, *et al.*, "Preparation of Ni powders through ultrasonic spray pyrolysis technique", Journal of Materials Research, 18(7), (2003), 1614–1622.

10. W.H. Suh, K.S. Suslick, "Magnetic and porous nano spheres from spray pyrolysis", Journal of American Chemical Society, 127 (34), (2005), 12007–12010.

11. K. Yu, et al., "Dispersed rod like nickel powder synthesized by modified pylol process", Materials Letters, 57 (24-25), (2003), 3992-3997.

12. F. Ebrahimi, Z. Ahmed, "The effect of current density on properties of electrodeposited nanocrystalline nickel", J. Appl. Electrochem., 33 (8), (2003), 733–739.

13. A. Agrawal *et al.*, "Recovery of nickel powder from copper bleed electrolyte of an Indian copper smelter by electrolysis", Powder Technology, 177, (2007), 133–139.

14. S. Komarnent et al., "Microwave hydrothermal synthesis of metal powders", J. Mater. Res., 10 (1995), 1687–1692.

15. M.S. Hegde et al., « Synthesis and chemical reactivity of polyol prepared mono-disperse nickel powders", Solid State Ionics, 93, (1997), 33-50.

AUTHOR INDEX
Ni-Co 2013

A

Adham, K. ... 213
Agarwal, S. ... 291
Agrawal, A. ... 291, 425
Alkan, M. ... 335, 365
Andrews, R. ... 315
Aspola, L. ... 315
Asselin, E. ... 107

B

Bednarski, T. ... 391
Boon, C. ... 177, 201
Bourget, C. ... 391
Bratt, M. ... 299
Burlet, C. ... 243

C

Cai, W. ... 117
Chatterjee, M. ... 353, 357
Chen, F. ... 221
Chen, J. ... 97
Chen, Y. ... 279
Clucas, J. ... 413
Cocalia, V. ... 391

D

DeLaat, R. ... 97
Demopoulos, G. ... 371, 379
Dhawan, N. ... 125
Doerfelt, C. ... 371
Dreisinger, D. ... 155, 255, 413
Drew, B. ... 231

E

Ewart, I. ... 191

F

Feldmann, T. ... 379
Ferron, C. ... 53
Fischmann, A. ... 391
Flynn, D. ... 231
Franzkowiak, A. ... 231
Fraser, R. ... 177, 201

G

Gillaspie, J. ... 75
Gluck, T. ... 155
Godiwalla, K. ... 291
Grassi, R. ... 143
Gregurek, D. ... 231

H

Haavanlammi, K. ... 325
Halim, M. ... 97
Hong, X. ... 221
Hughes, S. ... 315

I

Imeson, D. ... 255

J

Jana, R. ... 291
Janwong, A. ... 125
Jiang, T. ... 117
Johnston, T. ... 177, 201
Jyrkonen, S. ... 325

K

Karonen, J. ... 325
Kennedy, M. ... 299
Kishore, A. ... 357

L

Lakshmanan, V. ... 97
Li, C. ... 279
Li, G. ... 117
Li, H. ... 279
Li, W. ... 221
Li, Z. ... 405
Lijun, W. ... 405
Liu, B. ... 343
Louis, P. ... 243
Lu, J. ... 155
Luomala, M. ... 325

M

Matusewicz, R. ... 315

Mishra, D. 291, 425
Moats, M. 125, 191

N
Nam, C. 425
Nolet, S. 201
Nos, P. 299

O
Ounpuu, M. 255

P
Palmu, L. 269
Park, K. 425
Prasad, S. 353

R
Randhawa, N. 291
Rao, M. 117, 353, 357
Reiter, V. 231
Ressler, A. 231
Riekkola-Vanhanen, M. 269
Robinson, D. 177
Robinson, J. 191
Robinson, T. 143
Roy, R. 97

S
Sahu, K. 291, 425
Sarkar, P. 353, 357
Shuling, L. 405
Song, C. 405
Spanring, A. 231
Sridhar, R. 97
Suikkanen, P. 325

T
Tang, N. 343
Tasyürek, K. 365
Taylor, A. 3

V
Vanbrabant, Y. 243
Verbaan, N. 255
Vethsodsakda, T. 125

W
Wang, Q. 221
Wang, S. 191
Weaver, M. 299
Weight, D. 25
White, D. 75, 143
Wiggett, S. 391
Wu, C. 279

X
Xu, Q. 343

Y
Yesberg, J. 201
Yilmaz, O. 335
Yücel, O. 335, 365

Z
Zhang, B. 221
Zhang, Z. 279
Zhenping, C. 405
Zhi, Q. 117
Zuniga, M. 107

SUBJECT INDEX
Ni-Co 2013

A

Acid 413
Acid Mist 143
Agglomerate Size Distribution 125
Arsenic 371
Asbolane 243
Atmospheric Leaching 75, 413
Ausmelt 315

B

Bacteria 269
Base Metals 97
Bioleaching 269
Boleo 155
Brazing Alloys 365
Bulk Concentrates 325

C

Ca-ferritic Slag 231
Calcium Sulfate 379
Carbothermic Process 365
Caron Plant 413
Chloride Hydrometallurgy 379
Co-precipitation 371
Cobalt 155, 213, 291, 413
Cobalt Based Superalloy 353
Cobalt Oxides 243
Cobalt Powder 425
Cobalt Slag 53
Contacts 177
Coolant 299
Copper 143, 213, 255, 291
Corrosion Resistance Test 231
Cr-based Alloys 335
Creep 357
Crushed Ore Agglomeration 125
Crystal Growth 379
Current Density 191
Current Measuring 201

D

Dehydration 279
Democratic Republic of Congo 243
Direct Nickel Flash Smelting 325
Directional Solidification 343, 357

E

Effluent Solutions 371
Electrical Conductivity 125
Electrorefining 201
Electrowinning 143, 155, 191, 201
EMEW 191
Equiaxed 357
Extraction Chromatography 405

F

Fatigue 357
Ferric 371
Ferrous 371
Ferrous Sulphate 107
Flotation 255
Fluidized Bed 213
Furnace 299

G

Grain Size 353

H

Heterogenite 243
High Purity Cobalt 405
Hydrochloric Acid 379
Hydrometallurgy 269
Hydrothermal Reduction 425

I

IGT Blade 343
Intercell Contact Bar 177
Intermetallics 335

J

Jarosite Precipitation 75

K

Kinetics 117

L

Laterite 97, 221, 391
Laterite Ore 279
Laterites 107

Leaching 97, 107, 117
Levextrel Resins.. 405
Limonite.. 107, 279
Lithiophorite 243
Lixiviant Regeneration............................... 379
Low-grade.. 221

M

Magnesia-chromite Bricks 231
Metallothermic Process.............................. 335
Mineral Phase Transformation 279
Mixed Chloride 97

N

Ni-base Superalloy.................................... 357
Ni-based Alloys 335, 365
Ni-Co Separation....................................... 405
Ni-doped Goethite..................................... 117
Nickel......... 107, 143, 191, 221, 255, 291, 315,
.... 391,413
Nickel Boron... 365
Nickel Laterite 75, 125
Nickel Powder... 425
Nickeliferious Goethite 117
NorthMet... 255
Numerical Simulation 343

O

Outotec.. 315

P

Parallel Computing 343
Permeability.. 125
PGM.. 325
Pilot Scale Smelting.................................. 291
Platinum.. 255
PLATSOL-TM... 255
Poor Contact 201
Pressure Acid Leaching............................... 75

R

Reactive Crystallization 379
Recycling Cobalt... 53
Reduction Smelting................................... 291
Reduction-magnetic Separation................. 221
Refractories... 231
Removal.. 371
Resistance... 177
RHI... 231
Roaster 213
Rotary Kiln.. 231

S

Safety 299
SALT... 413
Saprolite 413
Sea Nodules.. 291
Short Circuit.. 201
Sintering... 279
Solvent Extraction 97, 391
Spent Batteries 53
Spent Catalysts 53
Stability.. 371
Stillwater Mining Company 231
Strength.. 357
Sulfuric Acid 117
Sulphide Ores 269

T

Tensile Properties 353
Thermo-Mechanical Treatment 353
TSL 315

Z

Zinc 143

MIX
Papier aus verantwortungsvollen Quellen
Paper from responsible sources
FSC® C105338

If you have any concerns about our products,
you can contact us on
ProductSafety@springernature.com

In case Publisher is established outside the EU,
the EU authorized representative is:
Springer Nature Customer Service Center GmbH
Europaplatz 3, 69115 Heidelberg, Germany

Printed by Libri Plureos GmbH
in Hamburg, Germany